Introduction to Human Factors

This textbook comprehensively introduces readers to the cutting-edge field of human factors psychology, offering real-world examples illustrating how experimental findings can be used to improve the design of tools and environments that we use every day. The revised second edition provides updated text, examples, pedagogical boxes, and references.

Features:

- Showcases pedagogical boxes that end with thought questions to encourage student processing and application of the material.
- Includes instructor materials such as PowerPoint slides, activities, and exam items to facilitate teaching for instructors who are new to the course.
- Presents theoretical and practical implications of applying psychology to design with the use of examples.
- Discusses anthropometric tools, anthropometric data collection methods, hand biomechanics, and hand tools.
- Highlights diversity and inclusion with applications to the special population section in each chapter.

Introduction to Human Factors is an ideal read for senior undergraduate and graduate students in the fields of ergonomics, human factors, and psychology.

Online teaching resources accompany this textbook, including PowerPoints, a test bank with answers, and an end-of-chapter questions and answers key for the instructors.

Introduction to Human Factors

Applying Psychology to Design

Second Edition

Nancy J. Stone, Alex Chaparro, Joseph R. Keebler,
Barbara S. Chaparro and Daniel S. McConnell

CRC Press
Taylor & Francis Group
Boca Raton London New York

CRC Press is an imprint of the
Taylor & Francis Group, an **informa** business

Front cover image: Artem Varnitsin/Shutterstock, Frame Stock Footage/Shutterstock, insta_photos/Shutterstock, Scharfsinn/Shutterstock.

Second edition published 2025
by CRC Press
2385 NW Executive Center Drive, Suite 320, Boca Raton FL 33431

and by CRC Press
4 Park Square, Milton Park, Abingdon, Oxon, OX14 4RN

CRC Press is an imprint of Taylor & Francis Group, LLC

© 2025 Nancy J. Stone, Alex Chaparro, Joseph R. Keebler, Barbara S. Chaparro, Daniel S. McConnell

First edition published by CRC Press 2017

Library of Congress Cataloging-in-Publication Data
Names: Stone, Nancy J., author. | Chaparro, Alex, author. | Keebler, Joseph R., author. |
Chaparro, Barbara S. (Barbara Stewart), author. | McConnell, Daniel S., 1969- author.
Title: Introduction to human factors : applying psychology to design /
Nancy J. Stone, Alex Chaparro, Joseph R. Keebler, Barbara S. Chaparro,
and Daniel S. McConnell.
Description: Second edition. | Boca Raton, FL : CRC Press, 2025. |
Includes bibliographical references and index. | Identifiers: LCCN 2024028770 (print) |
LCCN 2024028771 (ebook) | ISBN 9781032848853 (hbk) | ISBN 9781032370149 (pbk) |
ISBN 9781003515463 (ebk)
Subjects: LCSH: Human engineering.
Classification: LCC TA166 .S76 2025 (print) | LCC TA166 (ebook) |
DDC 620.8/2—dc23/eng/20241114
LC record available at https://lccn.loc.gov/2024028770
LC ebook record available at https://lccn.loc.gov/2024028771

ISBN: 978-1-032-84885-3 (hbk)
ISBN: 978-1-032-37014-9 (pbk)
ISBN: 978-1-003-51546-3 (ebk)

DOI: 10.1201/9781003515463

Typeset in Sabon
by codeMantra

Contents

8 Motor skills and control

12   Future trends in human factors   319

Preface to second edition

We are honored to have this opportunity to write a second edition of our book. Our goals for the first edition were to write a book that was comprehensive in its coverage of human factors and accessible to a wide audience of undergraduate learners; a book that provided examples of the application of human factors science to multiple settings and populations, including special populations; and a book filled with pedagogical features that can assist students and educators. Thank you to the students, faculty, and other reviewers who provided us with feedback on the book and to let us know we were on the right path to achieving our goals.

Therefore, we kept the same **approach** and **pedagogical features** as we had in the first edition. Updates to the second edition included the inclusion of more recent literature and databases; new, updated, and modified figures and tables; new and revised pedagogical boxes; and clarified writing for greater readability, as well as the deletion of dated material or examples. We also updated our **supplemental materials**, which include basic PowerPoint slides, activities, and test items.

Acknowledgments

We are grateful for the opportunity to revise the book. Some revisions are based on the insightful feedback from Shawn Doherty, as well as other colleagues and graduate students who gave suggestions for improving the book.

Author contributions

Nancy and Alex again served as our coordinating team and worked closely together. Nancy kept us on track and revised five chapters: Introduction to Human Factors, Research Methods, Anthropometry and Biomechanics, Environmental Design, and Human Error: Causes and Prevention.

Alex revised his four chapters: Vision, Tactile, and Olfactory Displays; Audition and Vestibular Function; Attention, Memory, and Multi-Tasking; and Decision-Making. He also revised the chapter on Future Trends in Human Factors with Joe.

Not only did Joe assisted with revisions to the chapter on future trends but also he reviewed all chapters after Nancy and Alex reviewed them. Joe's insights were extremely helpful in ensuring our chapters were complete and reflecting our goals for this text.

Barbara revised her chapter on Methods of Evaluation and reviewed some chapters.

Daniel did a tremendous job revising his chapter with all new images and updated content in the Motor Skills and Control chapter.

Preface to first edition

February 17, 2017

Increasing the safety of transportation systems, developing safer toys for children, making business processes more efficient, and reducing the challenges faced by individuals with disabilities are just a few examples of how human factors can benefit society. Traditionally, human factors have focused on designing tools and devices that are safer while also reducing the stress on the body, designing interfaces for pilots and drivers that are more intuitive and easier to use, and applying results in a military environment; however, the applications of human factors reach far beyond these settings. Whether you intend to be a nurse, teacher, pilot, engineer, or human factors specialist, familiarity with human factors and ergonomics principles can benefit you in your professional and personal life. The purpose of this book is to familiarize students in the field of human factors and ergonomics that impacts so many aspects of their personal and professional lives. Human factors training has been offered almost exclusively at the graduate level; however, we feel there is a need for a textbook aimed at undergraduates that exposes them to the field of human factors and how human factors are relevant to their school, home, leisure, and work environments.

We believe there is a need for a comprehensive, yet accessible, undergraduate text on human factors. Although there are a few undergraduate human factors texts available, our intent was to write a book that would be broadly accessible even to students who are not undergraduate psychology majors. Additionally, we sought to write a book that could be used by an instructor in departments that do not have a graduate program in human factors. A comprehensive text covering the breadth of the field of human factors would be a valuable resource for these instructors.

To expand our reach to a greater number of undergraduate students, we wrote this junior/senior level introductory but comprehensive text, with the only prerequisite being introductory psychology. Although the typical student in a human factors course majors in psychology or engineering, students from many other majors and colleges (e.g., social work, education, and business) will find the possible applications of the material to a variety of settings useful.

Approach

Our approach is to present major theories, principles, concepts, and methodologies that offer an overview of the breadth of the field of human factors. Existing texts target a graduate student audience and rely heavily on military and aviation applications of human factors. In contrast, we emphasize applications of human factors to everyday events such as the use of cell phones, driving, and functioning in our educational and living environments. This approach allows students to appreciate the relevance and value of human factors to the design of tools and environments they use daily.

Given the applied focus of human factors, there is also a strong emphasis on the applications of human factors science. Not only are the students exposed to applied examples, but also the theory, concepts, and principles that justify the application. It is essential that students gain an appreciation that the application is informed by the research findings.

We also wish to demonstrate that human factors can enhance the quality of life for many user populations. Human factors can address issues that are specific to special populations such as children, the elderly, or the disabled. Therefore, we address the applications to special populations in every chapter.

Key pedagogical features

The main highlights of this text will be (1) the use of real-world events to which undergraduates can relate, (2) critical thinking boxes, (3) the application of human factors, (4) exercises, and (5) designing for special populations.

1 The use of real-world events is important so undergraduate students can relate to the situation or event. These examples will be drawn from personal experiences as well as from current events. Each chapter opens with a vignette that describes a situation or situations that tie into the chapter material. This vignette and other examples are integrated within each chapter.
2 Critical thinking boxes occur throughout each chapter whereby relevant material is presented about a person, a topic, or an event. These boxes end with critical thinking questions to assist students in reflecting on and applying the material.
3 We also use applied examples throughout the book to demonstrate the applications of human factors to everyday events.
4 We developed a set of exercises at the end of each chapter to help students apply their new knowledge to specific situations or conditions.
5 Finally, each chapter addresses the application of the theories, principles, or concepts to special populations (e.g., children and disabled persons).

Supplemental materials

We created supplement materials that link to each chapter. In particular, we developed basic PowerPoint slides, activities, and test items.

Acknowledgments

We wish to thank Taylor and Francis, and Cindy Carelli, in particular, for giving us this opportunity to put forth what we consider to be a strong pedagogically based textbook on human factors.

Author contributions

Writing a comprehensive textbook about the diverse field of human factors requires teamwork. Nancy and Alex served as our coordinating team and worked closely together. Nancy, as the lead author, was in many ways the taskmaster (in a good sort of way!), keeping us on track. In her spare time, she wrote five chapters: Introduction to Human Factors, Research Methods, Anthropometry and Biomechanics, Environmental Design, and Human Error.

Alex, often the mediator, recruited Joe and Dan to join the team and contributed his expertise in his four chapters: Vision, Tactile, and Olfactory Displays; Audition and Vestibular Function; Attention, Memory, and Multi-Tasking; and Decision-Making. He also co-wrote with Joe the chapter on Future Trends in Human Factors.

We were lucky to bring Joe on board to not only co-write the Future Trends chapter with Alex but also serve as our reviewer for all the other chapters. Joe's insights and vast knowledge of the field were extremely helpful in ensuring our chapters were complete and reflected our goals for this text.

Barbara is an expert in User Experience, so it only made sense to have her write the Methods of Evaluation chapter; she uses and knows these evaluation methods well. In earlier drafts of the text, she also helped in reviewing initial chapters and establishing the format of chapters.

Then, there is Daniel who teaches more students in a semester than most of us teach in years, but nevertheless found time to share his expertise in the Motor Skills and Control chapter. Dan's great insights about teaching and the needs of teachers assisted our development of the supplemental materials.

About the authors

Nancy J. Stone received her Ph.D. in Experimental Psychology from Texas Tech University. She created and taught the undergraduate Human Factors course at Creighton University. She is a Fellow of the Human Factors and Ergonomics Society and currently serves on the Executive Council. She served as the Educational Technical Group (ETG) Program Chair and the ETG Chair. She also served on and then became Chair of the Education and Training Committee of the Human Factors and Ergonomics Society. Her involvement in human factors education led to her invited article on human factors education in the Special 50th Anniversary issue of *Human Factors*. Her research is in the areas of environmental design, teamwork, and student learning.

Alex Chaparro received his Ph.D. in Experimental Psychology from Texas Tech University and completed a postdoc at Harvard University in the departments of Psychology and Applied Sciences. He is a professor in the Department of Human Factors and Behavioral Neurobiology at Embry-Riddle Aeronautical University. He has taught human factors courses at the undergraduate and graduate levels and is a member of the Human Factors and Ergonomics Society. His research concerns the effects of distraction and aging on driving performance.

Joseph "Joe" R. Keebler received his Ph.D. in Applied/Experimental Human Factors Psychology from the University of Central Florida. He currently serves as an associate professor of human factors at Embry Riddle Aeronautical University. His work is aimed at experimental and applied research, with a specific focus on training and teamwork in military, medical, and consumer domains. He has partnered with multiple agencies and institutions in his career, with most projects aimed at the implementation of human factors in complex, high-risk systems to increase safety and human performance. This work includes over 100 publications, book chapters, proceedings papers, and presentations.

Barbara S. Chaparro received her Ph.D. in Experimental Psychology from Texas Tech University. She is a professor in the Department of Human Factors and Behavioral Neurobiology and head of the Research in User eXperience (RUX) Lab at Embry-Riddle Aeronautical University. Her research interests include the study of factors that influence the user experience (UX) of products and systems, the investigation of usability assessment methods, and the efficacy of augmented/mixed reality devices and applications.

Daniel S. McConnell received his Ph.D. in Sensory Psychology from Indiana University Bloomington and completed a postdoc at the Institute for Research in Cognitive Science at the University of Pennsylvania. He is currently a senior lecturer in psychology and co-director of the Technology and Aging Laboratory at the University of Central Florida. His research focuses on the visual control of reaching, the analysis of human movement kinematics, and human motor performance in the context of technology use.

Chapter 1

Introduction to human factors

Chapter Vignette

It is Friday afternoon in Los Angeles, and many individuals are beginning their long commute home. People board a late afternoon commuter train in anticipation of the start of their weekend. As with many train routes, this route has sections of track that are shared by trains traveling in opposite directions. The engineer operating the train must watch for stop signals along his route to determine whether it is necessary to stop for another train or clear to proceed. On September 12, 2008, the train cruised past a stop signal and shortly thereafter struck head-on into a freight train, killing 25 passengers and injuring more than 130 others. The engineer was texting on his cell phone during this critical time and missed the stop signal. Unfortunately, there was no warning signal alerting the engineer that he ran a stop signal. An appropriately designed alarm at the right time could have prevented this deadly accident.

In May 2010, a stock trader is entering sales into his computer. Upon completing the order, he must enter the number of shares to sell. He wishes to sell a number in the millions, so he is supposed to enter "m." Instead, he enters a "b"—which stands for "billion." By submitting this billion dollar sale, the trader initiates a large number of subsequent sales, creating chaos in the market and drastically affecting stock prices. The market quickly drops by about 1,000 points. Although not a deadly accident, some system safeguards could have been developed to reduce the likelihood of this error recurring. Having an alert and prompt to acknowledge such a large sale and providing a clearer understanding of the operator's decision would potentially have averted the ensuing financial fallout from the accident.

During the high-peak summer vacation months, many families are driving to and from their vacation destinations. To capitalize on this increase in traffic, some cities redesigned their street signs so drivers could more easily navigate the cities' main attractions, thereby increasing the cities' tourism.

Each of the above scenarios offers an example of a situation that could (or did) benefit from human factors, the application of psychological science to the design of environments that fit the capabilities and limitations of the human user. We can broadly classify human factors science into the study of human cognitive abilities and limitations; human physical abilities and limitations; and the design of devices, environments, and systems for human use. Properly designed environments and tools address human cognitive and physical capabilities and limitations. For example, one contributing factor to the train wreck incident could be related to displays and the presentation of information. The engineer apparently did not see the stop signal, a visual display. If a warning signal, possibly an auditory display, was activated, it might have caught his attention in time to stop the train.

Other aspects of human cognition besides perception include attention, decision-making, and memory. The train engineer might have been distracted by texting, drawing his attention from the primary task of operating the train. Similarly, road signs can be rede-signed to facilitate driver understanding, thereby reducing the distraction of sign reading and allowing the driver to focus on the primary task of driving. In the case of the trader, it is possible that the trader was talking to someone, not focusing on what he typed, making a mistake, a big mistake. Similarly, decision-making might deviate from the ideal because of systematic biases or because the decisions are based on information whose reliability or accuracy is suspect. Perhaps the train engineer believed he could text and maintain a look-out for the stop light at the same time. On the other hand, memory issues could be at play if the engineer did not remember the layout of that particular track. On a positive note, drivers might navigate the city routes more easily with new street signs.

In addition to cognitive abilities, understanding the capacity of human operators to produce certain types of motor movements for a given task is also critical. This includes consideration of the biomechanical limits of operators (i.e., strength, speed, and limits of their reach) and limits on performance (i.e., speed and accuracy). The design of the keyboard the trader was using may have been based on data including the size of hands and fingers, the reach of the individual fingers, coordination of multiple digits on the same hand, and accuracy of the movements that inform decisions including the optimal size of the keys, their placement, and distribution between the hands. Perhaps these factors affected the trader. It also could be the case that the trader has inadequate motor control, making him a poor typist.

Other human factors issues involve the design of the environment. Although effective visual displays or signs can help people navigate through an environment, the design of the city itself could also help. Understanding how the layout of an area affects people's ability to navigate through the space is just one aspect of environmental design that can reduce errors and make people more efficient wayfinders. In addition, we must con-sider how people interact with the environment. Perhaps the trader error reflects the high-paced nature of the job.

As you learn more about human factors, you should have fun identifying the various issues you encounter. As you can see from the above examples, human factors can be applied to just about everything around us. Unfortunately, you will notice that human factors prin-ciples are often not applied. Designers might not be aware of human factors principles and the extensive scientific knowledge and research literature that can inform design. In turn, designers might not involve users in the design process or might fail to consider users' needs early in the design process. These examples underscore how poorly designed equipment or environments can lead to devastating results including high cost, death, and destruction of property. Failure to consider human factors in design can lead to errors and decreases safety and/or efficiency. Yet, in the case of the redesigned city signs, the application of human fac-tors can enhance an individual's ability to do the tasks necessary to safely operate and utilize the system, such as the ability to find their way to the various city attractions.

Once environments, equipment, or devices are redesigned using human factors principles to reduce errors or injuries, the design often appears to be common sense (Howell, 2003). While human factors may appear tantamount to the application of common sense, this could not be farther from the truth, as evidenced by the plethora of poorly designed software interfaces, websites, and cell phones we encounter on a daily basis. Design by "common sense" assumes that users have commonly shared experiences and derive the same conclu-sions from these experiences, which obviously is an untenable suggestion. By applying human factors design principles, evaluative techniques, and methods, we can create more efficient and safe tools, tasks, and environments that users find pleasing and that reduce errors and often increase safety.

1.1 CHAPTER OBJECTIVES

After reading this chapter, you should be able to:

- Describe what human factors is.
- Describe what systems are and how they relate to human factors.
- Explain the history of human factors.
- Communicate the importance of systems theory.
- Identify human factors applications.
- Describe the education necessary to work in the field of human factors.
- Identify the different work settings in which human factors specialists work.

1.2 WHAT IS HUMAN FACTORS?

1.2.1 The application of psychological science

Human factors, as a field, is highly inter- and multidisciplinary. For someone studying psychology, **human factors** is an area of applied psychology in which psychological science, theory, and knowledge are used to design the environment to fit the limits and capabilities of human users. The "environment" is defined broadly and includes physical and social aspects (Barker, 1968; McGrath, 1984). The physical aspects include anything you can sense such as sound, visual information, motion, and the amount of space you have. The social aspects include the conditions of the situation that give us the sense that we should act a particular way. For example, we generally have a sense that particular behaviors are required in a classroom such as being on time, taking notes, and participating in group discussions. On a test day, there is also the additional time demand or pressure to perform well, which can impact how we cognitively process information. Similarly, an environment might be too dark, noisy, and crowded for reading, but ideal for socializing with friends. Therefore, how we perceive the physical and social aspects of the environment influences us. Given this definition, the environment could include an airplane cockpit, a car, your apartment or dorm room, a kitchen, a public library, a work environment or office space, or an outdoor space.

The environment also includes an assortment of items that the individual uses. To ensure that humans can effectively and safely fly planes, drive cars, and use equipment, computers, and cell phones, we must consider the limits of human functioning as well as the capabilities of the user. The limits and capabilities of humans involve the psychological, and often the physiological and physical, aspects of the user. For example, human factors considers the user's ability to perceive visual information (see Chapter 3) such as traffic signs or auditory information (see Chapter 4) such as sirens. Sometimes, weather or lighting prevents us from accurately perceiving these signs. Similarly, we do not always perceive or hear sirens depending on the noise level around us, where our attention is focused, or whether we are wearing headphones. The design of the environment, then, must include the application of theories of attention, memory, multi-tasking (see Chapter 6), and perception to ensure the users receive the information conveyed in the signs or sirens. That is, the driver must know when one must stop or if there is an emergency vehicle passing. In the case of the California train crash, an auditory alarm could have drawn the engineer's attention from texting and alerted him to the situation at hand.

Other limitations we might consider include the individual's ability to decipher the meaning of displays regardless of whether it is visual, auditory, or in some other mode. How individuals process information (see Chapter 6) and make decisions (see Chapter 7) influences their interpretation of the environment and their resultant behaviors or actions. For example, task interruptions or attempts to perform multiple tasks at the same time (i.e., driving and talking on a phone) can degrade performance on the individual tasks and affect the person's awareness of signs or hazards and memory of task-related information. Might this possibly explain how the trader entered "b" instead of "m" when making the sale? Human factors applies our understanding of cognition, memory, and information processing to the design of environments to reduce these potential information processing errors and enhance performance.

The acquisition and processing of information represents the basis of further action by the operator. These actions typically involve motor responses where the operator must activate a control like a steering wheel, mouse, or brake. Performance decrements occur when the motor demands exceed the capability of the user. Such a mismatch between demands and operator ability might arise when human abilities are not considered in the design. Design demands might exceed the operator's physical capabilities when the demands require a maximum force greater than the operator can generate, a higher level of precision than the operator can perform (see Chapter 8), or an inappropriate distance or movement to reach and activate a control (see Chapter 9).

Human factors is also an applied field because research is conducted not only to understand why some accidents occur but also to understand and discover ways to design the environment or system in order to reduce these accidents and improve performance. If individuals are turning on the wrong stove burner or not noticing that the car in front of them has stopped or slowed, accidents are more likely to occur. Research (see Chapter 2) and methods of evaluation (see Chapter 5) can help identify why people turn on the wrong burner or ways to improve the detection of braking vehicles. This knowledge can then be used to design more intuitive controls or vehicle signal lights. For example, the placement of a third brake light in a centered, high-mounted location was found to increase the likelihood that the driver of the following car would brake when the brake lights of the car in front were activated (Sivak et al., 1981). These findings led to a federal safety standard that required the placement of a third brake light on all vehicles, which is centered at the top or bottom of a vehicle's rear window. Figure 1.1a demonstrates the third brake light, which you can compare to no brake lights in Figure 1.1b.

1.2.2 Human factors has an engineering emphasis

Human factors also has an engineering focus that emphasizes the design of tools, equipment, or processes that function as users expect them to function; achieve the desired outcomes; and interact with related tools, equipment, or processes (Parsons, 1999). Let's consider the situation when you use your computer to write a paper. You expect that each key stroke will

(a)   (b)

Figure 1.1 A demonstration of the (a) centered, rear-mounted third brake light compared to (b) no brake lights. (© Nancy J. Stone.)

produce on the screen the letter, number, or character represented on the keys when you depress them; the backspace key will delete text as it moves backward; and the up arrow key will move the cursor up the page. As you write your paper, you might import graphs or pictures. The word processing program should allow you to download graphs or pictures from the Internet or other software. Further, users have expectations that they can format the paper and add embedded graphs and pictures without requiring specialized training. When the computer does not work as expected or is difficult to understand, the user is likely to make mistakes and get frustrated. We can determine if the user's experience (UX) will be good or bad through usability testing, an evaluation method (see Chapter 5) that can assess whether a computer system works as people expect it to work.

Human factors' relationship to engineering is so strong that human factors used to be referred to (and still is by some) as human factors engineering or engineering psychology. Other names often heard in relation to human factors are cognitive engineering and ergonomics. **Ergonomics** is also an applied field that strives to reduce fatigue and discomfort by the effective design of the workplace environment and equipment. Human factors is the term used predominantly in the United States, while ergonomics is more often used in European countries (Grether, 1968; Howell, 1985). To be inclusive, the Human Factors Society changed its name in 1993 to the Human Factors and Ergonomics Society. Although some individuals might refer to themselves as human factors engineers, the term "engineer" is generally reserved for those trained as engineers, not those trained as psychologists. Therefore, individuals not trained as engineers, psychologists in particular, often refer to themselves as human factors specialists.

1.2.3 Other areas related to human factors

Besides the relationship with engineering, human factors is also related to and draws upon theories and concepts from computer science, architecture, biological sciences (Howell, 2003), and medicine (see Figure 1.2). The reason for this overlap is the common interest each field has with reducing error by improving the design of equipment and user-equipment interfaces employed within the respective fields. This is one reason why human factors education often resides in a variety of departments, but generally in either psychology or engineering departments (Howell, 2003).

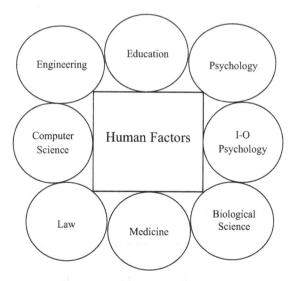

Figure 1.2 The eclectic nature of human factors. (© Nancy J. Stone.)

Human factors also shares a great deal of history with industrial-organizational (I-O) psychology, as both are applied fields; however, there are some key differences as well. Although both fields use the systems perspective (discussed below), the level of focus differs (Howell, 2003). Human factors generally focuses on the human–machine system. Human factors specialists seek to optimize the interaction between humans and machines whereby human users can interact with machines with minimal error and greater efficiency. In contrast, I-O psychology tends to focus on the larger organizational system. I-O psychologists attempt to optimize a worker's performance on a variety of tasks, while working alone or in groups as well as within various organizational structures.

Human factors and I-O psychologists also differ in their approach to increasing performance. I-O psychologists perform a job analysis in order to understand what knowledge, skills, and abilities are needed for a job. In turn, this information can be used to develop selection, training, and performance management systems. With these systems, the I-O psychologist can identify who is the best fit for the job, determine what skills need to be trained, and appropriately evaluate how well the individual is performing. Simply put, I-O psychologists attempt to find the best person for the job and then develop that person into a better performer with training or performance feedback.

Although some human factors specialists engage in selection and training, human factors specialists often begin with a task analysis. A **task analysis** identifies the required movements, actions, sequence of actions, and thought processes required to complete a task. These data are used to determine how best to design the job to fit the limitations and capabilities of the person. For example, a human factors specialist might redesign a job that makes it easier for people to understand a display (e.g., a stop sign, X-ray, or altimeter), to process job tasks more quickly, or to perform the necessary physical activities. Likewise, the job could be redesigned for someone with visual or auditory impairments or someone in a wheelchair.

Given the differences between I-O psychologists and human factors specialists, (see Box 1.1) how would professionals from these two fields approach the problem of inefficient employees at a meat packing plant? Imagine that the company has noticed a decline in the plant's output and products are not distributed to stores on time. An I-O psychologist is likely to look at issues of training—do the employees have the appropriate knowledge and skills to perform the job—or selection—have we hired people who are best suited for this job. In contrast, a human factors specialist is likely to evaluate the tools—do the employees have the proper cutting tools and equipment, are the packing materials easy to use, or the environment—do the employees have to work in unnatural or fatiguing positions such as working with raised arms.

BOX 1.1:   I-O VERSUS HF ASSESSMENTS OF THE TRAIN WRECK

In the opening vignette, there were a number of possible causes of the train wreck: the engineer (1) was texting, (2) ran a red light, and (3) apparently, had no way of knowing that he had passed the stop signal. An I-O psychologist might suggest a new selection program to identify individuals who are more conscientious and vigilant. Conscientious individuals might be less likely to text and perhaps more vigilant of their surroundings. Next, the I-O psychologist might recommend revising the training program. The new training could include a review of proper protocol (e.g., no texting while traveling in certain areas or at certain speeds), a study of the rules (e.g., when does one train stop while another train passes), as well as practice controlling the train under various simulated conditions (e.g., bad weather or high speeds).

Although helpful, selection and training programs do not address all problems. Once an individual is well trained, certain processes become faster and automatic (see Chapter 6).

This might give the train engineer the perception of having more "down time." It is also possible that the stop signal was faulty or that the sun or angle of the light made it difficult to see. Finally, there was no warning signal alerting the engineer to the problem. It would be extremely difficult to select or train someone to notice this particular error.

The human factors specialist might recommend installing a warning signal to indicate that a stop signal is near. Also, if the view of the signal is obstructed, it could be possible to create a backup signal sent electronically from the stop signal to the train. If the warning and signal are missed, it could be possible to have an alarm indicating that the train passed the red stop signal. If all of these changes were implemented, it would be necessary to revisit the selection and training programs for this "new" job.

As you can see, the work of I-O psychologists and HF specialists overlaps as well as complements each other. **Critical Thinking Questions:** Consider the market crash and the redesign of city street signs from the opening vignette. What are the issues an I-O psychologist and an HF specialist are likely to identify in these situations? How would these issues be resolved by an I-O psychologist and HF specialist?

1.2.4 Applications to special populations

Given that human factors is an applied science that can reduce error and increase the ease of use of equipment and environments, it can address the needs of special populations. Many of the early applications of human factors principles and methods focused on military personnel, especially the design of aircraft cockpits, while studying a population that was almost exclusively male. It became apparent that it was not possible to generalize data from adult males to children and women. Over the years, the application of human factors has broadened considerably to include a wide range of consumer products. In turn, human factors specialists have an increased recognition of the design challenges associated with accommodating men, women, and children, as well as users who might have physical (e.g., individuals in wheelchairs) and cognitive limitations.

Consider what might be the design challenges if our train engineer in the opening story was "color blind" and could not discriminate between a red and green light. How might a new design address this issue? What if the train engineer or trader had physical limitations that might affect how the job tasks were performed? Could new designs of the job remove or reduce these effects? If so, what impact might these job changes have on the selection and training of train engineers or computer users? Because human factors can address the many needs of special populations, we will address these applications in each chapter.

1.3 HISTORY OF HUMAN FACTORS

Now that you have an idea of what human factors is, let's consider how the profession of human factors evolved. The history of human factors may be traced back to the beginning of humankind, when our ancestors fashioned tools out of animal bones (Christensen, 1987). Yet, considerable growth of human factors began at the start of the industrial revolution (see Figure 1.3) and evolved according to the phases of the industrial revolution: the age of machines (1750–1870), the age of power (1870–1945), and the age of machines for minds (since 1945) (Christensen, 1987). Because the history of human factors was greatly affected by WWII, we will consider the history of human factors before WWII, which includes the age of machines and the age of power, and after WWII, the age of machines for minds.

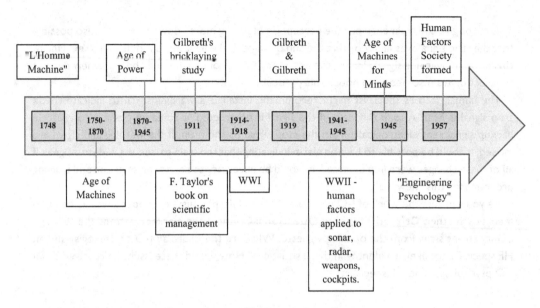

Figure 1.3 Historical timeline of human factors. (© Nancy J. Stone.)

1.3.1 Before WWII

The publication of La Mettrie's 1748 book, L'homme Machine, in which he compared humans to machines, marked the beginning of the Age of Machines (1750–1870) (Christensen, 1987). There was a realization that it was possible to study and learn about human behavior, performance, and proficiency by studying machines under similar conditions. The Age of Power (1879–1945) marked the advent of powered machinery in transportation, industry, and agriculture, as well as the classic time and motion studies conducted by Frederick W. Taylor and Frank B. Gilbreth (see Figures 1.4 and 1.5, respectively).

In the **time and motion studies**, Taylor and Gilbreth used the scientific method (see Chapter 2) to document physical movements made by the worker in the execution of a task. This information was used to identify ways the tasks could be reorganized, which would eliminate unnecessary movements, improve efficiency, and reduce fatigue. Taylor found that planned breaks could optimize performance for a particular job. Taylor also evaluated the design of shovels. He identified the size and shape of shovels that were best for shoveling different types of material. Taylor referred to his work as **scientific management**, the application of the scientific method to understand reasons for or causes of a worker's high or low level of productivity, a concern for management, and wrote about it in his book *Principles of Scientific Management* (1911).

An important result of Taylor's work was the recognition that because of individual differences there is no single best solution or design (Christensen, 1987). Even though there might be a better way or a best way to design things for the "ideal" user, human factors specialists must consider all potential users. Many designs (e.g., consumer products) should consider a wide range of users that involves special populations including children, women, older adults, or individuals with disabilities. Specific applications (e.g., airplane cockpit) will require designs for a narrower range of users. The final design should be the best for the particular population of users.

Taylor was not the only one interested in time and motion studies. Gilbreth, a mechanical engineer who was Taylor's contemporary, also conducted time and motion studies,

Figure 1.4 Frederick W. Taylor. (From Public domain, https://commons.wikimedia.org/w/index.php?curid=4548998.)

Figure 1.5 Frank B. Gilbreth. (From Public domain, via Wikimedia Commons https://commons.wikimedia.
org/wiki/File:Frank_Bunker_Gilbreth_Sr_1868-1924.jpg.)

specifically of bricklaying. Gilbreth's goal was to eliminate motions that were unnecessary, caused fatigue, and wasted human energy (Gilbreth & Gilbreth, 1919). He invented the scaffold, which eliminated the unnecessary motion of stooping by placing brick and mortar conveniently next to the wall where the bricklayers were working! Even though an old profession, Gilbreth was able to eliminate more than 67% of the bricklayers' motions after his analysis. Gilbreth's recommendations, derived from his motion studies, more than tripled the number of bricks set by a bricklayer (Gilbreth, 1911).

Besides his assessment of bricklayers, Gilbreth and his wife, Lillian, an industrial psychologist, argued for better standardization in the medical field. Gilbreth argued that a hospital was less efficient than most factories. In particular, the Gilbreths studied the motions during surgery (Gilbreth, 1916) and discovered that surgeons would look away from the patient to look for and then pick up the needed instrument from a nearby tray, losing valuable time. After their evaluation of this process, Gilbreth recommended standardizing the layout of surgical tools as well as the tools themselves, reducing the amount of time to find and retrieve a tool (Gilbreth, 1916). Having the surgeon request a tool, the surgeon can monitor the open wound while the nurse finds and places the tool in the surgeon's hand.

1.3.2 During and after WWII

Even though the influential work of Taylor and Gilbreth began in the early 1900s, the field of human factors became firmly established as a result of WWII (see Figure 1.3). The impetus for this growth was the large number of errors associated with the use of military equipment, specifically the high number of aircraft accidents (Grether, 1968). Individuals trained in experimental psychology were recruited to apply their knowledge of sensation, perception, and learning to the design of equipment such as sonar, radar, weapons, and cockpits to reduce errors (Howell, 1985). An alternative and often quicker solution was to have industrial psychologists select and train individual users better, but this did not eliminate errors caused by faulty designs (Grether, 1968).

The end of WWII marked the beginning of Christensen's (1962, 1987) third phase—the age of machines for minds (1945 to present day). This was also when experimental psychology applied to military environments evolved into the field of engineering psychology (Grether, 1968; Howell, 1985). Because experimental psychologists were seeking to determine the laws of behavior, they focused on solving applied problems while also attempting to develop the underlying theory of why the problem arose (Wickens & Hollands, 2000).

To make environments such as cockpits and other military situations safer, experimental psychologists conducted tests of human performance; sensation and perception; and memory, learning, and, later, cognition within various situations to determine what variables were related to poor performance. Because of the strong research orientation in engineering psychology, several research laboratories were formed to study engineering psychology issues (Grether, 1968). The general focus was on aviation, such as the design of cockpits or air traffic control systems. Other areas of research included the design of radar and sonar (Roscoe, 1997). The knowledge gained from these studies enhanced our understanding of the fundamentals of human performance, which could be used for other designs (Howell, 1985) such as tools, equipment, and systems.

By the late 1940s, human factors became a field of its own. The field became known as **human factors engineering,** a broader term that subsumed engineering psychology (Howell, 1985). One event that marked the beginning of the human factors engineering field was the founding of the Psychology Branch of the Aero Medical Laboratory headed by Paul M. Fitts (Grether, 1968; Summers, 1996). Other events occurred in 1949 that signaled the establishment of the human factors field. In particular, the Ergonomics Society, originally called the Ergonomics Research Society, was formed in Britain (Howell, 1985) and Alphonse Chapanis, Wendell Garner, and Clifford Morgan published their text, *Applied Experimental Psychology: Human Factors in Engineering Design* (Grether, 1968), the first book on human factors (Sanders & McCormick, 1993).

Further growth of the human factors field during the 1950s led to a monumental year in 1957. During 1957, the journal *Ergonomics* was first published; the Human Factors Society was formed in the United States; and the Society of Engineering Psychology was founded, which became Division 21-—Applied Experimental and Engineering Psychology—in the American Psychological Association (APA) (Howell, 1985).

Human factors' growth can also be attributed to significant world events. For example, in 1957, the launch of Sputnik by the USSR led to increased funding for the investigation of human space travel (Grether, 1968). In 1979, the accident at the nuclear power plant on Three-Mile Island (a partial meltdown at a nuclear power plant partly attributed to malfunctioning equipment and operator error in reading the equipment's output) focused attention on operator and supervisory control of complex systems. Other important events included the accident at the Chernobyl nuclear power plant in 1986 (a nuclear reactor in the former Soviet Union released radioactive gases into the atmosphere), the Challenger accident (1986) (the space shuttle exploded shortly after take-off due to poorly designed O-rings), and the Columbia accident (2003) (a second space shuttle accident that occurred upon entry into the earth's atmosphere due to a damaged exterior tile). The events of September 11, 2001 led to more human factors work to improve the effectiveness of airport screenings. Additional events, such as the Christmas day terror attack (2009) (a passenger allegedly attempted to detonate explosives that got through security on a flight from Amsterdam to Detroit, Michigan), suggest the need for further human factors research and application.

The growth of human factors can be linked to several individuals involved in the formation of various professional groups related to human factors (Parsons, 1999). Two individuals stand out as leading figures in the field of human factors—Alphonse Chapanis (see Box 1.2) and Paul M. Fitts (see Box 1.3). Chapanis was a prolific researcher and writer, contributing to the knowledge base of human factors. Fitts conducted seminal work in motor control and was one of the organizers of the Society of Engineering Psychologists and Division 21 of the American Psychological Association (Parsons, 1999).

BOX 1.2:   ALPHONSE CHAPANIS: 1917–2002

Alphonse Chapanis (see Figure 1.6) is a name synonymous with human factors and ergonomics and is considered one of the founders of the profession. Chapanis completed his graduate degree at Yale in 1942. A year later, he was commissioned into the Army Air Corps and worked as an aviation research psychologist (Krueger, 2002). During his time in the Army, he was involved in efforts to reduce pilot error. Some of these errors were attributed to the fact that many of the levers or controls were identical in appearance and were positioned near each other in the cramped cockpits. Pilots occasionally found themselves inadvertently retracting the plane's wheels when they meant to change the setting of the aircrafts' wing flaps after landing.

Figure 1.6 Alphonse Chapanis. (From S&T Archives, Ernie Gutierrez Photographs, R:1/172/15.)

Chapanis proposed modifying the control handles so they could be easily identified and each handle could be readily differentiated from the others when gripped. The control for the plane's landing gear was replaced with a rubber knob (see Figure 1.7), while the control for the wing flaps was replaced with a wing-shaped piece. These modifications, now referred to as *shape coding* (see Chapter 8), greatly reduced the confusion between the two controls and inadvertent activation of the wrong control (Roscoe, 1997), reduced pilot error, and are still used today in all types of aircraft.

Figure 1.7 Chapanis' use of a wheel knob on an aircraft's landing gear control. (Adapted from Varnav, CC0, via Wikimedia Commons. https://commons.wikimedia.org/wiki/File:A10A_cockpit.jpg.)

Chapanis also made important contributions to a number of areas including the design of displays and the allocation of functions between people and machines. In 1946, Chapanis joined the Psychology Department at Johns Hopkins University where he continued his research and trained about 30 graduate students (Krueger, 2002). Chapanis had over 100 publications, including a posthumous publication in 2004. He also served as president of the American Psychological Association's Division 21 (Applied Experimental and Engineering Psychology) from 1959 to 1960, president of the Human Factors Society from 1963 to 1964, and president of the International Ergonomics Association from 1976 to 1979 (Krueger, 2002). **Critical Thinking Questions**: Can you explain why adding a rubber wheel and a wing-shaped piece to the controls helped reduce pilot errors? If you are not familiar with a cockpit, consider a car's dashboard or controls. How might these displays or "dials" be improved to reduce error?

BOX 1.3:   PAUL M. FITTS: 1912–1965

From the end of WWII in 1945 until 1949, Lt. Col. Fitts (see Figure 1.8) oversaw the psychology branch of the Army Air Forces Aero Medical Laboratory. In 1949, Fitts left military service for an academic position at Ohio State University in Columbus (Roscoe, 1997). He attracted many students, often veterans, to his research program in aviation psychology. In 1958, he moved to the University of Michigan. During his academic career, Fitts was president of the American Psychological Association's Division 21 (Applied Experimental and Engineering Psychology) from 1957 to 1958 and president of the Human Factors Society from 1962 to 1963. He also supervised numerous students who entered the field of aviation psychology. In fact, Fitts at Ohio State and Chapanis at Johns Hopkins were two of the three researchers (Alexander Williams was at the University of Illinois) who had some of the most prominent military and academic aviation psychology laboratories in the late 1940s and 1950s, graduating the majority of the aviation psychologists at the time. Many of Fitts' students found work in the aviation industry, which was a major employer in the 1950s.

Figure 1.8 Paul M. Fitts. (From Paul M. Fitts, portrait, Professor of Psychology, HS15405, University of Michigan, News and Information Services, Bentley Historical Library, University of Michigan.)

Unfortunately, Fitts died from a heart attack in 1965 at the age of 53 (Roscoe, 1997). This was a great loss to the profession because in fewer than 20 years, Fitts published more than 70 articles. His research focused on motor control, information processing, eye movements by pilots during instrument landings, and aviation equipment design issues. His greatest impact was in his study of human motor control that led to the development of Fitts' law, a mathematical model that describes how the time to reach a target, or movement time, is affected by the target's distance and size. Fitts' law is widely used to determine how large a target should be to ensure an operator can reach the control or pedal quickly and accurately (see Chapter 8). **Critical Thinking Questions**: If you were cooking dinner and something began to burn, you would want to turn off the burner quickly. Would you prefer to reach for a small or a large stove top control if you had to reach for it across a distance of three feet? If you only had to reach a few inches to operate the control and turn off the stove, how comfortable would you be with a smaller control than the one that is three feet from you? Do you think the control should be larger or smaller if you are closer to it? How might this impact the design of where stove controls are placed?

There were women who had significant impacts on the field of human factors, and Frank Durso (2013, 2014a, 2014b, 2014c) called a group of them the "First Ladies of HFES." Three of these women, Ruth Hoyt Cameron, Dora Jean Dougherty McKeon (Strother), and Gloria Indus Lauer, were impactful in the 1940s and 1950s. The third woman, Nancy J. Cooke, began impacting the field in the 1980s and continues to do so.

1.3.3 Current directions in human factors

The field of human factors has changed significantly in both breadth and scope over the last five to six decades. This is reflected not only in the society's name change, but also in the broad range of applications reflected in the work of human factors experts. While military and aviation applications remain a cornerstone of human factors research and practice, a diverse set of areas now receive the attention of human factors experts including computer systems; consumer products; software and website interface design; space flight; nuclear power plants and issues related to automation; air traffic control and safety; and command, control, and communication. The field also has been responsive to recent events with an increased emphasis on security matters including identifying means of improving the detection of weapons and hazardous materials in passenger luggage, airport security (see Figure 1.9), fighting terrorism, cybersecurity, and improving homeland security. Human factors specialists continue to find employment with the Department of Defense, National Aeronautics and Space Administration (NASA), Federal Aviation Administration (FAA), aviation manufacturers, and even the Food and Drug Administration (FDA), with the range of employment opportunities now considerably broader than ever before in the history of the science.

There is also a significantly increased emphasis on the design of medical systems such as the equipment (e.g., surgical robotics and electronic medical records), management, and environmental design issues (e.g., the layout of rooms or spaces used by older adults or individuals with disabilities). Human factors specialists also might work for companies to design safe toys for children. Others work in the field of forensic human factors, helping individuals resolve litigation related to poor design which has led to injuries or issues of product liability.

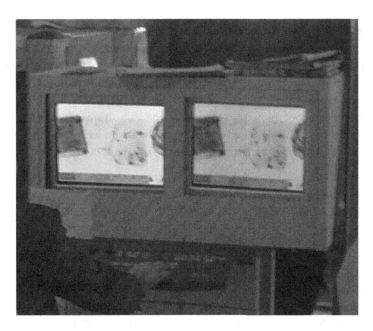

Figure 1.9 Airport security screening. (Adapted from User: Mattes, Public domain, via Wikimedia Commons. https://commons.wikimedia.org/wiki/File%3AVTBS-luggage_screening.JPG.)

Human factors is also being applied to educational systems. For example, human factors specialists are investigating better ways to present web-based instruction and how to enhance the design of the learning environment. And, of course, human factors is applied to our technological world—designing more user-friendly and efficient internet applications, websites, computers, and cell phones. Chapter 12 covers future trends in human factors in greater detail.

As the human factors profession has evolved, human factors specialists have played an increasingly larger role in the design of tools, devices, and environments. One major focus of human factors research and application concerns the person–machine interaction, also known as **human–computer interaction** (HCI). HCI focuses on how users interact with computers, how the design of the interface—both hardware (i.e., mouse, trackball, and keyboard) and software (i.e., the computer program)—aids or hinders the user's ability to accomplish the tasks. The application or lack of HCI principles is readily apparent in our interactions with computers, cell phones, and home-use medical devices. Box 1.4 describes the problems identified in blood pressure monitors that are used at home.

BOX 1.4:   HOW EASY ARE HOME-USE BLOOD PRESSURE MONITORS TO USE?

Using two evaluation techniques, hierarchical task analysis and heuristic evaluation (see Chapter 5), Abdusselam Selami Cifter (2017) assessed the usability of wrist cuff blood pressure monitors (BPMs) sold in Turkey, especially for people with various limitations (e.g., limited mental or physical capabilities). The three BPMs had the greatest number of violations (although the violations were minor) for the universal design principle of "perceptible information." That is, the PBMs needed improvements in how information was presented given potential user sensory limitations (e.g., limited vison or hearing), in creating legible essential information, or in giving appropriate feedback.

In addition, the three brands of BPMs operated differently in how one needed to position one's arm and body for an accurate measurement. Finally, although there were few major violations, the major violations occurred for "tolerance for errors," which means the devices did not prevent errors well or help someone to recover from an erroneous action. **Critical Thinking Questions**: What design changes could help make BPMs perceptible to someone with vision limitations, hearing limitations, or tactile (think about the buttons that need to be pushed) limitations? What type of visual or auditory information could help a user, especially someone with sensory or cognitive limitations better understand if an action was an error (or might be an error)?

HCI and other design processes emphasize the user experience (abbreviated as UX) and how designs might better adhere to the expectations and experience of a wider range of users. By applying a **user-centered design** (see Chapter 5) philosophy, each stage of the design process emphasizes how the proposed design meets the users' needs, expectations, and abilities by including the user in the design process. This is an **iterative process**, a repetitive cycle involving design, evaluation, and redesign before arriving at a final design. For an example of an iterative process, see Case Study 1. Unfortunately, human factors specialists often are consulted late in the iterative design process after important design decisions have been made without properly considering the user. At this late time, the human factors specialist might only have an option of tweaking the design and/or position of knobs and dials. It is for this reason that the human factors field has often been derisively called by some the "knobs and dials" profession (Christensen, 1987; Grether, 1968). The involvement of human factors specialists earlier in the product development cycle increases the likelihood that the design will meet the needs and expectations of the intended user population.

As the influence and acceptance of human factors have grown, human factors specialists have had a greater impact on individuals' lives. Currently, the fastest growing subspecialty within human factors concerns healthcare issues. Although human factors has been applied to aviation with considerable success, it was not until 1996 that representatives from the healthcare industry were brought together to talk with human factors specialists about patient safety (Leape, 2004). In 2000, the National Academy of Sciences Institute for Medicine (IOM) published a report on human error in the health profession (Kohn et al., 2000). This report disclosed that human error resulted in many negative outcomes, such as the amputation of the wrong limbs and overdoses. In fact, their estimates suggested that there were nearly 7,000 medication errors annually and nearly 98,000 deaths per year due to medical errors (Kohn et al., 2000).

Medication errors remain a concern as earlier estimates may have underestimated the frequency of medical errors by tenfold (Classen et al., 2011). The cost of an estimated 200,000 deaths and 10 million missed workdays associated with medical errors is estimated to total nearly 1 trillion dollars a year (Andel et al., 2012). Human factors specialists are working to address problems within health systems, including procedures and processes, the design of appropriate drug labels so the appropriate medications are administered, and the use of technology in healthcare environments.

1.4 THE IMPORTANCE OF SYSTEMS THEORY

As mentioned above, human factors is a science that studies systems. According to General Systems Theory (von Bertalanffy, 1950), or just **systems theory**, a system involves inputs, processes, and outputs that are highly interdependent. Sometimes systems are referred to by the acronym IPO. Inputs feed into a process, which produces outputs (Inputs→Process→Outputs). In turn, the outputs become inputs to the same or different process. Let's consider the 2010 stock market problem reported in the opening vignette. At some point, based on stock information (input), the trader decided to make the sale (process). To make the sale, the trader entered the keystrokes (output) to specify which processes the computer will do next. As he typed the information into the computer, the screen displayed this information, which provided additional input to the trader. (The discussion of the computer's input, process, and output is being intentionally omitted. You will understand why in the next paragraph). Once the trader submitted the sale, this action changed the information displayed on other trader's computer screens that was then processed, which led to additional behaviors—more stock trades.

Because human factors is concerned with the human user, the inputs, processes, and outputs tend to be viewed from the human user's perspective. Therefore, the inputs are the sensory inputs to the human user. The process stages occur when the user interprets the inputs, which represent **human information processing**—the cognitive processes such as perception, memory, learning, decision-making, and problem-solving. Once a decision is reached, the user then reacts to the input by making a response often in the form of a motor act (i.e., moving a steering wheel, pressing the brake pedal, or pressing a key). That, in turn, is an output. Applying this system to a cell phone, the ring is an auditory *input* to the user; the *process* involves the user's interpretation or identification of the ring and whether to answer the phone. If the decision is to answer the phone, then the user must control the phone by pushing some button (the *output*) to answer the phone.

Systems are classified as closed-loop or open-loop systems. A **closed-loop system** includes continuous feedback and control. For example, as you are driving, you must sense (mostly visually, but possibly auditorily or tactilely) whether you are staying on the road. As you look through the windshield, you see the road's path and your car's location relative to the road's path. If it appears as though you are going to drive off the road, you make corrections by steering the car to stay on the road. Thus, the *input* is the view of the road and the perception of where your car is on that road; the *process* may be the decision that the road is curving and that your current direction is not following that curve. That means that you need to *control* the car by turning the steering wheel to stay on the road. The *feedback* is the new visual input of the car's location relative to the

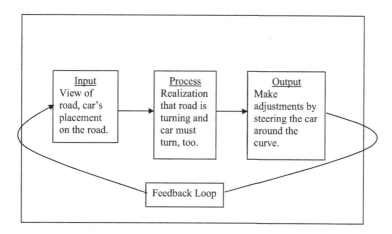

Figure 1.10 Example of a closed-loop system. (© Nancy J. Stone.)

road. Therefore, the feedback loop contributes other input to you. See Figure 1.10 as an example of a closed-loop system.

In contrast, an **open-loop system** does not have the immediate feedback loop allowing for corrections once an action is initiated. For example, once you throw a ball, you can watch the trajectory, but you cannot alter that particular ball's path. In this open-loop system, you can make adjustments for the next throw, but you cannot continue to make adjustments once the ball is thrown. The iterative design process can be considered an open-loop system. A design is created and then tested. Based on the test results, the design is modified and retested. This cycle would continue until the design achieved the desired outcome. In a closed-loop system, though, you can make continuous and immediate corrections such as when you are driving a car.

There are two other important components of systems. First, systems have a goal or set of goals and, second, the components are interrelated. Let's first consider the system of just you and the car when driving. In this case, your goal may be to get from school to home in an efficient and safe manner. If you increase your driving speed because you want to get home more quickly, this will impact your visual input as street signs pass by you more quickly, affecting your decision-making abilities because you will not have as much time to make decisions. Therefore, in a closed-loop system, your output (increased speed) will impact your inputs, such as your visual perception of the driving environment.

When considering the total driving environment, your output will also become the input for other drivers. Other drivers and pedestrians are likely to notice your changed speed. Given this input, other drivers and pedestrians might respond by speeding up, slowing down, changing lanes, or getting out of the street. So, when considering systems, remember that all parts are interrelated, and a change in one part is likely to have an impact on another part, even if the impact is small.

A common analogy for understanding the interrelatedness in systems is the "ripple effect" caused by throwing a stone into a pond. The ripples are largest near the point where the stone entered the water and become smaller as the ripples move out and away from this location; however, the ripples continue for quite a ways. The effects of a car accident can ripple through a road system affecting other travelers miles away. This relates to zero-, first-, and second-order systems, as discussed in Chapter 8. Zero order systems have a direct relationship between an input and output (e.g., braking in a car). First- and second-order systems reflect reductions in direct control due to more linkages and, hence, delays between an input and an output (e.g., stopping a freight liner).

Finally, it is important to realize that systems contain other systems or subsystems making the system complicated or complex. Although the terms are often used interchangeably, a complicated system has many subsystems, but we have an understanding of how these subsystems work together as a whole. Complex systems, on the other hand, are not fully understood or some of the subsystems are unknowable. In a highway system, there are the subsystems of each driver-and-car system. Each of these systems makes up a larger system of interacting cars. The important point to remember is that there can be inputs, processes, and outputs at various levels (see Box 1.5).

BOX 1.5: IDENTIFYING ENVIRONMENTS AS SYSTEMS

Any environment with which you interact can be viewed as a system. For example, a kitchen is a system. In a kitchen, the goals are usually to store or prepare foods in a safe manner. There are various inputs such as food, people, and an assortment of equipment such as the stovetop, oven, microwave, and blender. In addition, each piece of equipment conveys information (input) to the human user about whether the equipment is working, such as on/off lights, dials, and the sound some equipment makes such as the blender. The process includes the decision-making and information processing of the actual preparation of various foods. The outcome, hopefully, is the successful preparation of something edible. The inputs, processes, and outputs are interrelated. When cooking, if the proper on/off signals are not available, then the cook might believe the oven is off when it is on and the food continues cooking or someone gets burned. Also, if individuals do not understand the purpose of different tools and their proper use, there could be severe consequences. Furthermore, depending on the cooks' outputs, foods might not be handled properly or in a timely manner, which could lead to spoiled or tainted foods that could cause illness. **Critical Thinking Questions**: Can you identify another goal of a kitchen? What would be the inputs, processes, and outputs of the kitchen system for this other goal? Identify another system such as your school's library, your work environment, or a computer lab. What are the goals of this system? To meet these goals, what are the necessary inputs, processes, and outputs?

1.5 CASE STUDIES: EXAMPLES OF HUMAN FACTORS APPLICATIONS

Throughout our discussion of the definition of human factors, the history of human factors, and systems, there has been a constant message: human factors applies psychological theory and knowledge in order to reduce error. A more in-depth discussion of error and human reliability is presented in Chapter 11, but the following are examples of the applications of human factors to reduce error, increase safety, or both.

1.5.1 Case study 1: portable defibrillators

Over 1,000 individuals in the United States experience a sudden cardiac arrest every day, and they have a greater chance of survival if they receive an electrical shock within 5–7 minutes of experiencing the arrest (Suri, 2000). If an electric shock is given but the conditions are not right, this could have detrimental effects. Further, as you can imagine, trying to save someone's life could be a bit nerve-racking, especially for the untrained person (see Figure 1.11). Therefore, when designing a portable defibrillator (a device that administers an electrical shock to one's heart to return the heart to a normal beat pattern), the researchers' goals were to create a device that a wide variety of users in a wide variety of environments could use.

Because few laypersons are familiar with defibrillators, the researchers at Heartstream wanted to create a portable defibrillator that individuals could be confident in using and that could be used indoors or out, by young or old people, and by individuals untrained in how to use a professional defibrillator (Suri, 2000). To begin, the design team observed and talked with the experts who used defibrillators in their professional life. The particular devices were studied, as well as the process used in storing or handling the equipment. Based on these preliminary data, the design team was able to develop an initial, but incomplete, prototype.

Next, task and user analyses were performed on the proper use of the equipment. This step allowed the team to identify potential users and how they interacted with the equipment before redesigning the prototype. The new design included automatic functions to eliminate problems

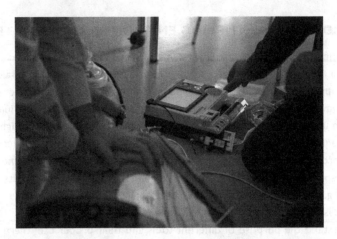

Figure 1.11 An example of a defibrillator in use. (From Rama, CC BY-SA 2.0 FR <https://creativecom-mons.org/licenses/by-sa/2.0/fr/deed.en>, via Wikimedia Commons. https://commons.wikimedia.org/wiki/File:CPR_training-02.jpg.)

users encountered when using the equipment. For example, users often became confused as to when to perform certain functions. Therefore, the defibrillator was designed to determine various vital signs of the patient and if a shock would be appropriate, as well as to provide audio and visual prompts to guide users through the process to ensure appropriate use (Suri, 2000).

With another redesign of the prototype, along with brainstorming sessions, a number of the design issues were resolved. Five different models were developed based on these data and tested with individuals who had basic CPR training. This initial test of the five models identified that users had problems in determining where to begin with the equipment. Based on these tests, the design team created a compact defibrillator that is packaged in red (a color code for emergency), that begins with audio prompts, and has audio and visual feedback (i.e., cognitive aids for decision-making) that ensures that the user is proceeding with the task correctly. The final product was approved by the U.S. Food and Drug Administration in 1996 for general use (Suri, 2000). Today, defibrillators can be found in numerous places where many people gather (e.g., shopping malls, gyms, airports, and schools), and they are designed for anyone to use.

Based on this example, can you identify other processes or tools used in the medical profession or by people in general, such as medical devices for tracking blood sugar levels, computer systems, or ATMs, that can cause confusion for users? What do you consider to be some of the major problems? Are the problems related to inputs, processes, or outputs? Can you identify where the problems are and make suggested changes? Finally, explain whether the design process for the defibrillators was iterative and whether the process was user-centered.

1.5.2 Case study 2: communicating instructions to individuals with low levels of literacy

Individuals with low levels of literacy are likely to have difficulty reading and comprehending instructions. Using diagrams or pictures that convey the message (i.e., "a picture paints a thousand words") in instruction manuals (i.e., dual coding, Paivio, 1971, 1991) could increase communication comprehension; however, the images must be tested for effectiveness. This is the case for farmers in Uganda, called smallholder farmers, as their land includes a small number of acres. The smallholder farmers provide most of the food (70%), and there is pressure to increase their productivity using interventions such as new technology that reduces the need for human labor (Kisaalita & Sempiira, 2022).

To implement the new technology, smallholder farmers must understand how to use the technology based on the written instruction manuals. Most of the smallholder farmers have low levels of literacy, and some have less than a fifth-grade education, whereby the farmers have a difficult time understanding written instructions. Hence, there was a desire to develop pictograms for the manuals to assist with the communication of the appropriate instructions. With no available electricity, the focus was on two devices used by the smallholder farmers. The EvaKuula is powered by biogas and wind to keep milk cold and fresh, and the hand-operated churn, IzeChurn, for making ghee (Kisaalita & Sempiira, 2022, p. 18).

To begin this process, key messages were identified in the instruction manuals and then a local artist created simple sketches as pictograms for each message. The artist made some modifications based on feedback from the researchers. Male and female smallholder farmers who were not visually impaired and had at least a fifth-grade education were randomly selected to participate in identifying the meaning of the pictograms. Only eight of the 20 pictograms met the criteria for understandability. To improve the pictograms, a focus group of four individuals who participated in identifying the meaning of the pictograms provided feedback on the misinterpretations of the pictograms that were not understood and made suggestions on what would be more meaningful in a pictogram. Using this feedback, the artist modified the pictograms that were not understood in the first round.

All pictograms (those understood in the first round and the modified ones) were then presented to a second group of randomly selected smallholder farmers (male and female, who were not visually impaired and had a minimum of a fifth-grade education). With the user's (smallholder farmers') input, 14 of the pictograms were identified as understandable. These pictograms can now be used to better convey the instruction manual information that will enhance the productivity of the smallholder farmers who had lower levels of literacy. Explain why many of the original pictograms created by the local artist and reviewed by the researchers did not effectively communicate the message. What might have been the reason for having a local artist draw these pictograms and the benefits of asking the smallholder farmers for their feedback and suggestions for changes? Can you think of situations in which well-educated individuals would benefit from well-designed pictograms?

1.6 EDUCATION AND EMPLOYMENT OF HUMAN FACTORS SPECIALISTS

1.6.1 Education

After taking this course, you might be interested in further pursuing a human factors education. If you visit the Human Factors and Ergonomics Society (HFES) web site (www.hfes.org) and go to the Education & Career Resources page, and then Academic Programs, you will find a list of undergraduate programs. Many human factors jobs, though, require education beyond the bachelor's degree, usually at the master's or doctoral level. A link to graduate programs is also available on this page. This listing includes programs predominantly in psychology and engineering.

Because the research areas of human factors specialists have grown, the specific undergraduate training for each person might vary. Given human factors' close relationship to experimental psychology, though, taking courses in cognitive psychology, physiological psychology, sensation and perception, statistics, research methods, memory, and learning will provide you with a good foundation. Depending on one's interests, courses in engineering or computer science could be helpful as well.

1.6.2 Accreditation, certification, and continuing education

Accreditation indicates that a particular human factors program has applied and met the requirements of the HFES Accreditation Review Committee (ARC). An ARC review ensures that students are receiving the proper coursework, research, and work experiences necessary to have the qualifications to work as a human factors specialist. To become accredited, a program must apply for accreditation and be approved by the ARC.

In contrast, human factors specialists may pursue *certification*. Not every human factors specialist is certified, but one argument for certification is to protect consumers from individuals who market themselves as human factors specialists, but lack the appropriate knowledge and skills (Hopkins, 1995). There are short courses, workshops, symposia, professional meetings, and seminars available to help individuals prepare for certification. Once certified, individuals may write CPE, CHFP, or CUXP after their names to indicate they are certified professional ergonomists, certified human factors professionals, or certified user experience professionals, respectively.

Professional CPEs, CHFPs, or CUXPs are recertified every 5 years to retain certification. To receive recertification, these professionals must demonstrate that they have been practicing in the field, attending relevant professional conferences, and/or receiving appropriate continuing education. You can find more information about the certification and recertification processes on the Board of Certification in Professional Ergonomics (BCPE) website (https://bcpe.org/).or on the HFES website under Related Organizations.

1.6.3 Work settings

Human factors specialists find employment in various settings including academia, industry, healthcare, research labs, and the government/military (Howell, 1985). Analysis of job openings posted by the HFES placement service (e.g., Anderson et al., 2005; Moroney, 2007; Schoeling et al., 2001; Voorheis et al., 2003) revealed significant changes in the types of industries seeking human factors specialists across time (see Table 1.1). These changes in types of companies reflect the changes in industry. For instance, during the boom of the dot-com industries at the turn of the century, 33.2% of the openings came from companies in the internet industry. After the collapse of the dot-com industry in 2000, industries that began to advertise openings included the consulting, government/military, and medical fields, which reflected the growth of consulting businesses, safety issues after the events of 9/11, and the cost and safety concerns in medicine.

In July 2023, a review of the HFES Job Board (found under Career Resources on the HFES. org website) found 144 positions posted. Companies such as Apple, Northrop Grumman, Leidos, Medstar Health, and Cherokee Nation Businesses posted several positions. These listings can change daily. With continued advancements in technology, such as artificial intelligence, robotics, unmanned aerial vehicles, and autonomous vehicles, the industries seeking

Table 1.1 Rank order by frequencies (and percent) of type of industry with greatest number of job openings

Year and number of placements posted	Top six industry types requesting HF specialists each year at the HFES job placement center
2001 (220 positions)	Internet (33.2%), Hardware/Software (10%), Consumer Products (7.3%), Risk Management (7.3%), Computer Software (6.4%), Aviation/Aerospace (5.9%)
2003 (117 positions)	Consulting (28.2%), Government/Military (12.8%), Medical (11.1%), Aviation/Aerospace (9.4%), Internet (8.5%), Transportation (6.8%)
2005 (92 positions)	Consumer Products (16.3%), Military (15.2%), Research & Development (14.1%), Consulting (14.1%), Engineering (12.0%), Medical (9.8%)
2007 (124 positions)	HF Consultant (23.4%), Healthcare/Medical (14.5%), Ergonomics (9.7%), Consumer Products (8.9%), Aviation/Aerospace (6.5%), Product Safety/Liability (5.6%)

human factors specialists might change; however, they will seek individuals with a strong foundation in human factors science. It is often the case that human factors specialists readily apply their knowledge and skills to new situations or environments. For example, another growing human factors area is educational research. Educational research requires expertise in the areas of usability testing and software design as well as environmental design, learning, memory, and cognition. Human factors expertise can be applied to the design of systems for human versus computer-generated grades, multimedia classrooms, multimedia designed for different aged learners, and distance education. Therefore, even if the industry types change, human factors knowledge and skills can be applied to a variety of environments.

As you might have noticed, the last entry in Table 1.1 is from 2007. To better understand the education needs of human factors practitioners based on the types of jobs available, Esa M. Rantanen and William F. Moroney surveyed current graduate students, young professionals (graduated within the past 5 years), and employers as to necessary skills and competencies needed on the job. The first report on these surveys appeared in 2011. Across all three surveys (Moroney & Rantanen, 2012; Rantanen & Moroney, 2011, 2012), there was an indication that there was a strong need to focus on communication skills within interdisciplinary teams, basic research methods and statistical skills, and abilities to translate theory learned in school to the actual application on the job. Further, these results suggest that the core competencies specified by the Board of Certification in Professional Ergonomics (https://bcpe.org/pathway-to-certification/) are a good model for the education of human factors practitioners. Regardless of the job or industry, employers are seeking individuals who can conduct quality research to identify solutions to real-world problems.

1.7 SUMMARY

Human factors applies psychological theory and principles to the design of environments with the intent to reduce error and increase safety. Originally, experimental psychologists applied their knowledge of sensation and perception, learning, and memory to decrease errors in military environments. Currently, human factors is applied to the design of everyday equipment, tools, and environments. In addition, human factors can address the needs of individuals from special populations.

Several other disciplines are related to the human factors field such as engineering and I-O psychology. The strong relationship with engineering exists because human factors specialists are interested in designing equipment, tools, and environments that are compatible with the capabilities and limitations of human users. Furthermore, these fields apply the systems perspective to their approaches, as does I-O psychology. I-O psychology and human factors both seek to enhance productivity and safety on the job but often approach problems from a different perspective. I-O psychologists attempt to create a strong workforce through better selection, training, and performance management of the workers. Although some human factors specialists work on selection and training, human factors specialists generally attempt to design a job that fits the capabilities and limitations of the workers, thereby accommodating a more diverse workforce.

One reason human factors applications can help create a safer work environment is through the use of a systems perspective—the notion that all parts of the system are interrelated and affect each other. According to systems theory, inputs to the system go through some type of process that creates outputs. These outputs then become inputs either for this same system or for another. Feedback, which is the output of the system and loops back into the system as input, creates a closed loop and allows for continuous control and adjustments within the system, helping to reduce errors. Many of the human factors systems considered are conceptualized as closed-loop.

Because humans have always been seeking ways to improve functions, the beginnings of human factors might be traced back to the beginning of time. Yet, human factors evolved into its own profession just after WWII. Human factors still has a strong military influence; however, the field has since diversified and now is involved in the design of space, land, sea, and air travel systems; computer and other communication systems; home, work, and educational environments; airport security systems; and equipment, tools, and environments for use by all individuals, including special populations.

Individuals interested in becoming a human factors specialist can pursue an undergraduate education in human factors, but a student with a master's or doctoral degree will be better prepared to enter the profession. With the appropriate training, the human factors specialist can work in private consulting, industry, military settings, or academia.

LIST OF KEY TERMS

Closed-loop system
Environment
Ergonomics
Human–computer interaction
Human factors
Human factors engineering
Human information processing
Industrial-organizational psychology
Iterative process
Open-loop system
Scientific management
Systems theory
Task analysis
Time and motion study
Usability
User-centered design

SUGGESTED READINGS

Casey, S. M. (1993). *Set Phasers on Stun: And Other True Tales of Design, Technology, and Human Error*, Santa Barbara, CA: Aegean.
   This book presents captivating stories that describe true incidents that could have been avoided with proper human factors.
Casey, S. M. (2006). *The Atomic Chef: And Other True Tales of Design, Technology, and Human Error*. Santa Barbara, CA: Aegean.
   This is the second book of captivating stories by Steven Casey about true incidents of error that often led to injury or death, but could have been reduced or eliminated with the application of human factors principles.
Stuster, J. (Ed.). (2006). *The Human Factors and Ergonomics Society: Stories from the First 50 Years*. Santa Monica, CA: Human Factors and Ergonomics Society. https://www.hfes.org/Portals/0/Documents/HFES_First_50_Years.pdf?ver=2020-09-10-143021-533&timestamp=1599766230529.
   This book provides short articles and stories about human factors and the society in the first 50 years. The authors in this edited work were part of and experienced the evolution of HFES.
Roscoe, S. N. (1997). The adolescence of engineering psychology. In S. M. Casey (Ed.). *Human Factors History Monograph Series*, Vol. 1. Santa Monica, CA: Human Factors and Ergonomics Society. Retrieved 9/24/2024: chrome-extension://efaidnbmnnnibpcajpcglclefindmkaj/https://cms.hfes.org/Cms/media/CmsImages/adolescence.pdf.

This article gives a good, concise overview of the beginning of engineering psychology, which started near the end of WWII.

Meister, D. (1999). *The History of Human Factors and Ergonomics*. Mahwah, NJ: Lawrence Erlbaum Associates.

Here is a book that can give you more details about the history of human factors and ergonomics.

Sanders, M. S., & McCormick, E. J. (1993). *Human Factors in Engineering and Design* (7th ed). New York: McGraw-Hill.

This book is an excellent reference book and is used as a resource text for individuals studying for the certification exam.

CHAPTER EXERCISES

1. Human factors is the application of psychological science. First, identify a psychological phenomenon or finding (e.g., the Zeigarnik effect). Now, give an example of how this information would or would not be considered human factors.

2. Given the following systems, (1) identify the goals, inputs, process, and outputs, (2) describe how these different parts of the system are interrelated, and (3) describe how the system would be affected if one or more of these parts or functions changed:
   a. Highway system
   b. Air transportation system
   c. Elevator
   d. Bus system
   e. Rest rooms
   f. Offices
   g. Parking lots

3. Identify various events that occurred during WWII and explain how they helped the development of the human factors profession.

4. If you were having trouble entering phone numbers into your cell phone, explain how a human factors specialist is likely to approach the problem compared to an engineer or an I-O psychologist.

5. As a human factors specialist working during WWII, describe the different types of research you would likely conduct and explain how the results could be applied to everyday use.

6. What are some of the reasons we need to consider special populations? Give an example of how a lack of consideration of special populations is likely to create errors or decrease safety for the users.

7. Compare and contrast the terms engineering psychology, ergonomics, and human factors.

8. Describe where human factors specialists are likely to work and the types of work they might do.

9. Consider everyday things you use. Identify what might be considered a human factors problem and describe how you might use human factors to design something that works well.

10. Human factors is an adaptable field. What makes the human factors specialist capable of applying one's knowledge and skills to a variety of areas (e.g., law, medicine, or travel)?

REFERENCES

Andel, C., Davidow, S. L., Hollander, M., & Moreno, D. A. (2012). The economics of health care quality and medical errors. *Journal of Healthcare Finance,* 39(1), 39–50.

Anderson, T. J., Bakowski, D. L., & Moroney, W. F. (2005). Placement opportunities for the human factors engineering and ergonomics professionals in industry and government/military positions. *Proceedings of the Human Factors and Ergonomics Society Annual Meeting, September 2005,* 49(7), pp. 788–792. https://doi.org/10.1177/154193120504900710

Barker, R. G. (1968). *Ecological Psychology.* Stanford, CA: Stanford University Press.

Bederson, B. B., Lee, B., Sherman, R. M., Herrnson, P. S., & Niemi, R. G. (2003). Electronic voting system usability issues. *Proceedings of the CHI Annual Meeting,* April 5–10, 2003, pp. 145–152.

Berkun, M. M. (1964). Performance decrement under psychological stress. *Human Factors,* 6, 21–30.

Byrne, M. D., Greene, K. K., & Everett, S. P. (2007). Usability of voting systems: baseline data for paper, punch cards, and lever machines. *Proceedings of the CHI Annual Meeting,* April 28–May 3, 2007, pp. 171–180.

Christensen, J. M. (1962). The evolution of the systems approach in human factors engineering. *Human Factors,* 4, 7–16.

Christensen, J. M. (1987). The human factors profession. In G. Salvendy (Ed.), *Handbook of Human Factors,* Chapter 1.1, New York: John Wiley & Sons, pp. 3–16.

Cifter, A. S. (2017). Blood pressure monitor usability problems detected through human factors evaluation. *Ergonomics in Design,* 25(3), 11–19.

Classen, D. C., Resar, R., Griffin, F., Federico, F., Frankel, T., Kimmel, N., Whittington, J. C., Frankel, A., Seger, A., & James, B. C. (2011). 'Global Trigger Tool' shows that adverse events in hospitals may be ten times greater than previously measured. *Health Affairs,* 30(4), 1–9. doi: 10.1377/hlthaff.2011.0190

Conrad, F. G., Bederson, B. B., Lewis, B., Peytcheva, E., Traugott, M. W., Hanmer, M. J., Herrnson, P. S., & Niemi, R. G. (2009). Electronic voting eliminates hanging chads but introduces new usability challenges. *International Journal of Human-Computer Studies,* 67(1), 111–124. doi: 10.1016/j.ijhcs.2008.09.010

Durso, F. (2013). Finding eve. *Human Factors and Ergonomics Society Bulletin,* 56(12). https://www.hfes.org/Portals/0/Documents/First%20Ladies%20of%20HFES.pdf?ver=2020-11-09-123830-387.

Durso, F. (2014a). Whirly-girl. *Human Factors and Ergonomics Society Bulletin,* 57(2). https://www.hfes.org/Portals/0/Documents/First%20Ladies%20of%20HFES.pdf?ver=2020-11-09-123830-387.

Durso, F. (2014b). Top girl. *Human Factors and Ergonomics Society Bulletin,* 57(6). https://www.hfes.org/Portals/0/Documents/First%20Ladies%20of%20HFES.pdf?ver=2020-11-09-123830-387.

Durso, F. (2014c). Making change. *Human Factors and Ergonomics Society Bulletin,* 57(10). https://www.hfes.org/Portals/0/Documents/First%20Ladies%20of%20HFES.pdf?ver=2020-11-09-123830-387.

Gilbreth, F. B. (1911). *Motion Study. A Method for Increasing the Efficiency of the Workman.* New York: D. Van Nostrand.

Gilbreth, F. B. (1916). Motion study in surgery. *Canadian Journal of Medicine and Surgery,* 40, 22–31.

Gilbreth, F. B., & Gilbreth, L. M. (1919). *Fatigue Study. The Elimination of Humanity's Greatest Unnecessary Waste. A First Step in Motion Study.* New York: MacMillan Company.

Grether, W. F. (1968). Engineering psychology in the United States. *American Psychologist,* 23, 743–751.

Hancock, P. A., & Hart, S. G. (2002). Defeating terrorism: what can human factors/ergonomics offer? *Ergonomics in Design,* 10(1), 6–16.

Harris, D. H. (2002). How to really improve airport security. *Ergonomics in Design,* 10 (1), 17–22.

Hopkins, C. O. (1995). Accreditation: the international perspective. *Human Factors and Ergonomics Society Bulletin,* 38(2), 1–5? How site bulletin?

Howell, W. C. (1985). Engineering psychology. In E. M. Altmaier & M. E. Meyer (Eds.). *Applied Specialties in Psychology,* pp. 239–273, Chapter 10. New York: Random House.

Howell, W. C. (2003). Human factors. In W. C. Borman, D. R. Ilgen, & R. J. Klimosky (Eds.). *Handbook of Psychology: Industrial and Organizational Psychology,* Vol. 12, pp. 541–564, Chapter 21. Hoboken, NJ: John, Wiley, & Sons.

Kisaalita, W. S., & Sempiira, E. J. (2022). Development of pictograms to communicate technological solution instructions (Labeling) among low-literacy users. *Ergonomics in Design*, 30, 17–29.

Kohn, L. T., Corrigan, J. M., & Donaldson, M. S. (Eds.) (2000). *To err Is Human: Building a Safer Health System*. Washington, DC: National Academy Press.

Krueger, G. (2002). Alphonse Chapanis 1917–2002. *Human Factors and Ergonomics Society Bulletin*, 45(10), 1 & 4.

Leape, L. L. (2004). Human factors meets healthcare: the ultimate challenge. *Ergonomics in Design*, 12(3), 6–12.

McGrath, J. E. (1984). *Groups, Interaction, and Performance*. Englewood Cliffs, NJ: Prentice-Hall.

Miller, D. P., & Swain, A. D. (1987). Human error and human reliability. In G. Salvendy (Ed.). *Handbook of Human Factors*, New York: John Wiley & Sons. pp. 219–250, Chapter 2.8.

Moroney, W. F. (2007). Placement opportunities for human factors engineering and ergonomics professionals. *Proceedings of the Human Factors and Ergonomics Society 51th Annual Meeting*, October 1–5, 2007. Baltimore, Maryland. pp. 516–520.

Moroney, W. F., & Rantanen, E. M. (2012). Student perceptions of their educational and skill needs in the workplace. *Proceedings of the Human Factors and Ergonomics Society 56th Annual Meeting*, October 22–26, 2012. Boston, MA. pp. 576–580. doi: 10.1177/1071181312561120

Paivio, A. (1971). *Imagery and Verbal Processes*. New York: Holt, Rinehart & Winston.

Paivio, A. (1991). *Images in Mind: The Evolution of a Theory*. New York: Harvester Wheatsheaf.

Parsons, H. M. (1999). A history of Division 21 (Applied Experimental and Engineering Psychology). In D. A. Dewsbury (Ed.), *Unification through Division: Histories of the Divisions of the American Psychological Association*, Vol. 3, American Psychological Association. pp. 43–72, Chapter 2. doi: 10.1037/10281-002

Rantanen, E. M., & Moroney, W. F. (2011). Educational and skill needs of new human factors/ergonomics professionals. *Proceedings of the Human Factors and Ergonomics Society 55th Annual Meeting*, September 19–23, 2011. Las Vegas, NV. pp. 530–534. doi: 10.1177/1071181311551108

Rantanen, E. M., & Moroney, W. F. (2012). Employers' expectations for education and skills of new human factors/ergonomics professionals. *Proceedings of the Human Factors and Ergonomics Society 56th Annual Meeting*, October 22–26, 2012. Boston, MA. pp. 581–585. doi: 10.1177/1071181312561121

Roscoe, S. N. (1997). The adolescence of engineering psychology. In S. M. Casey (Ed.). *Human Factors History Monograph Series*, Vol. 1. Santa Monica, CA: Human Factors and Ergonomics Society. pp. 1–9. chrome-extension://efaidnbmnnnibpcajpcglclefindmkaj/https://cms.hfes.org/Cms/media/CmsImages/adolescence.pdf.

Sanders, M. S., & McCormick, E. J. (1993). *Human Factors in Engineering and Design*. New York: McGraw-Hill.

Schoeling, S. E., Boliber, M. J., & Moroney, W. F. (2001). Placement opportunities for human factors engineering and ergonomics professionals in industry and government/military positions. *Proceedings of the Human Factors and Ergonomics Society 45th Annual Meeting*, October 8–12, 2001. Minneapolis/St. Paul, MN. pp. 768–772.

Sivak, M., Post, D. V., Olson, P. L., & Donohue, R. J. (1981). Driver responses to high-mounted brake lights in actual traffic. *Human Factors*, 23, 231–235.

Summers, W. C. (1996). 50 years of human engineering. *Human Factors and Ergonomics Society Bulletin*, 39(3), 1, 3.

Suri, J. F. (2000). Saving lives through design. *Ergonomics in Design*, 8(3), 4–12.

Taylor, F. W. (1911). *The Principles of Scientific Management*. New York: Harper and Brothers.

von Bertalanffy, L. (1950). An outline of general systems theory. *British Journal of Philosophical Science*, 1, 134–165.

Voorheis, C. M., Snead, A. E., & Moroney, W. F. (2003). Placement opportunities for human factors engineering and ergonomics professionals in industry and government/military positions. *Proceedings of the Human Factors and Ergonomics Society 47th Annual Meeting*, October 13–17, 2003, Denver, CO. pp. 908–912.

Wickens, C. D., & Hollands, J. G. (2000). *Engineering psychology and human performance*. Upper Saddle River, NJ: Prentice Hall.

Chapter 2

Research methods

Chapter Vignette

Bill was talking with some friends about his grandfather, who seemed to be depressed. His grandfather lives in a retirement home where he has his own room, but there are also many shared spaces such as the kitchen and living room. Bill was trying to figure out a simple but effective way of helping his grandfather feel better. One of his friends, Frank, thought Bill should paint his grandfather's room yellow because it is a bright color that Frank feels lightens his own mood. His other friend Geena was concerned that yellow was perhaps too bright, and that it might get annoying. She recommended painting the room green and placing various plants around the room to make it feel like a park or forest. Because Geena feels so much better whenever she is outdoors and in a place like a park, she believed this type of environment would make Bill's grandfather happier. Although Frank agreed that the color green along with a multitude of plants could be nice, he felt strongly that Bill should paint his grandfather's room yellow. Frank suspected that the color yellow might positively affect Bill's grandfather's mood. Bill decided to try Frank's idea and paint his grandfather's room yellow. Geena asked Bill to let them know how his grandfather reacts to the yellow. As her own grandparents sometimes seem to fall into a rut, Geena is interested in knowing how this works.

For a short time after painting his grandfather's room yellow, Bill's grandfather's mood seemed to improve, but this was only temporary. Slowly, Bill's grandfather returned to his original depressed mood. As Bill was concerned for his grandfather, he thought he would try Geena's idea next. What did he have to lose?

The scenario described above raises questions that can be addressed through research. It is possible that Bill's grandfather's depressive symptoms could be affected by room color, but it could be the change in color itself and not the specific color that alters his mood, or something else entirely. As Bill's grandfather's mood improved after the painting of the room, this suggests that the change could have been due to the color, the change in color, or just having Bill there and being able to talk to him. In addition, because the change in mood was temporary, this could suggest that Bill's grandfather adapted to the color or he was not as happy once Bill no longer came around to paint. At this point, Bill does not know what caused the change in his grandfather's mood. Just because Bill's friend thinks the color yellow makes him happy does not necessarily mean that yellow will make Bill's grandfather happy or that the color yellow generally makes people happy. What Bill needs to know is how the color yellow influences people with mild depression. This requires research on the impact of color on people's moods. Understanding how most people, especially people who are mildly depressed, respond emotionally to the color yellow would be a better indication of how his grandfather might react or be affected.

DOI: 10.1201/9781003515463-2

The questions raised in the situation described above are concerned with how individuals will react to, think about, and interpret particular situations. Environmental design solutions based on our intuition of how we *think* people will react, or how we *want* them to think about or respond to them, are less likely to result in effective solutions. Instead, we need research data to understand how most users react to the environments in order to create the best design.

2.1 CHAPTER OBJECTIVES

After reading this chapter, you should be able to:

- Describe what the scientific method is.
- Compare and contrast basic and applied research.
- Define study, experiment, and quasi-experiment.
- Define the terms related to conducting a research project.
- Define reliability and validity.
- Describe the difference between descriptive and inferential methods.
- Understand what types of ethical behaviors are required for research.

2.2 WHAT IS THE SCIENTIFIC METHOD?

The **scientific method** is a systematic, unbiased, and objective process of acquiring knowledge. If the methods are systematic, unbiased, and objective, then the scientific knowledge that is gained should be amoral (Arnoult, 1972). Amoral results reflect the truth and are unbiased by any desired outcome or preconceived notion.

To be amoral when employing the scientific method, the researcher should be skeptical and seek empirical evidence. A skeptical researcher questions and critically evaluates data, and the research methods used to ensure that all aspects of the scientific process being implemented are appropriate. Skepticism helps ensure that science is not based on intuition and biases but on the facts acquired through the scientific process. This process is often referred to as the *empirical method* because we gain knowledge through direct observation.

Because we use the scientific method, human factors specialists determine better designs for equipment, environments, and tools that are easier to use, reduce errors, and make the equipment, environment, or tool safer. Sometimes this approach is casually dismissed as being common sense. The fact that users are confronted daily by software that is not intuitive or tools that are difficult to use and can result in accidents and injury are testament that this view is in error. If human factors is just common sense, why are poorly designed tools, environments, and processes so common? The application of sound methodological principles in design and evaluation of environments, equipment, and tools increases the likelihood that the products and environments will work well, are less likely to cause errors, and are intuitive to the user, which Donald Norman (1988) calls "in the world" knowledge (see Chapter 11). Therefore, we need to understand the scientific method and the challenges of applying the scientific method in our discipline.

One difficulty human factors specialists experience when applying the scientific method is that we study people who are complex, differing in many ways, and behaving or responding differently due to learning, aging, and situational factors (Chapanis, 1992). While our own experiences may offer some insight into the thoughts, motivations, and abilities of other users, we must base design recommendations on science, not intuition or gut feeling, as our friends in the chapter-opening vignette opted to do.

2.2.1 Hypothesis testing

Implementation of the scientific method involves the process of testing hypotheses. A hypothesis is a prediction made by the researcher that is developed from observation and previous research, and it is normally based on theory. Further, hypotheses must be testable statements. For example, Frank's hypothesis would be that yellow is a color that makes people happy. Similarly, Geena hypothesizes that a green room with many plants creates a relaxing area that also makes people happy. In contrast, the statement "Invisible people live among us" is not testable and would not be a good hypothesis because there's simply no way to prove or disprove that invisible beings are around us.

While Frank, Geena, and Bill developed their hypotheses based on personal experiences, normally scientists would need to review the literature for evidence consistent with or contrary to their hypotheses. Further, a literature review would tell the researcher whether relevant theories already exist that address the causal relationship between affect and the properties of color. A theory is a set of testable assumptions or suspected relationships among variables that attempt to explain a certain event or behavior. Therefore, a hypothesis or set of hypotheses could be established based on a theory. For example, if individuals experiencing seasonal affective disorder (SAD), or depressive symptoms during the winter months (Lee et al., 1998), feel depressed due to reduced light levels, one might hypothesize that yellow could make someone happier because the color yellow reflects more light than many other colors.

After hypotheses have been established, an appropriate method and means of analysis are determined, data are collected, and then the scientist evaluates the data and tests the hypotheses. If we hypothesized that yellow makes people happy, this hypothesis might be written more formally as, "It is expected that individuals exposed to the yellow room will report higher levels of positive affect than individuals in a neutral or other colored room." We expect that the data will support our hypothesis (also known as the *alternative hypothesis*). Yet, we cannot be sure that something is true with just one test of our (alternative) hypothesis. We might have been lucky with that one test. Rather, we can establish that something is not true with just one test. To do this, we test the *null hypothesis* that we hope to reject. The null hypothesis would state that there is no difference between individuals' moods when in a yellow or any other colored room. If the null hypothesis can be rejected, then we can say that the data support the alternative hypothesis—the color yellow tends to make people happier.

A failure to reject the null hypothesis could be due to a variety of factors including a lack of an effect, a lack of sensitivity in our measurement to find an effect, or a flaw in the research. In our research, we might have used the "wrong" yellow color, or the participants might not have been exposed long enough to the yellow room to demonstrate a change in mood. Therefore, we never accept the null hypothesis. In other words, the lack of an effect does not necessarily indicate that the relationship in question does not exist. Instead, we would report that the null hypothesis was not rejected or the alternative hypothesis was not supported. This logic is similar to the guilty or not guilty outcomes of a court trial. If someone is found not guilty, this does not indicate innocence, but only that there was insufficient evidence to establish guilt.

If we reject the null hypothesis, which indicates that yellow makes people happier, we should identify a *parsimonious* explanation of the results. When interpreting results, it is always best to defer to the simplest interpretation of the data. Complex interpretations are not likely to be the best or correct interpretations. Finally, if the experimental process was appropriate, the data should be *replicable* by other researchers.

2.3 BASIC VERSUS APPLIED RESEARCH

When conducting research and considering what our research questions or hypotheses will be, the focus could be on basic or applied research. **Basic research** is conducted for the sake of seeking knowledge or understanding. In addition, basic research helps us develop theory by controlling many real-world factors in order to focus on the specific question of interest. Researchers might investigate individuals' visual acuity (see Chapter 3) under low levels of lighting, the cognitive processes used to negotiate one's way around a new building (e.g., wayfinding, see Chapter 10), or how individuals solve complex problems (see Chapter 7). In basic research, the knowledge of how the visual system works, how individuals develop cognitive maps, or how individuals solve problems, for example, is the outcome desired.

It is possible, and often the case, that basic research findings are later used to solve applied problems. For instance, an understanding of visual acuity could be applied to vehicle dashboard design (see Figure 2.1) to ensure all labels and numbers are legible during night driving, knowledge about cognitive map development could be applied to the design of buildings to create more intuitive hallway layouts, or information about how individuals solve problems could be applied to the design of computer games to create a more logical or challenging game. The original intent of basic research, though, is to understand the underlying mechanisms, not to solve an applied problem.

In contrast, **applied research** is conducted in order to address a real-world problem. Researchers could investigate the impact of various room colors on individuals' moods and mood changes to enhance mental health, the impact of various signs or landmarks on the ease of learning how to navigate a new environment, or the impact of cell phone use on safe driving behaviors. As an applied field, a human factors design might benefit from basic research, but according to Alphonse Chapanis (1991), "the only kind of research that qualifies as human factors research is research that is undertaken with the aim of contributing to the solution of some design problem" (p. 3). As an example, the investigations of visual adaptation were basic research, but the results were applicable to human factors in the design of equipment people use (Chapanis, 1991). Therefore, human factors research solves problems or reduces errors that cause problems at work or in everyday life.

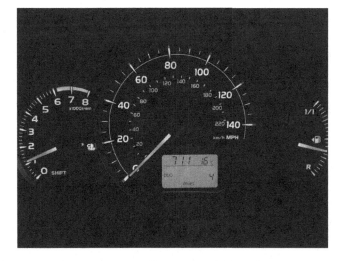

Figure 2.1 Illuminated car dashboard. (From Pixabay.)

2.4 TERMS AND CONCEPTS

When considering the use of color in an environment, there are several different perspectives that can be taken or questions that can be asked such as: How many unhappy people live in rooms that are not yellow? Is the color yellow more strongly related to happiness than the color blue or green? Are people who work or live in yellow colored environments happier than people who live in rooms of different colors? Depending on the question, a researcher will conduct either a study or an experiment.

2.4.1 Studies, experiments, and quasi-experiments

Although people often use the terms "study" and "experiment" interchangeably, they do have different meanings to researchers. A **study** refers to situations where there is no manipulation of the experiences people encounter. Administering questionnaires to people at the mall or in a safe place after a natural disaster (as shown in Figure 2.2) would be a study because you are not manipulating or controlling what types of experiences the participants might have. Similarly, one could study how often individuals who report being happy live or work in a yellow colored room. In an **experiment,** the researcher controls the experiences of the participants. In the case of an experiment involving room color, color is a manipulated variable whereby some participants might be asked to work in a yellow room while others work in a green room to determine how the different colors affect their moods.

A type of study that appears to be an experiment, but does not have the same level of manipulation of variables as an experiment, is called a quasi-experiment. In a **quasi-experiment,** although individuals have different experiences, these differences are not controlled or manipulated by the experimenter. For example, if you were to compare individuals from different universities, you would find that the students at University A have different experiences than students at University B; however, these students self-selected their university, and hence, their experiences. Similarly, if you survey individuals and ask them what their room color is, you can group these individuals by room color, but you did not randomly

Figure 2.2 An example of a study. A study can include survey research, which can occur under various conditions and at a multitude of locations. (U.S. Navy photo by Photographer's Mate 2nd Class Elizabeth A. Edwards [Public domain], via Wikimedia Commons.)

assign them to these room colors. These individuals may have selected the color yellow for a variety of reasons. These differences could reflect different personality types that are related to levels of happiness, which is a good basic research question. Although we might not know what these differences are, we must nevertheless be aware that these differences are likely to exist. Applied researchers interested in the impact of room color on performance are less likely to focus on the underlying mechanisms of how the color impacts mood. They are more likely to focus on the impact of color on various behaviors and cognitions, but will want to make sure the effect is not due to individual differences related to the self-selection of color.

For the following discussions, it will be made clear when a term or concept applies only to studies or only to experiments. The terms research or research projects will be used to describe an investigation, whether it is a study or an experiment.

2.4.2 Populations and samples

When conducting research, whether it is basic or applied, we collect information from a group of participants or a sample to study, understand, and make inferences about the population. Let's say you are interested in studying the impact of talking on a cell phone on driver reaction time. The focus of this experiment is on drivers, which could include anyone who is 16 years of age and older. The **population** is the larger group of individuals you are attempting to understand. Your end goal is to make inferences about, or generalize your findings to, this larger group of individuals. In this case, it would be all individuals of driving age.

For your experiment, it is not feasible to evaluate every driver's performance mainly due to time and money constraints. Therefore, we evaluate a **sample**, or a subset, of the population. Ideally, we want to select a set of participants using random sampling, which reduces the chance that the group will vary in some systematic way from the broader population. When using **random sampling**, every person in our population, in theory, has an equal probability of being selected, ensuring that our sample is more representative of the population than if we do not use random sampling. As we often study individuals at institutions such as universities, an Army base, or nursing homes, not every individual in the population of interest has an equal probability of being selected for our research project. We must be aware of these limitations and report descriptions of our sample so that other researchers can appropriately evaluate who was studied and to whom the results apply. Because of these potential limitations, the sample can impact the external validity of the research, which is discussed later in the chapter.

Because we are often restricted by who is available to participate in our research, it is important to consider different types of samples. A *stratified sample* attempts to collect data from various subgroups of the population, such as men and women or people of different racial groups, whereby the sample contains subsamples that represent the same proportion of individuals as found within the population. A *cohort* sample would contain people born in a similar time period, such as the "baby boomers" who were born between 1946 and 1964. Finally, a *convenience sample*, as the name implies, means that the sample is selected because the participants are readily accessible to the researcher and little or no attempt is made to ensure the sample is representative of the population.

Once we determine the type of sample we will use, we need to select our sample. If we want to compare reaction times of individuals using cell phones to those individuals who are not using cell phones, then random selection is preferred, but not likely feasible. If researchers at Big-Time University ask students to participate in their driving experiment, then the researchers will be selecting drivers from a subsample of that student population. Undergraduate college students, in general, and Big-Time University students, in particular, might not be representative of the larger population of drivers because they tend to be younger (17–23 years of age), more educated, and affluent. Therefore, instead of using random selection to determine our sample, student volunteers should be randomly assigned to the two driving groups, with and without a cell phone.

Although random sampling deals with ensuring that the sample is reflective of the population, there is an even more important consideration when assigning participants to various experimental conditions. This is referred to as random assignment. **Random assignment** occurs when we use a method such as flipping a coin, rolling a die, or using a computer-generated randomized list that determines to which experimental group or condition the participants will be assigned. The rationale is that students have an equal probability of being placed in any of the experimental groups. If you are studying driving behavior in which some individuals are driving with a cell phone and others are driving without a cell phone, the sample of students in the "driving with" group that represents the population is expected to be similar to the sample of students in the "driving without" group. When the groups are assumed to be similar at the beginning of the experiment, we can argue that differences in driving performance are due to the experimenter's manipulation; otherwise, the cause of our results would be unclear.

Imagine recruiting participants for your driving study from students who attend two classes, a morning class that consists of younger traditional undergraduates and a late afternoon class that targets older, non-traditional students who work during the day. Because of convenience, students in the morning class are assigned to the "driving without" and students in the afternoon class are assigned to the "driving with" condition. If we were to find that the "driving without" group has faster reactions times than the "driving with" group, it is not possible to determine whether this outcome is because the participants in the "driving without" group reacted faster simply because they are younger or because they were not using a cell phone at the same time.

Besides determining how to select individuals for your research, you might wonder how many people you should include. The required *sample size* to see an effect is an important consideration and is affected by the effect size. An **effect size** is essentially the impact of the manipulation. In your driving experiment, you might have your participants drive a simulated route that has low, medium, or high task complexity, regardless of whether they are "driving with" or "driving without" a cell phone. If you expect (based on previous research) that the difference between low, medium, and high task complexity will have a large impact on the number of errors drivers make, then the effect size is expected to be large. If the expected effect size is large, then one can use a smaller sample. If the effect size is expected to be small, then the sample size needs to be larger in order to achieve a meaningful statistical result.

It is important that the sample does not become unnecessarily large. Mathematically, large samples help us achieve statistical significance because we reject (or do not reject) the null hypothesis based on the calculated statistic, which is influenced by the sample size. We might reject the null hypothesis simply because the sample size is large, not because there is a true difference. With extremely large samples, results might be statistically significant, even with a small effect size, but these results are essentially meaningless and not practical. Therefore, we should design our research project to recruit the fewest number of participants possible, but with a large enough sample to see an effect if it actually exists. We use **power analysis** to calculate the sample size needed given the expected effect size, which is estimated from the literature. We do not want to waste people's time and energy as well as resources if we could achieve a similar result with fewer participants.

In summary, whether you send individuals a survey, ask participants to come to a laboratory to complete a task, or evaluate workers on the job, each of these participant groups makes up your sample. In turn, we use the results from the sample to understand the population.

2.4.3 Independent and dependent variables

Variables change or vary, taking on a range of values across individuals. An **independent variable** is the variable that we manipulate in an experiment such as whether the participant uses a cell phone or not (i.e., "driving with" or "driving without"). It is "independent" because it is a freely selected variable that is not influenced by other variables in the research project. We choose what the independent variable will be and the aspects of it we will be studying. If we are researching the impact of color on mood to inform Bill as to how to help his grandfather, we might have people perform a task in rooms of different colors we selected. In this case, room color would be our independent variable. We might also want the participants to perform a task that has various levels of difficulty whereby task difficulty would be a second independent variable.

The **dependent variable** is the outcome measure we are evaluating. In our color-and-mood experiment, mood would be our dependent variable. As we also have individuals performing tasks of different levels of difficulty, a second dependent variable could be performance level. We would expect our mood and performance measures to be dependent on the color of the room and/or the level of task difficulty.

As color is often impacted by the amount of lighting, it is possible that the amount of daylight in the room could serve as a covariate. A **covariate**, such as the amount of daylight in the room, interacts with the independent variable affecting the relationship between the independent and dependent variables. The amount of daylight covariate could remove the effect of daylight, clarifying the relationship between room color and mood or room color and performance level.

The independent variable is sometimes referred to as the *predictor*, a variable used to predict the dependent variable. Similarly, the dependent variable is sometimes called the *outcome measure*, *criterion measure*, or *criterion*. A **criterion** is a set standard or measure of performance used to evaluate the effect of the variable of interest, which in this case is the color of the room. Although some people use the terms independent and dependent variables within studies, the terms predictor and criterion measure are more common. Normally, only the terms independent and dependent variables are used in experiments.

Let's reflect back a moment on quasi-experiments. Remember, quasi-experiments are studies that appear to be experiments but are not true experiments because they lack a manipulated independent variable. Imagine researching the relationship between a student's sex, the university the student attended, and the starting salary. Sex and the university attended by the student are independent variables, but they are not manipulated. Although we can analyze the data as if sex and university attended were manipulated and this was a true experiment, it is critical that we interpret the results appropriately given that we did not manipulate sex or which university the participants attended. We can only report relationships between sex or university attended and the dependent variable. We cannot say that being a certain sex or attending a certain university <u>caused</u> the observed effects on the dependent variable. It is possible that another variable not identified yet could be the cause of this result, which is known as the **third variable problem**. What conclusions might you make based on the graph in Figure 2.3? Box 2.4 and the supplement in Section 2.9 provide additional discussions about the proper interpretation of correlation coefficients.

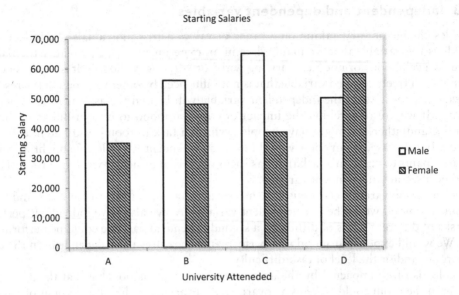

Figure 2.3 What conclusions can you make? Based on this chart, what conclusions can you make about salaries women make or salaries women make if they attend certain schools?

BOX 2.1:   WHAT ARE THE INDEPENDENT AND DEPENDENT VARIABLES?

Individuals who forget to turn off the stove top create a fire and burn hazard for themselves and others. Your research project for your human factors course is to determine ways to reduce the number of times individuals forget to turn off the stove top burners. Assume that a stove top has the capability of sensing changes in pressure (e.g., pressure from stirring or a pan placed on the burner), indicating when the burner is being used. If the stove senses that the burner is still on, but not being used, the appliance will start beeping after a certain time, signaling that the stove is still turned on. From your review of the research, you find that an intermittent beeping tone tends to get people's attention more quickly than a steady tone. Based on this information, you decide to evaluate three different beeping rates to determine which rates are more effective in signaling that the burner is turned on. In your experiment, individuals are randomly assigned to a slow, moderately fast, and fast beeping tone. To measure their reaction time, you ask them to respond once they hear the tone by reaching for and turning a dial as if they are turning off a burner on a stove. **Critical Thinking Questions**: What is the independent variable? What is the dependent variable? What might be a possible second independent or dependent variable?

2.4.4 Operationalizing your variables

Regardless of the type of research we conduct, it is important to operationalize our variables. To **operationalize** a variable is to describe the variable in concrete terms so that it is clearly understood how the variable is being measured. This not only allows us to ensure that we are measuring what we intend to measure, but also allows other scientists to replicate the research. In our cell phone experiment, besides having individuals drive with or without using a cell phone, we could include a second independent variable—the complexity of the driving environment. If we expose research participants to a low, medium, or high complexity situation, a well-operationalized variable would make it clear how low, medium, and high complexity are defined and created. A low complexity environment might be defined as one that has only one additional car besides the driver's car on a five-mile flat and straight stretch of highway. In contrast, the high complexity condition might be defined as ten cars along with the driver's car on a five-mile curvy mountain road.

BOX 2.2:   OPERATIONALIZING YOUR VARIABLES

In the driving experiment described in the text, how might you operationalize "accident?" When we think of accidents, we normally think of a car hitting something else. It is still an error, though, if we accidentally drive off the road and onto the curb or if we drove too close to other cars. When driving at high speeds, it is important to have a larger gap between cars to avoid accidents, but that large gap does not always occur, which creates a dangerous situation. Therefore, it is important to determine if the research is focusing on just accidents or the broader scope of driving errors. It is possible to consider a range of driving behaviors from driving poorly, such as on the sidewalk, to near misses, to actually crashing into another vehicle. These activities would all be classified as errors, but only the last event (i.e., crashing) would be considered an accident (see Chapter 11 for a discussion on errors and accidents). To clarify what an "accident" is in your study, it would be necessary to specify your definition and how accidents will be measured. Well-operationalized variables allow us to compare research findings across an assortment of projects and inform our understanding of the generalizability of results. **Critical Thinking Questions**: How would you operationalize "accidents?" In our stove burner experiment in Box 2.1, how might you operationalize slow, moderately fast, and fast beeping sounds?

2.4.5 Variable reliability and validity

When measuring your variables, whether independent or dependent, predictor or criterion, qualitative or quantitative, the variables must be reliable and valid. **Reliability** refers to the consistency of the measure. When using an electronic timer to measure reaction time in our cell phone and driving experiment, a reliable timer would record similar times in repeated trials for an individual who has a consistent reaction time. **Validity** assesses whether the measure is truly measuring or tapping what you want to measure. A personality measure is not a valid assessment of reaction time, but a stopwatch is. Also, if you are attempting to assess the value of some characteristic or behavior such as mood or reaction time, the measure will not be valid if it cannot reliably determine the value of this characteristic or behavior. Therefore, a measure must first be reliable in order to be valid.

2.4.6 Assessing reliability and validity

There are a variety of ways to demonstrate reliability and validity. Many methods for determining reliability and validity use the correlation statistic (see Section 2.9 for a review), which measures the strength of the relationship between two (or more) variables. We will focus on the correlation between two variables. As a general guideline, the correlation coefficient for reliability scores should be positive and greater than or equal to 0.70. It is often desirable to have reliability scores of 0.90 or higher. If the correlation coefficient is negative, then the variables are inversely related (i.e., as the values of one variable increase, the values of a second variable decrease). For reliability, inversely related variables reflect a lack of reliability, as this suggests that as one variable is rated more positively the other scale is rated more negatively. In scale development applications, inversely related variables generally indicate that one of the variables should be reverse-scored to establish reliability. Below we focus on the conceptual aspects of assessing reliability and validity, not how to calculate the correlation coefficients.

One of the most common measures of reliability is **test–retest reliability**, a comparison of scores or values obtained using the same measure at two different times. A high test–retest reliability indicates that the scores are fairly similar at both test times, and values of 0.90 or higher are often desired. If you assess individuals' likelihood of using cell phones while driving at two different times, you would expect high scorers on the first administration to have high scores on the second administration when there is no intervention to alter cell phone use. Similarly, we would expect low and moderate scorers to have low and moderate scores, respectively, on the second administration. Low test–retest reliability indicates that there is no consistency between the scores measured at time 1 and time 2.

Another common type of reliability includes **inter-rater reliability**, which measures the degree of rating similarity between two or more raters. If you have individuals rating behaviors or judging performance, you want to assure that the raters are using the same criteria for their ratings and judgments, which is reflected in a high and positive inter-rater reliability score.

Shifting now to the assessment of validity, the intent is to determine if the measures are truly measuring what they are intended to measure. There are four ways to collect evidence of validity: face, content, criterion, and construct. Face and content validity evidence is based on individual assessments and do not require the calculation of a correlation. A tool that has **face validity** *appears* to be appropriate for the variable being assessed. For example, most individuals will agree that having participants talk on a cellular phone while navigating a road course on a driving simulator would be an appropriate way to assess the impact of cell phone use on driving behaviors. Face validity is nice to have and might impact how involved participants are in the research project, but other measures of validity are needed.

Evidence of **content validity** also does not require a calculation; it is based on expert opinions. If you created a short survey to evaluate individuals' moods, you would have mood experts, or *subject matter experts (SMEs)*, review the items to ensure those items tap the appropriate content. The main requirement is that the person who creates the tool should *NOT* be the same person who assesses the content validity. If we created the items, we tend to be biased in thinking the items are good. Someone who did not create the items can be more objective.

Unlike face and content evidence of validity, both criterion and construct evidence of validity require the calculation of a correlation coefficient. The objective of **criterion-related validity** is to determine how well the measurement tool relates to some criterion measure. Often, the desire is to use the tool to predict the criterion, known as *predictive validity*. Let's say you create a questionnaire to assess a person's "openness" to colors (i.e., the degree to which individuals like a variety of colors). After individuals complete the questionnaire, you would have them experience rooms of different colors and you would record their moods. If individuals with high color "openness" were generally happy, but individuals with low color "openness" were generally sad in colored rooms, your questionnaire would be considered a valid predictor of mood in colored rooms, assuming the correlation coefficient was sufficiently strong (i.e., generally greater than 0.30).

Finally, **construct validity** provides evidence that your tool relates to a known construct such as personality. As an example, you might consider your "openness to color" questionnaire to be a short personality test. To test this, you could compare the scores on your questionnaire with scores on a known measure of personality (i.e., the construct). If there were a measurable relationship, a strong positive correlation between these scores, the questionnaire data would demonstrate construct validity. If the questionnaire is not correlated with the measure of personality, there is no construct validation and the questionnaire is measuring something other than personality.

Validity scores also should be positive, but they tend to be lower than reliability scores. A validity coefficient of 0.30 or higher can be considered extremely strong, depending on the context.

BOX 2.3: DETERMINING RELIABILITY AND VALIDITY

If we use the "openness to color" scale in our color experiment, first we would want to verify that the scale is reliable. To assess reliability, we could administer the scale to a group of individuals and then administer the scale to the same individuals several weeks later. The length of time between the two administrations of the scale cannot be too short or there is the possibility of individuals remembering how they responded previously and familiarity with the items can lead to the *practice effect*. If the time between administrations is too long, there is the possibility that something could impact an individual's level of openness. If the scores are similar on the two occasions, we would have a strong, positive relationship reflecting a reliable measure. After reliability is established, we want to determine if the scale is valid. We are most likely interested in evidence of criterion-related validity in order to predict the criterion (i.e., mood) based on one's predictor score (i.e., level of openness to color). To assess criterion-related validity, we need to identify the criterion measure that will be predicted. Once the criterion measure is determined, we can measure the predictor and criterion at the same time or at different times. If we measure the predictor and the criterion at the same time, this is *concurrent validity*. Measuring the predictor at time one and then measuring the criterion a little later assesses *predictive validity*. **Critical Thinking Questions**: Given our mood and color experiment, what could be the criterion? How would you measure the validity of our "openness to color" scale? How would you measure reliability and validity for the variables in Box 2.1 or Box 2.2?

2.5 Descriptive versus inferential methods

Recall that the purpose of human factors is to help ensure that the design of an environment meets the capabilities and limitations of human users, thereby reducing errors or increasing safety and ease of use. The purpose of psychological research is to increase our understanding of human behavior and cognition to enhance the quality of life. To do so, we need to understand the type of research being conducted, which varies depending on our research goal. The goals of science are to describe, predict, and explain. Our ultimate goal is to control, or apply what we know, to improve the situation. Do not think of control as a negative term. Stoplights are created to control our behavior, but in a positive and safe fashion. Similarly, human factors research attempts to determine how to "control" factors within our lives that might otherwise reduce our reliability, increase our errors, or make our environments unsafe.

To control behaviors, we need to understand what causes the behavior. If we can describe behaviors (e.g., Grandpa normally burns the toast), then we can predict behavior (e.g., Grandpa will burn the toast this morning, just watch!). Although we can predict grandpa's behavior, we might not be able to explain it. Grandpa might burn the toast because he forgot it was in the toaster, he could not see that the toaster settings were set too high, he could not smell the burning toast, or he needs a new toaster. Only after we can explain why grandpa burned the toast can we determine how to "control" the environment or behavior. If it is a visual issue, giving grandpa tips on how to remember the bread is in the toaster is not going to be helpful.

Similarly, if we observe that air traffic controllers tend to commit more errors toward the end of their shifts, we only have an understanding of what is related (errors and time on shift), but we do not know the true cause of the errors. The true cause of errors could be related to fatigue, sleepiness, anticipation of getting off work, or some other yet-unknown variable. Based on this relationship, we could remove all fatigued or sleepy controllers in an attempt to create a safer environment. After removing all fatigued controllers, we might find that the remaining controllers still experience errors. Only with research that explains and helps us understand the cause of behavior can we work to control our environment. Fatigue might cause the errors, or it is possible that fatigue can lead to attention deficits, which cause the errors. If this is the case, fatigue is related to errors, but only through the relationship to attention deficits.

To understand what variables are related and what explanations are plausible given a certain research question, we need to understand the different types of research methods. If you are unsure of the statistical aspects discussed next, a brief review of basic statistics is in Section 2.9. Below we discuss the different types of research methods and data that one might collect or have to interpret.

2.5.1 Descriptive methods

Descriptive methods are used to collect data that identify and define the situation and can be used to make predictions, which correspond to the first two goals of psychological research (describe and predict). These *descriptive data* include measures of how many, how much, or how often and are typically reported as frequencies. Baseball fans are familiar with descriptive data, as baseball, compared to all other sports, has the greatest number of recorded descriptive statistics, including runs batted in, home runs, errors, innings pitched, and strikeouts. Just like baseball fans, human factors researchers want to understand the situation. In our driving simulation, we might record descriptive data such as how many men and women participated, age of participants, how many times the individuals were distracted or crashed, how much time the individuals spent talking on the "phone," and mean time to complete a road course. Common descriptive statistics include measures of central tendency or the "average" (e.g., mean, median, or mode) and the dispersion of data (e.g., range or standard deviation). Review the supplemental quick review (Section 2.9) if these terms are not familiar.

One of the more common means for collecting descriptive data is through observation. Using the *observation method*, researchers watch individuals perform a job or task such as actually driving or performing in a driving simulator. The data collected during observation can be recorded as a narrative. Because people have different writing styles and can describe the same situation quite differently, a content analysis should be conducted on any narrative data.

In performing a **content analysis**, individuals identify which written or oral comments are similar and group them together. Then, the researcher, but possibly the individuals sorting these comments, assigns these categories a label that reflects the overall meaning of the category. For example, participants in our color research project might say, "I feel very comfortable," "This is a pleasant room," and "It seems like I have been in here only a few minutes." It is possible that the first two comments would be grouped together as reflecting a sense of comfort, whereas the third comment might be grouped with others that reflect a perception of time. Researchers often report the category label and the frequency of comments. To avoid the issues related to narrative responses and the need for a content analysis, some form of *checklist* or *scoring sheet*, which lists various behaviors (e.g., swerved to avoid crash, changed lanes to avoid crash, or crashed), can be used to record each time a behavior occurs. When researchers observe and tally the behaviors that occur, we must ensure that our raters have similar ratings. To check for similar ratings, we conduct a measure of inter-rater reliability.

Reviewing records is another means of collecting descriptive data. *Records* include any recorded data such as length of time to process a customer, a daily work log, or procedure or equipment manuals. Procedure or equipment manuals can give us information about expected processes, which can be compared to the actual process observed. Assuming confidentiality is not an issue, records are usually easily accessible and contain a great deal of descriptive data.

Besides observation or records, *surveys* or *questionnaires* are also a common means for collecting descriptive data. Surveys may be conducted by phone, in person, via email or the web, or by regular mail. See Chapter 5 for ways to construct a survey.

Once collected, descriptive data need to be organized in a useful way. We might present the data in a frequency distribution table or calculate a correlation coefficient (see Section 2.9). Correlations inform us as to how strong the relationship is between two variables and can help us make predictions. If the correlation between practice or training and accidents is strong and negative, we would predict that individuals with more training would have fewer accidents than individuals with less training.

BOX 2.4:   CORRELATION VERSUS CAUSATION

A study evaluated the relationship between an individual's self-reported driving behavior and actual data on driving behavior (Arthur et al., 2005). They found a positive correlation between the number of self-reported crashes and actual crashes ($r = 0.43$) and a positive correlation between self-reported tickets and the actual number ($r = 0.37$) of tickets received. These correlations indicate that individuals who reported more crashes and tickets also tended to have more crashes and tickets. As you might guess, these correlations do not suggest that reporting more crashes *caused* more crashes or reporting more tickets *caused* them to receive more tickets. Rather, the individuals were somewhat accurate in recalling the number of crashes and tickets. What if we measured the drivers' personality instead of asking for their self-reported information? Let's say we found that conscientiousness was negatively correlated with number of accidents and openness to experience was positively correlated with the number of tickets received. We would not change how we interpret the results just because we changed the variables in the study. **Critical Thinking Questions:** How might you interpret these results between personality and driving behavior? What other descriptive data could you collect in this line of research? What would be the expected correlation and respective interpretation of the result?

2.5.2 Inferential methods

In contrast to descriptive methods, **inferential methods** are used to explain relationships and infer cause-and-effect relationships. Continuing our previous example, if we know that individuals with more practice or training tend to have fewer accidents, we can calculate how many accidents a person is expected to have if the person has 6 hours of practice. A **regression** analysis calculates a line of best fit to the data and is used to make these predictions, but the line of best fit does not explain the relationship between the two variables of interest. We also could test the question, "Does spending time in a yellow room make people happier than spending time in other colored rooms?" If we find that people who are randomly assigned to yellow rooms compared to individuals in other colored or neutral rooms report higher levels of happiness, then we could conclude that yellow rooms increase people's level of happiness.

When using the **experimental method**, the researcher manipulates the independent variable. The manipulation of the independent variable involves assigning participants to an *experimental group*, the group that will be exposed to the treatment you are trying to test, such as a yellow room. The *control group* is the comparison group, such as those individuals who are assigned to experience a neutrally colored room. In our research on cell phone use and driving, the independent variable "cell phone use" has two levels, "driving with" and "driving without." The "driving with" level is considered our experimental group and the "driving without" group would be considered the control group because this group would not experience the use of cell phones. In contrast, the independent variable "complexity" does not have a "no complexity" group; rather the groups differ according to magnitude (low, medium, high) of complexity.

It is possible to create a matrix that represents a crossing of both of our two independent variables (see Table 2.1). Individuals who participate in our experiment will be randomly assigned to one of the cells (A–F), also known as groups or conditions. Individuals randomly assigned to cell A will be in the "driving with" group and will experience a low complexity situation when in the driving simulator. Individuals in cell B will also be in the "driving with" group, but will experience a medium level of complexity in the driving simulator.

After we have collected all the data for each condition, we will evaluate the main and interaction effects for this experiment using analysis of variance (ANOVA). A **main effect** is concerned with the result on the dependent variable for a single independent variable, such as cell phone use and whether there is a difference in reaction time between the "driving with" and "driving without" groups. The **interaction effect** informs us as to whether the impact of driving with or without using a cell phone leads to different reaction times depending on the level of complexity. It is possible that reaction time increases steadily for cells A, B, and C (in Table 2.1), but that the increase in reaction time across cells D, E, and F is not very great, as demonstrated in Figure 2.4. This would indicate a possible interaction because complexity appears to have a greater effect for individuals driving while using cell phones than for individuals driving without using them. A good way to "see" an interaction is to draw lines from the mid-points of the top of each bar for each condition (driving with— A, B, C; driving without—D, E, F). If the lines are not parallel, an interaction is likely.

When using the inferential method, we could use a **between-subjects design** whereby different participants are assigned to each group or condition. With this design, different participants would be exposed to a low, medium, or high complexity situation and to either the "driving with" or "driving without" conditions. We would compare responses "between" individuals. In contrast, a **within-subjects design** (sometimes referred to as a *repeated measures design*) would have the same participants experience all the different conditions of the independent variable. In this case, the results are from "within" the same participants. In our driving experiment, it might make sense to have the same participants experience the low, medium, and high complexity situations (in random order). We could design an

Table 2.1 A data matrix for cell phone use and task complexity

| Cell phone use | Complexity | | |
	Low	Medium	High
Driving with	A	B	C
Driving without	D	E	F

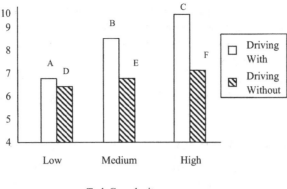

Figure 2.4 An example of an interaction effect.

experiment where different participants experience the "driving with" condition than those participants who experience the "driving without" condition (between-subjects), but the participants experience all three complexity conditions (within-subjects) regardless of the cell phone use condition. This is known as a **mixed design**, a combination of between (driving with or without a phone) and within (complexity level) designs.

Using within-subjects designs reduces error variance whereby we need fewer participants; however, we need to be aware of how the first trial affects the second or subsequent trials. If our mixed design had participants drive with and without a cell phone at their assigned complexity level, the participants will likely learn something or gain experience about the driving situation on the first trial that positively or negatively affects performance on subsequent trials. If we **counter-balance** the conditions by having half of the participants experience the "driving with" condition first and the other half experience the "driving without" condition first in each of the low, medium, or high complexity conditions, this will help offset any learning or practice effects.

We also need to be aware of confounding. **Confounding** occurs when we change the levels of the independent variable and a systematic change occurs in another variable. Going back to our room color and mood project, the change in color might also affect the amount of reflected light in the room. If we reject the null hypothesis for room color, we cannot be sure if color caused the difference or if it was the amount of reflected light. We could test this by comparing rooms that differ in color but are equally bright and have the same reflectance.

BOX 2.5: HYPOTHESIZING AFTER THE RESULTS ARE KNOWN

When formulating hypotheses, an investigator bases the hypotheses on what is currently known given the published research literature. This means that hypotheses are developed *before* collecting data. It is important to have good hypotheses that can be supported by the data, as research generally does not get published when there are no significant findings. Unfortunately, some researchers rewrite hypotheses because this makes the paper "flow" better or better fit the data collected. This practice of **H**ypothesizing **A**fter the **R**esults are **K**nown (HARKing, Kerr, 1998) is more widespread than it should be. In fact, it should not occur at all. Because statistical analyses are based on probabilities, which are determined from the hypotheses created without this new knowledge, changing the hypotheses *after* the results are known is similar to stating what the correct answer to an exam question is after learning the correct answer. Kerr (1998) discusses reasons why people might HARK, but argues that the costs of HARKing are harmful to science. HARKing also raises the issue of ethics, as discussed later in the text. **Critical Thinking Questions:** What do you think some of those harmful effects might be? What are some reasons for HARKing? Do you consider HARKing to be ethical?

2.5.3 Generalizability of the results

Once we analyze our data, we want to identify how our results apply to real-world settings. Because individuals selected in our samples are often college and university students, our data might not generalize well across all societies because our samples are WEIRD (Western, Educated, Industrialized, Rich, and Democratic, Henrich et al., 2010). Therefore, we should understand our population to determine if they are WEIRD and how that might impact our interpretations.

Even if we select from outside the college and university setting, generalization can be problematic. If we sample only 20-year-old males in our driving experiment, this limits the **generalizability**, or applicability, of our results to the broader population, which includes *all* individuals of driving age. A random sample will be more generalizable, whereas a convenience sample will generally have lower generalizability.

If your data are generalizable, then your project has **external validity**. One of the main problems for external validity is the sample, which usually is less diverse than the population as a whole. Experiments on the impact of room color on school-aged children might not generalize to Bill's grandfather's situation. If we want to learn about the appropriate colors to use in rooms for older adults, we need to sample from that population.

When determining who we will sample, we need to understand whether our results will generalize to individuals of different racial backgrounds, children, and the elderly. Obviously, for a driving experiment, children not of legal driving age would not be included in our population of interest. This is why it is critical that we are careful in describing our sample when we report our results, as the sample impacts to whom our results might apply. Until we conduct studies on different people under different conditions, or use **multiple operationalizations**, we should limit our generalizations to the population our sample represents.

Besides needing a representative sample to have generalizability, or external validity, the research project must have internal validity. **Internal validity** is achieved when our measures and procedure are reliable and valid. That is, the measures and procedure must be accurately and consistently testing or evaluating what you want to assess. When you place individuals in a driving simulator such as the one shown in Figure 2.5 with or without a cell phone and then give them various reaction time tests, are you really measuring the impact of cell phone use on reaction times? It would seem so.

Internal validity also can be affected by several other factors such as subject bias, experimenter bias, and demand characteristics. In the case of **subject bias**, the participants might be trying to determine the purpose of the experiment. When this happens, sometimes the individuals try to give the answer they think the experimenter wants. It is probably less common for participants to give answers they suspect are not desired by the experimenter, but still a possibility. In contrast, **experimenter bias** occurs when the researcher's own biases influence the results due to a reduction in objectivity. As an example, objectivity might be reduced if a coder is aware of the hypotheses and research conditions under which the data were collected. In this situation, the coder might "see" responses or behaviors that align with the hypotheses,

Figure 2.5 Does this driving simulator allow for internal validity. (Adapted from https://commons.wikimedia. org/wiki/File:Drunk_driving_simulator,_Montreal_by_CAA_of_Quebec.jpg.)

potentially favoring those responses or behaviors and biasing results. Experimenter bias generally arises due to natural processing biases, not evil intent. This is why at least one of the coders should be *blind* to (i.e., unaware of) the hypotheses and conditions.

Finally, **demand characteristics** are cues or clues within the research setting that suggest a certain response or behavior is desired. When one interviewer constantly smiles and nods approval while a second interviewer does not smile or nod, the demand characteristics differ. In the former case, the interviewer is encouraging responses, whereas the second interviewer is neutral, but possibly discouraging certain types of responses. These subtle differences can impact your research project or evaluation, reducing generalizability.

Generalizability is a concern in all types of evaluation of people's behaviors. A description of other methods of evaluation and ways of determining how people function with and within their environment are covered in Chapter 5: Evaluation Methods.

2.6 ETHICS

If the purpose of human factors research is to inform our decisions for designing environments to increase the quality of life, we must be careful not to degrade the quality of life in our search for answers during the research process. It is critical that we, as researchers, continuously review the ethical treatment of participants, including the use of human volunteers in human factors studies or experiments.

The methods of conducting research might appear minor or trivial to some individuals; however, these methods might raise concerns for others relative to the ethical treatment of research participants. We have ethical principles to guide our treatment of participants in research to avoid the pitfalls, and the biases or incidents of poor judgment that might lead to the unethical treatment of participants. One of these principles is to avoid conflicts of interest. **Conflicts of interest** arise when individuals have a vested interest or obligation to more than one objective, goal, or party. A researcher who experiences conflicts of interest might be negatively influenced whereby his or her judgment or interpretation of the data is not objective. For example, a researcher investigating the safety of cell phones whose research is sponsored by a cell phone company might feel pressure to minimize the importance of certain negative findings.

2.6.1 Nuremberg code and Belmont principles

The Nuremberg Code and Belmont Principles are critical components of the ethical principles for research. The establishment of the Nuremberg Code was a result of the Nuremberg trials (see Figure 2.6) in which Nazis were tried for the inhumane treatment of prisoners. The **Nuremberg Code** (*Trials of War Criminals*, 1949) articulates certain processes that must be followed in research such as the use of informed consent and allowing individuals to withdraw from research without penalty.

The development of the Belmont Principles was in response to the Tuskegee Syphilis Study (1932–1972). Because the researchers wanted to study the progression of the disease, they did not inform these Black men who had syphilis that they had the disease, nor did they provide treatment even after penicillin became a known cure in the 1940s (Jones, 1981). The underlying premise of the **Belmont Principles** (National Commission, 1979) is that researchers need to respect people and protect participants from harm.

2.6.2 Determining if a study or experiment is ethical

Determining if a study or experiment is ethical is a process of evaluating the benefits versus the risks to participants. Researchers must protect the participants from invasions of privacy and physical or psychological harm, and we must give participants sufficient information to give informed consent. It is ultimately the researcher's responsibility to ensure that the research is conducted ethically. Several guidelines and procedures assist us in determining if the research is ethical.

Figure 2.6 The Nuremberg trials (National Archives Identifier 540127).

2.6.2.1 Ethical principles of psychologists

Guidelines such as the *Ethical Principles of Psychologists and Code of Conduct* (American Psychological Association, 2017) help us determine what is or is not ethical. This *Ethics Code* includes principles on supervising students to professionalism to appropriate behavior with clients. The principles are constantly evolving and being updated because of incidents that occur. We will highlight only aspects related to research.

Of particular concern to research are the principles of competence, privacy and confidentiality, education and training, and research and publication. The *principle of competence* implies that researchers should conduct research only within the field for which they were trained. Based on the *principle of privacy and confidentiality*, it is also critical that the information collected from the participants is protected and the participants are not personally identifiable to people outside the research project.

Because research often occurs in educational institutions, students with various backgrounds and levels of expertise might become involved in research. The *principle of education and training* states that the supervising researcher, usually a faculty member, is responsible for supervising these students in the appropriate use of research methods and statistics, as well as in the appropriate ethical treatment of participants. Research students should learn how to keep confidential records, appropriately maintain confidentiality, and obtain informed consent according to the *principle of research and publication*. In addition, when reporting your results, you must write ethically by avoiding *plagiarism* (e.g., not giving appropriate credit to your sources of information), not using misleading statistics or misusing statistics, using the appropriate research design, and not manipulating the data to fit the desired results.

BOX 2.6:   ETHICAL DILEMMAS—FALSIFYING OR MISREPRESENTING DATA

In the research world, there are pressures to publish. To publish, a study or experiment needs to have interesting, if not statistically significant, results. If researchers falsify or misinterpret their data, these behaviors are unethical and produce misleading results. False results used by other researchers will influence their hypotheses, which could lead to further misdirected and inappropriate research. Similarly, if researchers withhold information, such as the negative effects of child restraints (and this is just a made-up example), this could have serious if not harmful or fatal results. To reduce these issues of research misconduct, researchers can use preregistration of their project. There are also ethical issues about confidentiality. When collecting data from individuals, you can learn quite a bit more about these individuals than they might voluntarily disclose (e.g., how quick they are to learn a particular task, potential biases, or various personality characteristics). It would be improper for us to discuss these personal aspects in any identifiable way. In addition, if someone we know asks us about how well a friend performed on the task, it would be unethical to disclose that information. Even though we are talking with a friend, we have an obligation to uphold the confidentiality agreement we made with the participants. **Critical Thinking Questions:** In either of our color or driving simulation experiments, what are some possible negative consequences if these data were falsified? What are some potential ethical concerns regarding the principle of privacy and confidentiality? How could preregistration reduce these ethical issues?

2.6.2.2 Informed consent

Researchers are ethically obligated to inform participants about the procedures of their study so that participants can make an informed decision about whether to participate based on the potential risks and benefits of the research. This is called **informed consent**. Unless you are assessing an educational program or using data that already exist in a database and no individuals can be identified with the data, your participants will need to give informed consent.

After giving informed consent, a participant still has the right to discontinue participation, even if the experiment has already started. This is where conflicts of interest might arise. A researcher might be tempted to try to talk the participant into staying because the researcher needs only one more participant to finish the study. Retaining an individual could be unethical because participants cannot be coerced to participate in research.

2.6.2.3 Risk/benefit ratio

If participants must give informed consent and they cannot be coerced to participate, people sometimes wonder how it is ethical to pay or recruit participants, or to use deception. Determining if an experiment or study is ethical is a judgment, and those judgments must weigh the risk or cost relative to the potential benefit gained.

The **risk/benefit ratio** is a subjective evaluation of what is gained versus what might be a cost to the participants and society. Research of a high quality is usually perceived to be worth the risk if the project can demonstrate the hypothesized outcomes. In some cases, the risk is too high regardless of the benefits, and the research would not be considered ethical. Therefore, the problem lies with determining risk.

Risk may include physical, psychological (mental or emotional), or social injury. As life is risky, risk is evaluated relative to everyday activities. Research that exposes individuals to situations or experiences such as driving, going to school, or playing video games are similar to their everyday experiences. Therefore, the likelihood of injury is not perceived to be greater than life itself and is considered *minimal risk*. If the risk level is greater than minimal risk, then the participants are *at risk*.

Experiments that put participants at risk include those that invade a participants' privacy by asking questions of an extremely personal nature (e.g., "Do you take drugs before driving?"). At-risk research may be conducted, but certain procedures must be followed. First, the data must be kept confidential. It is not necessary for data to be anonymous; however, with anonymous data, the possibility of determining who completed a particular questionnaire or last drove the simulator is minimized or eliminated. Collecting data anonymously is the best practice if there is no need to associate the data with a specific participant. If there is a need to collect the data over the long term, as with a repeated measures design, then it is necessary to have some type of identifier such as names or codes to track the data. Using names or some type of decipherable code is acceptable as long as the data remain confidential. That is, the data are only seen by the approved experimenters, no data are reported that could be linked to a particular individual to protect from an invasion of privacy, and the codes that link the names to the data are destroyed after all the data are collected and verified.

Further, researchers also should have a plan to deal with risk if something happens during data collection. If someone becomes extremely stressed in the high complexity situation in our driving experiment, it would be important to stop the experiment, debrief the individual, and excuse the individual from the study. If it appears as though the person is more stressed than what might normally be expected and the individual does not appear to be calming, you might escort the person to, or make the person an appointment at, a counseling center. The likelihood is small, but it could happen, so researchers need to be prepared.

2.6.2.4 Use of deception and debriefing

Another issue that puts research participants at risk is the use of deception. The use of deception is perceived by some to be extremely problematic, whereas others perceive it to be a necessity in order to conduct certain types of experiments. **Deception** occurs when individuals are deliberately misled, which could mean giving participants misinformation or misleading information, or even withholding information. Because of the latter definition, the use of deception is fairly common, but usually innocuous. For example, if you want to study the impact of room color on mood and you tell the participants that you are studying the impact of room color on mood, the participants are likely to exhibit or report the mood they *think* you want. To avoid this problem, it is often necessary to withhold information so that the data collected are not biased.

If deception is used, the experimenter must explain the true purpose and hypotheses of the study and allow participants to ask questions after the data are collected during a **debriefing**. The purpose of the debriefing is to reduce the participants' possible misconceptions of the study, to reduce any possible discomfort, and to remove any possible risk from deception.

2.6.2.5 Institutional review boards (IRBs)

Universities that conduct research have an Institutional Review Board (IRB) composed of researchers at the university as well as individuals from the surrounding community. The **IRB** is an advisory group that provides an external objective review of the proposed research. Hospitals and many companies that conduct research also have their own IRB.

Researchers must submit their research proposals to the IRB for approval, and IRB approval must be received before beginning any research project involving human participants. It is a Federal regulation, the law, for a researcher to acquire IRB approval. Further, to receive federal funding, a project must receive IRB approval. Even though the IRB "approves" the project, the researcher is ultimately responsible for the ethical conduct within the research project. If any study or experiment is in violation of or non-compliant with the federal regulations, *all* research at that institution can be shut down.

2.7 APPLICATIONS TO SPECIAL POPULATIONS

Human factors research is extremely diverse and can include investigations of computer desk use in schools on musculoskeletal disorders in children, the impact of cell color on prisoner behavior, or the usefulness of tactile or auditory displays in helping individuals with physical disabilities find their way through a town. Some groups of individuals are considered vulnerable groups. **Vulnerable groups** include individuals who have some type of limitation of either their capacity to give informed consent or in voluntariness, the extent to which they feel some pressure or obligation to participate (CITI, n.d.). Children, prisoners, pregnant women, individuals with physical disabilities, mentally disabled persons, and economically or educationally disadvantaged individuals are all considered to be vulnerable populations. The IRB will check to be sure you have taken special precautions to protect these vulnerable groups.

Other issues to consider are whether the research results will generalize to various groups of diverse individuals. Will the results of a study apply to African Americans, Hispanics, and Native Americans; to men and women; to children and the elderly; and to individuals with disabilities?

BOX 2.7:   WHAT DO WE KNOW ABOUT THE OLDER ADULT DRIVER?

Keep in mind that a great deal of research is conducted on the college student population. Therefore, the research findings might not always apply to all age groups. In particular, the driving study example used in the text might result in different results if the participants were older adults. Mouloua et al. (2004) operationalized "older" as 60 years of age or greater. They found that, although older adults were more careful in a simulated driving task than younger drivers, they tended to display greater degradations in attention. Therefore, if attentional demands are already high for the older driver, having them talk on the cell phone simultaneously could be catastrophic. Similarly, elderly adults (65 or older) in a driving simulation merged more slowly into traffic than younger drivers (26–40) when large trucks were on the simulated Dutch highway (de Waard et al., 2009). **Critical Thinking Questions**: What should be considered when attempting to generalize the results of our driving experiment? What other considerations about special populations might we need to consider?

2.8 SUMMARY

Before making adjustments to the environment by painting rooms various colors to reduce depression or prohibiting our elders from driving to reduce traffic accidents, you should use the scientific method to determine whether the data support these decisions. Because human factors research is conducted to answer real-life problems, it is applied research. In the search for these solutions or answers, human factors specialists might conduct studies, experiments, or quasi-experiments to investigate the problem. Either descriptive or inferential or both methods might be used.

Regardless of the type of research conducted, it is important to consider how the sample will be selected and how this will affect the generalizability of the results to the population. In addition, the variables must be operationalized well, reliable, and valid, and the research project itself must have internal validity.

Most importantly, it is critical that the researchers act ethically throughout the research process, which includes the collection of data, proper treatment of participants, data analysis, interpretation of the data, and the writing of the final report or article. In particular, research data must remain confidential, and participants usually must have the opportunity to give informed consent.

2.9 SUPPLEMENT: A QUICK REVIEW OF DESCRIPTIVE AND INFERENTIAL STATISTICS

This section is a quick review of basic descriptive and inferential statistics for individuals who have already taken a statistics or research methods course. Individuals who have not taken one or both of these courses should find this information helpful with basic statistical information.

2.9.1 Descriptive statistics

Descriptive data only describe the situation and people in our research. Common forms of descriptive statistics include frequencies, measures of central tendency, and measures

Table 2.2 A frequency distribution table

Sex	Number	Percent
Men	35	43.75
Women	45	56.25
Total	80	100.00

of dispersion. A *frequency distribution table* (see Table 2.2) allows us to see quickly how many of each type of person or situation exist.

Besides the frequency distribution, we can calculate the "average" or *central tendency*. We measure central tendency in three ways: the mean, the median, and the mode.

The *mean* is the arithmetic average. If five individuals responded to a survey item using a scale from 1 to 5 (very satisfied to very dissatisfied) and the responses were 1, 2, 3, 4, and 5, then the mean would equal the sum of all responses divided by 5, the number of respondents. In this case, the mean would be $((1 + 2 + 3 + 4 + 5)/5) = 3$ The *median* is the mid-point of the distribution. If your data set included the following responses: 4, 3, 4, 5, 2, 4, 4, 1, 5, 6, 7, a frequency distribution table can help you "see" the median (see Table 2.3). In this example, the median would be 3 because this splits your distribution in half—there are four values below and four values above 3. Finally, the *mode* is the most frequently occurring number. From Table 2.3, we see that there are five threes, which are easier to see when plotted in a histogram as shown in Figure 2.7. Therefore, the mode also would be 3. It is possible to have more than one mode in a data set. Sometimes the data sets are *bi-modal*, where there are two modes, *tri-modal*, which includes three modes, or *multi-modal*, where there are more than three modes.

Table 2.3 Frequency of response options

Response option	Frequency
1	1
2	3
3	5
4	3
5	1
Total	13

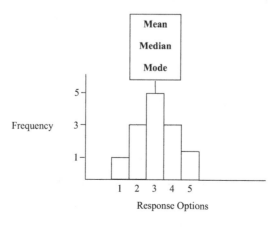

Figure 2.7 Histogram of response options.

Table 2.4 Frequency distribution table to compare differences in dispersion

Response	Data set 1 Frequency	Data set 2 Frequency	Data set 3 Frequency
1	3	9	0
2	9	10	0
3	26	12	45
4	9	10	3
5	3	9	2

Next, we normally want to know how much the data are dispersed. Even if two data sets have the same mean, the data could be distributed differently. In Table 2.4, the means are 3.00, 3.00, and 3.14 for Data Sets 1, 2, and 3, respectively. Although the means are identical or similar, the dispersions are quite different. The *range* is one measure of dispersion and is calculated by subtracting the smallest number from the largest number in the data set. Data Sets 1 and 2 represent a wide range of scores, with a range of $5 - 1 = 4$. There is a restriction in the responses in Data Set 3, as no one responded with a rating of 1 or 2, whereby the range for Data Set 3 is $5 - 3 = 2$.

As the range is a relatively crude measure of dispersion, we generally calculate the *standard deviation (s)*, which essentially measures, "on average," the distance of the data from the mean. Because Data Sets 1 and 2 have values from 1 to 5, the standard deviation will be greater than for Data Set 3, which has data ranging from 3 to 5 (see Table 2.4). Even though both means are 3.0 in Data Sets 1 and 2, the values in Data Set 2 fall farther from 3.0 because there are more responses at the lower and higher values than in Data Set 1.

We calculate the standard deviation with $s = \sqrt{\dfrac{\left(\sum x\right)^2 - \dfrac{\left(\sum x^2\right)}{n}}{n-1}}$. For the data in Table 2.4, the standard deviation is larger for Data Set 2 ($s = 1.37$) than Data Set 1 ($s = 0.93$) or Data Set 3 ($s = 0.45$).

We can also assess the relationship between two or more variables. This is correlation and, essentially, we measure how the variables co-vary. The *correlation coefficient (r)* is the calculated value and measures the direction (positive or negative) and strength of a relationship between two variables, such as between the number of minutes talking on the "phone" and the number of "accidents." When values on one variable, such as minutes on the

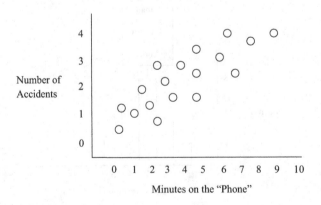

Figure 2.8 An example of a positive correlation.

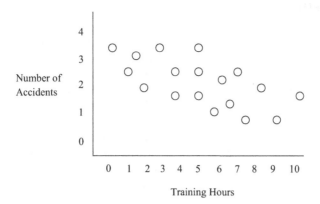

Figure 2.9 An example of a negative correlation

phone, increase and correspond to increases in values on a second variable, number of accidents, this represents a *positive correlation*, as depicted in Figure 2.8. For this positive correlation, we could say individuals who talk more minutes on the phone tend to have more accidents.

In contrast, a *negative correlation* exists when the values of the variables change in opposite directions. Let's say individuals in our driving study reported the number of hours of training to drive while talking on a cell phone. If the number of hours of practice increases while the number of accidents decreases, this would be a negative correlation, as shown in Figure 2.9. This negative correlation suggests that as training increases, the number of accidents decreases, which would be a good relationship to have.

The correlation coefficient may fall between negative one and positive one (i.e., $-1 \leq r \leq +1$), inclusively, representing negative and positive relationships, respectively. When the absolute value of the correlation coefficient is large (i.e., closer to either to -1 or $+1$), this reflects a strong relationship between the two variables. With a strong correlation, you can predict better the change in one variable as the other variable changes. Although the correlation of -0.90 is negative, it is a stronger relationship than a correlation of $+0.75$. If there is little or no relationship between the variables, the correlation value will be close to zero.

Correlation coefficients only tell us the strength and direction of the relationship between the two variables. We do not gain any information about causation and cannot conclude that practice causes fewer accidents. There could be other factors that influence both practice and the number of accidents. For instance, if the number of hours on a cell phone is positively correlated with accidents, it is possible that another variable such as carelessness is actually the cause. Careless people might be more likely to spend more time talking on their cell phones while driving. This is known as the third variable problem, which is discussed in the chapter.

2.9.2 Inferential statistics

Inferential statistics are the tests used to interpret and understand our data, as well as to determine if our manipulations have an effect on the dependent variable. There are various statistical tests including the t-test and analysis of variance (ANOVA). We use probability to determine whether a t-test or ANOVA has *statistical significance*. Statistical significance indicates that a highly unlikely result (based on probability and an assumed "true" null hypothesis) was observed, and the null hypothesis can be rejected.

LIST OF KEY TERMS

Applied research
Basic research
Belmont principles
Between-subjects design
Conflicts of interest
Confounding
Construct validity
Content analysis
Content validity
Counter-balance
Criterion
Criterion-related validity
Debriefing
Deception
Demand characteristics
Dependent variable
Descriptive methods
Effect size
Experiment
Experimental method
Experimenter bias
External validity
Face validity
Generalizability
Hypothesis
Independent variable
Inferential methods
Informed consent

Institutional Review Board (IRB)
Interaction effect
Inter-rater reliability
Internal validity
Main effect
Mixed design
Multiple operationalization
Nuremberg code
Operationalize
Population
Power analysis
Quasi-experiment
Random assignment
Random sampling
Regression
Reliability
Risk
Risk/benefit ratio
Sample
Scientific method
Study
Subject bias
Test–retest reliability
Theory
Third variable problem
Validity
Vulnerable groups
Within-subjects design

SUGGESTED READINGS

American Psychological Association. (2017). Ethical principles of psychologists and code of conduct (2002, amended effective June 1, 2010, and January 1, 2017). https://www.apa.org/ethics/code/.

This is the complete code of ethics psychologists are to follow, published in 2002 and amended in 2010 and 2016.

Heiman, G. A. (2002). *Research Methods in Psychology*. Boston: Houghton-Mifflin.

This is a good resource book for understanding research methodology and related statistics.

Jones, J. H. (1981). *Bad Blood: The Tuskegee Syphilis Experiment*. New York: The Free Press.

This book uncovers how the U.S. Public Health Service began and continued a 40-year study on untreated syphilis in black men in Macon County, Alabama, around the city of Tuskegee, even after the Nuremberg trials ended and penicillin became a known treatment.

National Commission for the Protection of Human Subjects of Biomedical and Behavioral Research. (1979, April). *The Belmont Report: Ethical Principles and Guidelines for the Protection of Human Subjects of Research*. Washington, DC: National Institute of Health.

If you are interested in learning more details about the Belmont Report, this is the full document. The Belmont Principles discuss in detail what informed consent is, how to assess risks and benefits, and proper ways of selecting individuals for research.

Trials of War Criminals before the Nuremberg Military Tribunals under Control Council Law No. 10,
  2, 181–182. (1949). Washington, D. C.: U.S. Government Printing Office.
  You will gain a more thorough understanding of the Nuremberg Code and why the IRB requires
  various procedures.

CHAPTER EXERCISES

1. You are studying the impact of room color on mood. You have college students come into a laboratory where there are two identical rooms, but one room is painted yellow and the other is painted pink. The students perform a vigilance task in either the yellow or pink room. After 30 minutes, the students' mood is assessed.
   a. Describe the population that is being studied.
   b. Describe the sample, how the sample should be selected, and the pros and cons of this type of sample. Explain how this sample is different from other types of samples.
   c. Describe the groups to which these results could be generalized.
   d. Write a possible hypothesis for this study.
2. You are investigating whether individuals truly understand how to use home-use blood pressure monitors.
   a. Design a research project you could conduct that would have high internal validity.
   b. Explain whether this is or is not a quasi-experiment.
   c. Explain how your sample might influence the results.
   d. Discuss what variables might be confounded.
3. For the experiment described in #1,
   a. What is the independent variable?
   b. What is the dependent variable?
4. Operationalize each of the following:
   a. Friendliness
   b. "Best" performer
   c. Depressed grandparent
   d. Clear and understandable instructions
   e. Legible street signs
   f. Safe environment (e.g., a kitchen or roadway)
   g. Older employees
5. For the following situations, explain how this could be a conflict of interest and how a conflict of interest could be avoided.
   a. You wish to study the impact of seating design on lower back pain. A large, well-known chair manufacturer is willing to fund your research.
   b. A leading drug manufacturer is willing to fund your research on the development of labels to help reduce the number of misread medicine labels in hospitals.
6. For the following situations, (1) explain how informed consent would be given, (2) explain whether there is minimal risk or the participants would be at risk, and (3) explain your decision on the risk/benefit ratio (i.e., should the IRB approve this research?).
   a. A researcher wants to investigate the validity of a new machine to install in vehicles for identifying if someone is too drunk to drive. The researcher wishes to include in the research individuals who have had at least one DUI (Driving Under the Influence or DWI: Driving While Intoxicated).
   b. In order to study the impact of ambient noise on hearing perception, the researcher wishes to study the individuals with hearing loss at a nearby nursing home.

    c. Faculty members ask their students to participate in a research project in which they will play a videogame. The game has the students "flying a mission" and "eliminating" the enemy. The students will get extra credit if they participate.

7. Which type of research, studies or experiments,
    a. Has greater **internal** validity? Explain.
    b. Has greater **external** validity? Explain.

8. You wish to test the hypothesis that individuals exposed to yellow rooms are happier than individuals exposed to green or blue rooms.
    a. Design a between-subjects experiment that would test this experiment. That is, identify the independent and dependent variables and explain what the participants would experience. Explain if you would have to use counter-balancing.
    b. Design a within-subjects experiment that would test this experiment. Explain if you would have to use counter-balancing.

9. For each of the following, (1) explain what the relationship means and (2) explain how the result could be applied.
    a. The correlation between attendance and performance in an Algebra class is 0.87.
    b. The correlation between age and cognitive alertness while driving is −0.45.
    c. The correlation between room temperature and depression level is 0.22.

10. For each of the following, explain how the result could be applied.
    a. The performance of individuals in the pink room was significantly slower than the performance of individuals in the yellow room.
    b. Individuals in the more complex driving simulation had significantly more accidents than individuals in the simpler driving simulation.

11. For the following, (1) determine if the process is ethical and (2) discuss why it is ethical or what you could do to make it ethical, if possible.
    a. Some individuals in your vigilance project continue to fall asleep, so you decide to sound an alarm every few minutes to keep them awake.
    b. Near the end of your data collection, you have some students who are frustrated with the project and want to quit. You persuade the students to continue and complete the project.
    c. Your results did not come out as you predicted so you decide to rewrite your hypotheses so that the paper will flow better.
    d. Some of your results are as you predicted, but others are difficult to explain. You decide to eliminate the confusing data.

12. Given the following frequency distribution table, determine the mean, median and mode.

Response option	Frequency
1	1
2	2
3	8
4	6
5	3
Total	20

13. Given the following scatterplots, estimate the correlation coefficient:

(a)

(b)

(c)

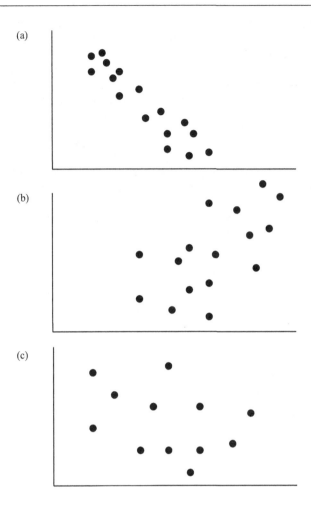

REFERENCES

American Psychological Association. (2017). Ethical principles of psychologists and code of conduct (2002, amended effective June 1, 2010, and January 1, 2017). https://www.apa.org/ethics/code/.

Arnoult, M. D. (1972). *Fundamentals of Scientific Method in Psychology*. Dubuque, IA: Wm. C. Brown Company Publishers.

Arthur, W. Jr., Bell, S. T., Edwards, B. D., Day, E. A., Tubre, T. C., & Tubre, A. H. (2005). Convergence of self-report and archival crash involvement data: a two-year longitudinal follow-up. *Human Factors*, *47*, 303–313.

Chapanis, A. (1991). To communicate the human factors message, you have to know what the message is and how to communicate it. *Human Factors Society Bulletin*, *34*(11), 1–4.

Chapanis, A. (1992). To communicate the human factors message, you have to know what the message is and how to communicate it: part 2. *Human Factors Society Bulletin*, *35*(1), 3–6.

CITI: Collaborative IRB training initiative – a course in the protection of human research subjects. (n.d.). Retrieved January 2002 from the University of Miami Web site: https://about.citiprogram.org/series/human-subjects-research-hsr/.

de Waard, D., Dijksterhuis, C., & Brookhuis, K. A. (2009). Merging into heavy motorway traffic by young and elderly drivers. *Accident Analysis and Prevention*, *41*, 588–597.

Henrich, J., Heine, S. J., & Norenzayan, A. (2010). The weirdest people in the world? *Behavioral and Brain Sciences*, *33*(2–3), 61–83.

Jones, J. H. (1981). *Bad Blood: The Tuskegee Syphilis Experiment.* New York: The Free Press.

Kerr, N. L. (1998). HARKing: hypothesizing after the results are known. *Personality and Social Psychology Review, 2,* 196–217.

Lee, T. M. C., Chen, F. Y. H., Chan, C. C. H., Paterson, J. G., Janzen, H. L., & Blashko, C. A. (1998). Seasonal affective disorder. *Clinical Psychology: Science and Practice, 15,* 275–290.

Mouloua, M., Rinalducci, E., Smither, J., & Brill, J. C. (2004). Effect of aging on driving performance. *Proceedings of the Human Factors and Ergonomics Society 48th Annual Meeting,* September 20-24, 2004, New Orleans, LA. pp. 253–257.

National Commission for the Protection of Human Subjects of Biomedical and Behavioral Research. (1979, April). *The Belmont Report: Ethical Principles and Guidelines for the Protection of Human Subjects of Research.* Washington, DC: National Institute of Health.

Norman, D. A. (1988). *The Psychology of Everyday Things.* New York, NY: Basic Books.

Trials of War Criminals before the Nuremberg Military Tribunals under Control Council Law No. 10, 2, 181–182. (1949). Washington, DC: U.S. Government Printing Office.

Chapter 3

Visual, tactile, and olfactory displays

Chapter Vignette

Steve knew that his vision was different from other people. He experienced difficulty discriminating between red and green signal lights at intersections and colored text on computer displays that others seemed to find easy. He learned to rely on a variety of subtle cues such as noticing that the green light usually was the bottom most of the three signal lights and was usually brighter than the red top-most light. He used these cues and others to pass his driving test and to negotiate intersections when he was driving alone. Later, an ophthalmologist tested his color vision after Steve reported his difficulty with traffic lights. Using the Ishihara color test that consists of different color dots some of which form numbers, Steve found that he could see a number in some of the color plates but not in others. He was diagnosed with a common form of color blindness—red-green color deficiency. A more apt description was that he had some form of "color limitation" not color blindness since he could tell the difference between blues and yellows—it was just the reds, green, and yellows that he had difficulty with. He continues to have trouble interpreting the LED lights used to indicate normal or abnormal states on printers and electrical devices, as well as deciphering colored maps and graphics in journals or news articles.

The challenges faced by Steve are an example of the issues that users confront using displays that do not match their abilities or that are viewed under environmental conditions that affect their visibility and legibility. Much is known about the sensory abilities of human operators, normal variation in these capabilities, and how these abilities vary depending on environmental conditions and across the life span that can inform the design of information displays. This chapter introduces the design of information displays based on our knowledge of the sensory capabilities of human operators.

3.1 CHAPTER OBJECTIVES

After reading this chapter, you should be able to:

- Describe the different classes of displays.
- Describe the different factors that should be considered in the design of effective displays.
- Describe basic visual functions like acuity, color, and contrast sensitivity.
- Describe the different types of cues that observers use to judge depth.
- Describe examples of tactile and olfactory displays.

DOI: 10.1201/9781003515463-3

3.2 WHAT ARE DISPLAYS?

A central issue in human factors concerns how to convey information efficiently and accurately to the user to support their behavioral goals. **Displays** are central to this process as they are the primary means by which users acquire information about a system including its current state and feedback regarding the effect of the user's inputs. Consider when you drive, the displays in the vehicle provide you information on a wide range of vehicle parameters including your speed, battery charge, fuel and oil level, and engine temperature. Symbols on the dashboard indicate whether the car is in drive, reverse, and whether you have secured your seat belt. An auditory display chimes reminding you of your failure to buckle up. Although employed less frequently, tactile displays such as the vibration of your cellular phone and olfactory displays like the odorant added to natural gas can also be used to convey information to the user. In this chapter, we focus on visual, tactile, and olfactory displays. Chapter 4 addresses auditory displays.

3.3 VISUAL DISPLAYS

There are many different types of visual displays, and it is useful to make a distinction between static and dynamic displays. **Static displays**, as the name implies, are displays that do not change such as stop or yield road signs, warning or caution labels on equipment, building names, and signs on women's and men's bathrooms. Dynamic displays are used to represent information that changes over time such as the signal strength on your Wi-Fi, the progress bar when copying a file, the speedometer, and oil temperature light in your car.

3.3.1 Static displays

The effectiveness of static displays is influenced by several factors that need to be considered when designing a display. Summarized in Table 3.1 is a list of these factors (Woodson 1981). Two of the main concerns regarding the design of a static display like a sign are **conspicuity** and **visibility**.

Table 3.1 Optimizing sign detection, recognition, and understanding

Conspicuity	A sign should be positioned so that it stands out prominently relative to other objects in the environment. Color, size, shape, flashing, and symbology can be used to increase conspicuity.
Visibility	Environmental conditions (day, night, rain, fog, etc.) change and a signs visibility under these different conditions should be maximized.
Legibility	Font, fore and background contrast, and text size should be selected to ensure legibility under different environmental conditions.
Intelligibility	A sign should make explicit what the hazard is and what may happen if a warning is ignored.
Emphasis	Symbols or words that denote "danger," "hazard," "caution," and "warning" should be the most salient feature of a sign.
Standardization	When possible, standardized symbols use by government agencies or industry should be employed.
Maintainability	Signs should be of materials that make them resistant to the effects of the elements (snow, rain, sun, temperature, etc.) and vandalism.

Adapted from Woodson (1981).

3.3.1.1 Conspicuity

Conspicuity refers to how well a sign attracts attention and can be enhanced using color and movement such as in the case of marquee lights and by placing the sign near where the viewer might be looking. In the case of vehicle warnings or caution lights, conspicuity is enhanced by placing them in the instrument cluster near the center of the field of view of the driver and using salient colors like red and yellow that are associated with warnings and cautions. Conspicuity is further enhanced by having the signal increase in brightness when activated and/or blink on and off. Conspicuity is also dependent on the environmental context as in the case where a sign appears in a cluttered visual scene like that shown in Figure 3.1, which makes the sign less conspicuous increasing an operator's time to find the relevant information.

3.3.1.2 Visibility

Visibility refers to how well a sign can be seen. The visibility of a sign is primarily dependent on its size, its distance from the viewer, and where the image falls on the retina. Both acuity (i.e., the ability to resolve small image details) and sensitivity to color differences are worse in the peripheral retina than in the fovea, the central region of the retina, where an image falls when you look directly at it. Hence, critical information that must be detected should be located near where the users gaze would normally be directed. The design of the sign should also consider whether it must be visible under day- and nighttime conditions. These issues are discussed in more detail later in the chapter.

3.3.1.3 Legibility

The use of text labels, part identification codes, and passwords depends on legibility, which is defined as the ability to distinguish letters, numbers, and symbols from each other. When using passwords that use combinations of numbers and letters, the choice of the font can make it difficult for the user to discern the number "1" from a lowercase "L" or a zero from the letter "O." Difficulty discriminating between alphanumeric characters can pose serious safety concerns as exemplified by an incident where unclear text on air traffic control display

Figure 3.1 Visual clutter can increase the time to find the relevant sign and slow a driver's response to road information. (By MPD01605 [Public domain], via Wikimedia Commons.)

was blamed for controller's errors in reading aircraft altitudes and destinations resulting in the controller directing the aircraft to the wrong airspace sectors. The air traffic controllers reported problems discerning between the numbers zero, six, and eight due to the small font size used on the displays (BBC-News 2002). Legibility is also dependent on both observer characteristics such as visual acuity and on stimulus properties including the shape of individual characters, the length of ascender and descenders of lowercase characters like the "p" and "h," stroke width, and the contrast of the text or figure relative to the background.

3.3.1.4 Readability

Readability concerns how text qualities including the spacing between characters or lines of text, line lengths, font size, uppercase letters, lowercase or mixed case affect how easily a reader can read the text. Use of all uppercase text slows reading performance (Forbes et al. 1950) and is not recommended when speed is an important design consideration or for long sections of text like paragraphs where it may affect intelligibility.

3.3.1.5 Symbolic

Symbols or pictographs are pictorial figures used to represent information. They have found wide use in modern graphical user interfaces and are an effective means of conveying information. They are most effective when used to depict a concrete object or action and are less effective in conveying complex or abstract concepts. Symbols are also advantageous because, unlike text, they do not need to be translated for persons unfamiliar with the language.

The design of pictographs represents a number of challenges because people must not only be able to readily identify the symbol, but also recognize its referent—the word, object, place, or concept designated by the symbol. While people may readily recognize a depicted object, they may not necessarily comprehend the referent. Consider the complex symbol shown in Figure 3.2. Can you identify what the complex symbol represents

Figure 3.2 Can you identify what this icon means? (©Alex Chaparro.)

Figure 3.3 Examples of icons used for camping. (By Tkgd2007 [Public domain], via Wikimedia Commons.)

and its referent? What should people do if they see this symbol? Hint: This pictograph was found on a ferry. Compare this to the camping icons shown in Figure 3.3. How many of the referents can you identify?

More detailed illustrations, sometimes accompanied by text, are used in pamphlets like the passenger safety and evacuation instructions illustrated in Figure 3.4 and other materials to convey more complex actions. This type of illustration is less ambiguous and can convey more complex information than simpler icons.

3.3.1.6 Text

Until recently, paper documents were the primary means of distributing information to users. With the widespread availability of personal computers, smart phones, and watches, we have come to rely on these devices to access media (newspapers, books, and magazines),

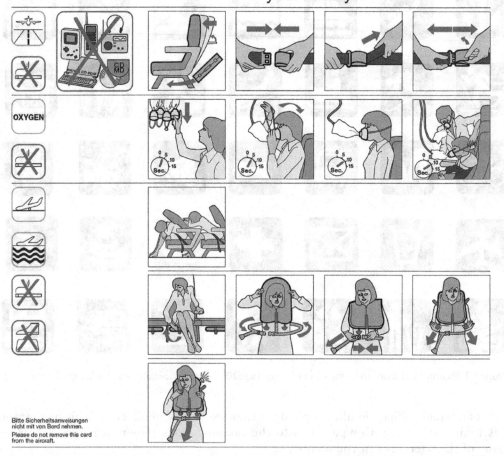

Figure 3.4 A copy of a pamphlet describing passenger safety and evacuation instructions. (Available via CC-BY-SA 3.0 from http://creativecommons.org/licenses/by-sa/3.0/de/deed.en.)

technical documentation (manuals and instructions), and communication (email and text messages). These devices pose challenges for the display of alphanumeric information given the small size of the display screen.

The readability of alphanumeric information can be maximized with careful consideration of several basic visual properties including the visual angle and contrast of the text as well as properties of the individual characters including their stroke width and the type of font used to display the alphanumeric information.

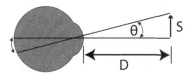

Figure 3.5 The calculation of the visual angle (θ) subtended by a stimulus takes into account both the distance of the object from the observer (D) and the physical height or width (S) of the object. (©Alex Chaparro.)

3.3.1.6.1 Visual angle

The size of alphanumeric characters is specified by their **visual angle** or the angle subtended by the stimulus on the retina rather than the characters physical width or height. The visual angle subtended by stimulus, shown in Figure 3.5, can be calculated using the equation shown below, where H corresponds to the height of the stimulus and D corresponds to the viewing distance expressed in the same units. The units for visual angle are minutes of arc or seconds of arc. A circle has 360 degrees and 1 degree of arc is equal to 60 minutes of arc, while 1 minute of arc is equal to 60 seconds of arc. For a comparison, a full moon subtends 0.5 degrees or 30 minutes of arc, and a person with good vision can discriminate details subtending 0.8 minutes of arc.

$$VA\,(\text{minutes}) = \frac{3{,}438 \times H}{D}$$

The advantage of specifying the dimensions of stimulus in terms of visual angle is that it accounts for the combined effect of a stimulus's size and an observer's viewing distance. After all, even a large stimulus can be difficult to discern if seen from a large distance and likewise a small stimulus is quite visible if seen up close.

3.3.1.6.2 Contrast

The visibility of static stimuli like alphanumeric characters depends on the luminance contrast of the foreground relative to the background. The black characters on this page are visible because they reflect less light than the white background. The contrast of a stimulus can be expressed using one of several equations including luminous contrast and the contrast ratio, where L_{max} and L_{min} are the maximum and minimum measured luminance of the visual pattern, respectively.

$$\text{Luminous contrast} = \frac{L_{max} - L_{min}}{L_{max}},$$

$$\text{Contrast ratio} = \frac{L_{max}}{L_{min}}$$

The recommended size and contrast of a stimulus depends on a number of factors including the criticality of the information, the viewing distance, observer characteristics (i.e., age and visual impairment), and ambient illumination levels. Adults with normal vision are reasonably tolerant of variations in contrast under normal viewing conditions (Legge et al. 1987). However, studies of reading show that performance is more dependent on contrast especially when the text is very small or viewed under lower illumination levels. Important from a design perspective is the fact that the effect of contrast can be offset to some degree by making the stimuli larger and similarly the effect of small stimulus size can be partially offset by increasing the contrast of the stimulus (see discussion of contrast sensitivity). It is not by chance that the Snellen acuity chart consists of high contrast black letters on a white background. Older adults may require both larger text size and greater contrast due to age-related changes in visual acuity and sensitivity to contrast which decline with age, as well as eye conditions like cataracts that degrade the quality of the visual image formed on the retina.

3.3.1.6.3 Stroke width

Alphanumeric characters vary in their aspect ratio (i.e., width-to-height ratio), or what typographers who design fonts call the **stroke width** (see Figure 3.6). The stroke width of letters depends on the font chosen to render the letter, and it is usually recommended that stroke width for black letters on a white background has a value of 1:6 to 1:8, whereas for white letters on a black background, a stroke width ranging between 1:8 and 1:10 should be used (Sanders & McCormick 1993). The greater stroke width for white letters on a dark background seeks to minimize the effects of irradiation where light reflecting off the white areas of the letter appears to bleed into the darker regions of the characters blurring the appearance of the letters and reducing legibility. Characters with a narrow stroke may be less legible when depicted against a high reflective background due to the effects of irradiation. The effects of irradiation can be ameliorated by increasing the characters' stroke width and by manipulating character features including the size of interior spaces of letter forms like "a," "o," and "e." See Box 3.1 for a description of a new font designed to improve the readability of highway signs.

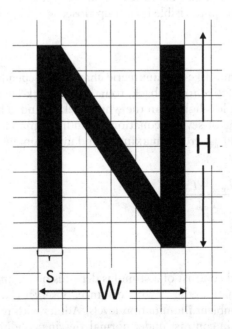

Figure 3.6 The calculation of stroke width-to-height and character width-to-height is expressed as the proportion of two dimensions, S/H (e.g., 1/5 = 0.20) and W/H (e.g., 3/5 = 0.60), respectively. (Available via a CC-BY-SA 3.0 from http://creativecommons.org/licenses/by-sa/3.0.)

3.3.1.6.4 Font selection

The fonts used to render alphanumeric characters belong to one of two categories: serif and sans serif, where serifs refer to the decorative features found at the end strokes of letter segments (Cheng 2005). The font chosen for this textbook is an example of a serif font. The words that follow "sans serif font" are printed using a non-serif font. There are many serif and sans serif fonts designed specifically for displaying large quantities of text. Serif fonts are widely used in printed materials like books, magazines, and newspapers, whereas sans serif fonts have found wide use on web pages where the lower resolution of computer display screens may make serif fonts more difficult to read.

BOX 3.1   IMPROVING THE READABILITY OF ROAD SIGNS

Have you ever considered the identity of the font used on interstate highways signs as you drive by? The signs may receive barely a thought from travelers yet they convey valuable and important information on the distance to destinations, the names of cross roads, the availability of eating establishments, lodging, fuel stations, and rest stops. In the United States, the font used on most highway signage goes by the little known name of Highway Gothic. It has been used for over 50 years, and while its use was approved by Federal Highway Administration, its adoption was more a matter of historical precedent rather than the outcome of principled testing and evaluation (Yaffa 2007).

Highway Gothic was known to have several shortcomings, one of the most significant being that at nighttime the text becomes illegible when illuminated by vehicle headlights. The interaction of the headlights with the highly reflective material used on the signs causes them to appear luminous and blurry, reducing the readability of the text (Garvery et al. 1998). This effect called halation affects all drivers, but it may be a particular concern for older drivers who are more susceptible to the effects of glare caused by halation.

A new font called Clearview has been introduced as a replacement for Highway Gothic. The font was developed by an environmental graphic designer and type designer. The size of the lowercase characters called the "x height" was increased as was the counter shapes. The counter shape refers to the size of interior space formed in the "o" and "p." The counter shape was believed to be too small in Highway Gothic accounting for the loss of definition when illuminated at nighttime by headlights. Illustrative examples of a sign in both highway gothic and the new Clearview font are shown in Figure 3.7.

Preliminary studies (Garvery et al. 1997) indicate that signs depicting the new Clearview font significantly increased the distances at which the signs are legible, which translates into more time for drivers to read and respond to road signs at nighttime or under poor environmental conditions. **Critical Thinking Question:** Can you identify examples of signs you have seen that are difficult to read? What properties of a sign contribute to its poor legibility choice of the font, text size, or contrast?

Clearview

Highway Gothic

Figure 3.7 Examples of the older Highway Gothic font and the new improved font called Clearview. (Clearview image available via https://commons.wikimedia.org/wiki/File%3AClearview_font. svg via Wikimedia Commons [Public domain]; Highway Gothic image available via https:// commons.wikimedia.org/wiki/File:Highway_Gothic_font.svg via Wikimedia Commons [Public domain].)

3.3.2 Dynamic displays

Many real-world tasks involve the monitoring of dynamic displays to obtain quantitative information concerning a variable's value, qualitative information such as the direction or rate of change, or to confirm that system parameters fall within normal bounds. In comparison to static displays, **dynamic displays** can change constantly or periodically and may be used to create situation awareness as in the case of an air traffic controller who must project into the future the locations of aircraft displayed on the computer screen given what he or she knows of the current speed, altitude, and phase of flight (climb, descent, and cruise) to avoid potential conflicts (Endsley 1995).

3.3.2.1 Analog and digital displays

Modern technologies afford the designer considerable flexibility in the selection and design of displays. Consider the all-important bathroom scale, which we approach with some trepidation and that displays weight using either an analog display consisting of a numerical scale and pointer or a digital display which announces your weight in clear and uncompromising digits. Shown in Figure 3.8 is an example of an instrument cluster with linear, circular, and semicircular displays. Different analog scales, so named because the pointer position is analogous to the value it represents, may be distinguished based on whether the scale is fixed, and the pointer moves or the pointer is fixed and the scale moves. The analog bathroom scale is an example of the fixed pointer moving scale, whereas an automobile speedometer is an example of the fixed scale moving pointer. Analog scales can also be categorized according to whether the display is linear, circular, or semicircular.

The choice of employing analog or digital displays depends on the types of tasks an operator must perform. For instance, digital displays are advantageous if an operator has to read exact numerical values but is less advantageous if the numerical values change rapidly. The rapid change may make reading the display difficult if not impossible, and it may take greater effort to identify trends in the values (e.g., are the values increasing or decreasing?). By contrast, one of the strengths of analog displays is that they can communicate spatial information, trends, and rate of change well, but not precise numerical values.

The design of analog displays should consider other factors including the design of the pointer, the scale range, scale markings including the size of the major and minor tick marks, as well as numerical scale progressions (i.e., the numerical difference between adjacent numbers on the scale). Several human factors texts (Woodson 1981; Sanders & McCormick 1993) offer detailed recommendations and guidelines regarding these design matters.

Figure 3.8 BMW instrument cluster illustrating the different types of analog, digital, and semicircular displays. (Image available via a CC -BY-SA 3.0 from https://commons.wikimedia.org/wiki/File:BMW_320i_ E46_instrument_cluster_during_fuel_cut-off_(aka).jpg.)

3.3.2.2 Display location and arrangement

In many applications, operators are required to monitor a complex arrangements of displays and warning lights. Consider the dashboard of a car, as shown in Figure 3.8, and the many different types of displays it contains. There are four principles that guide the location and arrangement of displays in a panel: the importance of use, the frequency of use, the sequence of use, and functional organization (Wickens et al. 2004). The *importance of use* and *frequency of use* principles dictate that regions near the center of the visual field of the operator should be reserved for the most important and frequently used displays. Thus, the operators need to only make small eye movements to frequently monitored information and also ensures that the operator may more readily detect changes in readings or status information because it is near center of the visual field where vision is best. The *sequence of use* principle specifies that the physical arrangement of displays should match as closely as possible the order in which they are monitored or used by an operator in the execution of their tasks. The principle of *functional organization* states that displays illustrating different but functionally related information should be grouped together. The orderly arrangement of displays reduces both eye and head movements, as well as the time to acquire system information.

3.3.2.3 Quantitative displays

As any student can readily attest, textbooks contain many quantitative displays that illustrate the relationship between two or more variables. This relationship can be documented in either a table format or a graph form; however, the choice can have a significant effect on a reader's ability to interpret the data, as illustrated by the data shown in Table 3.2.

Compare the sets of data: can you visualize what the data would look like if plotted on a graph? As is readily obvious from the graphical representation as shown in Figure 3.9, this form of representation is a much more effective means of illustrating the data. Not only can the viewer readily recognize the differences between the sets of data, but they can also appreciate the global trends, including the rate of change (i.e., slope of the data) of each set as well as how individual data points may differ from these trends. The strength of tables lies in the ability of the reader to obtain a precise value which, depending on the task the operator must perform, may be an important consideration in the selection of an appropriate display format. The application of several simple guidelines listed in Table 3.3 offers a means of ensuring the effective communication of data in using graphs.

Table 3.2 Can you visualize what the graph of each pair set of numbers will look like?

Set A		Set B		Set C		Set D	
X_1	Y_1	X_2	Y_2	X_3	Y_3	X_4	X_4
10	8.04	10	9.14	10	7.46	8	6.58
8	6.95	8	8.14	8	6.77	8	5.76
13	7.58	13	8.74	13	12.74	8	7.71
9	8.81	9	8.77	9	7.11	8	8.84
11	8.33	11	9.26	11	7.81	8	8.47
14	9.96	14	8.1	14	8.84	8	7.04
6	7.24	6	6.13	6	6.08	8	5.25
4	4.26	4	3.1	4	5.39	19	12.5
12	10.84	12	9.13	12	8.15	8	5.56
7	4.82	7	7.26	7	6.42	8	7.91
5	5.68	5	4.74	5	5.73	8	6.8

Data from https://en.wikipedia.org/wiki/Anscombe%27s_quartet.

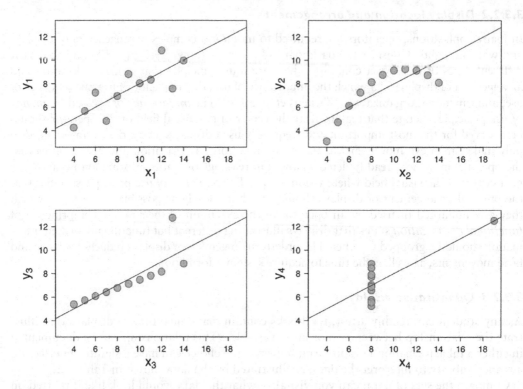

Figure 3.9 Graphs depicting the data from Table 3.1. (Image from https://commons.wikimedia.org/wiki/ File:Anscombe%27_quartet_3_cropped.jpg.)

Table 3.3 Guidelines for the design of effective graphs

Effective communication Using Graphs	
Maximize legibility	Maximize the legibility of graphical components (lines and symbols) by increasing their size, and contrast and selecting graphical elements are readily discriminable from each other. Some symbols are more readily confused, and confusions may be exacerbated when the image is degraded by photocopying or reduced in size.
Reduce visual clutter	Reduce visual clutter by eliminating all nonessential graphical elements. Modern software applications like word and excel allow the user to ornament their graphics by adding superfluous features like visual textures to bar graphs and adding dimensionality (3D shape cues) to a two-dimensional illustration. Edward Tufte (1983), a well-known proponent of good graphics design practices, has argued in favor of the related principle of maximizing the data to ink ratio. Basically, this principle states that as much data as possible should be displayed while minimizing the amount of ink used to display it. As is true in most cases, this holds within reason so long as the absence of graphical elements does not make the graphic difficult to understand.
Minimize cognitive cost	Minimize the cognitive cost of understanding a graph by placing descriptions on the same page as the graph, using meaningful labels or abbreviations for different sets of data, and placing data and labels that must be integrated close to each other so that they can be easily associated with one another. The latter point represents the application of the proximity compatibility principle which states that sources of information that must be integrated (i.e., a warning light and an associated display, a symbol and related label) for operator/user understanding should be placed proximal to one another. Where physical proximity is precluded by other constraints associations can be made explicit by using linking lines or using position mapping. Examples of these attributes are shown in Figure 3.11.

Adapted from Wickens et al. (2004).

3.3.2.4 Navigation displays and maps

The last type of display we will discuss is navigational displays. Travelers use a variety of navigational aids including hand drawn maps, a list of directions, and published navigation aids like maps. More recently, Global Positioning Systems (GPS) allow drivers to enter a new destination and the system prompts the drivers when to turn. Alternatively, drivers may access maps online, downloading maps and printing a "command list" which they can supplement with street views. Tools such as MapQuest and Google maps give a list of direction and associated maps.

Many of the same design concerns related to legibility and clutter discussed previously are also pertinent to maps. Labels and symbols must have sufficient contrast that they are discriminable and subtend a large enough visual angle that they are legible under the different conditions where they might be used. Color is frequently used in maps for ascetic purposes, but also more importantly to convey information. Edward Tufte (1990) outlined other uses of color including using color differences to indicate quantities such as temperature, using colors like blue and green to imitate features like lakes and forests on maps, and using different colors as a label to distinguish between different types of roadways on a map (e.g., state highway versus interstate highway). The use of color can pose a challenge because colored text can be difficult to read due to low contrast produced by certain foreground–background color combinations like yellow text or symbols on a white background. Color differences are also more difficult to discern under low illumination conditions (Pokorny et al. 2006) and for individuals with abnormal color vision.

The design of maps represents a balance between the level of detail and map size that allows the user to acquire the information they need. Electronic maps offer greater flexibility since the user can zoom in or out, and they can also declutter a map by choosing what information to display depending on the circumstances. The increased user customization offered by electronic map displays can, however, increase the cognitive workload for operators.

3.3.3 Sum-up of visual displays

We have discussed the main types of visual display and some of the principles that guide their design and use. While design principles and recommendations aid the human factors expert, they alone are often not sufficient as each application involves unique design considerations that are not addressed by design handbooks. Familiarity with the limitations and capabilities of human sensory abilities offers the human factors expert a basis for making informed decisions about the relative importance and applicability of alternative design recommendations. The following section offers a brief review of basic visual functions.

3.4 VISION

The color deficiency experienced by Steve described in the introductory scenario is not uncommon, approximately 8% of males experiencing some form of color deficiency (Padgham & Saunders 1975). In many cases, these deficiencies are a minor annoyance, but in certain situations, they can pose a risk. We will review the visual, tactile, and olfactory capabilities and limitations of human observers in turn. Considerations of these capabilities are critical to ensuring that operators can obtain the necessary information to perform tasks or support judgments.

3.4.1 The physical properties of light

The visible light that we see is a small part of the larger electromagnetic spectrum that includes X-rays, microwaves, and ultraviolet rays. The photons emitted by a light source and reflected off of surfaces differ in terms of their wavelength, which is expressed in nanometers where a nanometer is equal to a billionth of a meter. The eye is sensitive to light with wavelengths between 400 and 700 nm. The intensity of the light is determined by the number of photons that are absorbed by **photopigments** found in photoreceptors.

3.4.2 Structure of the eye

Figure 3.10 shows a schematic diagram of the eye. Light enters the eye through the transparent cornea. It is the curved surface of the cornea that accounts for two-thirds of the optical power of the eye. Optical power refers to the ability to bend or **refract** light. The light continues through the aqueous humor, the pupil, the lens, and the vitreous humor to form an image of the external world on the photoreceptors neural tissue lining the back of the eye. The aqueous and vitreous humors are transparent watery fluids that give the eye its shape.

The pupil diameter varies between a maximum of 8 mm under low illumination to a minimum of 2 mm under bright sunlight (Blake & Sekular 2006). A change in pupil diameter

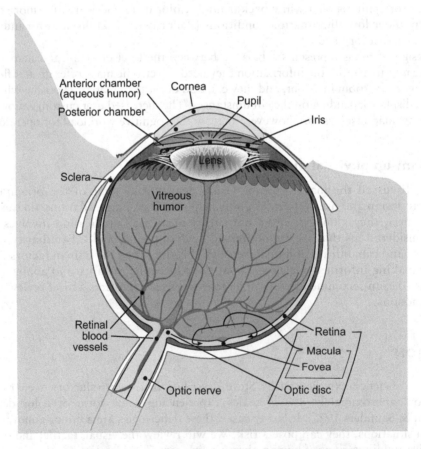

Figure 3.10 Schematic of the eye illustring the major strucutres of the eye and optic nerve. (Artwork by Holly Fischer. Image available via a CC-BY-3.0 license from https://ticket.wikimedia.org/otrs/index.pl?Action=AgentTicketZoom&TicketNumber=2013040410011627.)

regulates the amount of light entering the eye. The pupil dilates under low illumination allowing more light to enter the eye which is especially helpful under low illumination where photons are scarce and image quality is directly related to how many photons are absorbed by photoreceptors. At higher illumination levels where the photons are abundant, the pupils constrict, reducing the amount of light entering the eye.

Changes in retinal illumination produced by variations in pupil diameter are relatively modest given that retinal illumination, which is proportional to the area of the pupil, varies by about 16-fold and we are capable of seeing over a 40 billion-fold range in illumination (40,000,000,000 or 4×10^{10}; Rodieck 1998). This range extends from the dim star at night to the bright noon time glare of the sun on a ski slope. Also, because the pupils have a relatively slow response, they offer little protection when illumination levels change rapidly. Finally, pupil size is dynamic, varying in diameter in response to changes in emotional state (Võ et al. 2008) and cognitive workload (Hoecks & Levelt 1993).

The lens is the next important optical component of the eye accounting for about a third of the optical power of the eye. By making small adjustments in the refractive power of the lens through a process called accommodation, the visual image is brought into sharp focus on the retina. **Accommodation** involves altering the thickness of the lens by changing its shape. The ciliary muscles attached to the lens can be tightened or relaxed to alter the natural shape of the lens which is spherical. Accommodation is what allows you to keep an image of your finger in focus as you bring it closer to your nose.

The image formed by the cornea and lens is focused on the **fovea**, which forms a slight depression on the retina and supports high acuity because of the high density of photoreceptors found there. Signals from the photoreceptors are conveyed to the brain by neurons called ganglion cells, whose processes called axons come together at the optic disk forming the optic nerve that exits the eye. The optic nerve conveys information from the eye to areas of the brain in the form of electrical impulses. Unlike the fovea, the optic disk contains no photoreceptors; hence, it forms a **blind spot** in each eye that observers are not consciously aware of.

3.4.3 Duplex visual system

We have the ability to see under dim and high illumination conditions due in part to the fact that we have two distinct visual systems. One of these systems is called the scotopic or night visual system, which is mediated by rod photoreceptors. The rod photoreceptors are exquisitely sensitive to light and mediate our ability to see under low to moderate levels of illumination. The other system is called the photopic or day visual system, and vision is mediated by the cone photoreceptors, which function at higher illumination levels and are responsible for our color vision and good acuity. The existence of two functionally and anatomically distinct visual systems is called the **duplicity theory of vision**. Despite the fact that we associate seeing with the capabilities of the daylight visual system, retina is predominantly covered by rods—the roughly 120 million rods found in each retina greatly outnumber the roughly 6 million cones (Goldstein 2010).

The distribution of rod and cone photoreceptors across the back of retina is very uneven. The rods are totally absent from the central fovea, increasing in number and density as you move away from the fovea and reaching a peak density around 20 degree in the peripheral retina. The total absence of rods in the fovea means that a small dim object like a star cannot be seen if you attempt to fixate it. Instead, the ability to detect dim stimuli improves when you look away from the star so its image falls in the peripheral retina where the greatest density of rod photoreceptors is found. Similarly, the distribution of cones across the retina is also not uniform. The highest density of cones is found in the central fovea and drops rapidly with eccentricity.

3.4.3.1 Dark adaptation

We have all experienced the difficulty of seeing in a darken theater after coming in from outside. After some period of time, our eyes adjust, and we can see features that were previously not visible to us. The adaptation to dim illumination from bright illumination is called **dark adaptation** and reflects a shift in vision from the photopic to the scotopic visual system. During this transition from bright to dim illumination, we experience difficulty seeing in dim environments. To become fully dark adapted takes approximately 30 minutes; however, when we walk out of the dark theater into the daylight, it takes about 2 minutes to fully light adapt!

3.4.3.2 Detection thresholds of the rod and cone visual systems

The sensitivity of the rod and cone visual systems to different wavelengths of light is not uniform, but rather varies considerably depending on several different factors including the wavelength of light, where the stimulus is presented on the retina, and whether the observer is dark adapted. To measure the maximum sensitivity of the rod photoreceptors, for instance, the observer would dark adapt in a dark room for 30–40 minutes. Rather than looking directly at the test light, the observer is instructed to fixate a small dim red light. While fixating, the observer is instructed to report when they detect a dim test flash positioned so that it falls about 20 degrees in the peripheral retina where the rod photoreceptors reach their peak density and where rod sensitivity is greatest. We can also repeat the experiment to measure the maximal sensitivity of the cones by restricting the small test flashes to the rod-free area of the central fovea.

The lower curve labeled "Rods" in Figure 3.11 shows the type of data one would obtain for the rod photoreceptors using this experimental paradigm. The X-axis of the graph represents the wavelength of the test flash and the Y-axis represents how intense

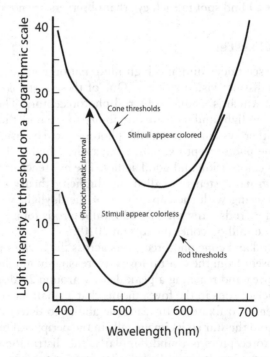

Figure 3.11 Threshold sensitivity of the rod and cone systems as a function of wavelength. (Adapted from Hecht and Hsia 1945.)

the test flash has to be so that the observer detects the flash on half of the test trials. There are several features of the data that are important. First, note that the curve for the rods is shifted downward indicating that the rods are considerably more sensitive to dim test flashes than the cones over much of the visible spectrum. Second, note that the maximum sensitivity of the rods and cones are different, with the rods being maximally sensitivity to light with a wavelength of 510 nm and the cones being maximally sensitive to a wavelength of 555 nm. Third, note that for wavelengths greater than 620 nm, the rod and cone curves nearly overlap. Thus, for these wavelengths, a stimulus that is just detectable by the rod visual system is also at detection threshold for the cone visual system. This fact has practical uses. For instance, soldiers use flashlights with red filters to read maps and other materials in the field at nighttime. This allows them to read smaller details on maps using the superior acuity of the cone visual system while not appreciably affecting dark adaption of the rods and reducing their sensitivity.

The region between the rod and cone curves is called the **photochromatic** interval that represents a range of light intensities where a light is sufficiently intense to be detected by the rods but too dim to be detected by the cones. The size of the photochromatic interval is largest at shorter wavelengths illustrating the greater sensitivity of the rod visual system for shorter wavelength light (<510 nm). As lighting levels decrease and we are using more rods to see, there is an associated change in the perceived hue of the stimulus. Imagine we measure an observer's absolute detection threshold or the dimmest light they can detect 50% of the time using a short wavelength 480 nm test light. When the stimulus is detected by the cone system, the participant can report detecting it, but they can also report its hue (i.e., blue) or color. However, once the light is so dim that it is detected solely by the rod system, participants report detecting the light, but it now appears achromatic (i.e., varying levels of gray). Thus, lights in the range of intensities defining the photochromatic interval that are detected by the rods appear colorless.

The shift from rod to cone vision is also accompanied by another interesting phenomenon called the Purkinje Shift, so named after a 19th century physiologist Johannes Purkinje (Anstis 2002), who observed that as illumination levels decline, red colored flowers appear darker and blue flowers appear brighter. This relative change in brightness is a direct result of the shift from the cones to rods, which are considerably more sensitive to short-wavelength light (450 nm light) than to long-wavelength light (650 nm light).

Sensitivity to color worsens as illumination levels decline from high photopic to mesopic levels. Observers report a preferential loss of sensitivity to blue and yellow under mesopic illumination levels. This was recently demonstrated in an experiment where subjects were asked to sort samples of color chips into groups that they would call "red, pink, orange, yellow green, blue, purple, and gray." While categorization of reds and oranges remained consistent, other color chip samples were assigned black, green, or blue-green (Pokorny et al. 2006). The same chips seen under scotopic illumination levels would all appear as varying shades of gray.

3.4.3.3 Cone spectral sensitivity functions

Unlike the rods, of which there is one type, there are three different types of cone photoreceptors. The three cones, called the Short-, Middle-, and Long-wavelength cones, are aptly named to emphasize the fact that they differ in terms of the portion of the visible spectrum they are maximally sensitive to. Figure 3.12 shows the individual spectral sensitivity of each cone type; the Short-, Middle-, and Long-wavelength cones and the rod photoreceptors. The short-wavelength cones are more sensitive to light of shorter-wavelength than either the

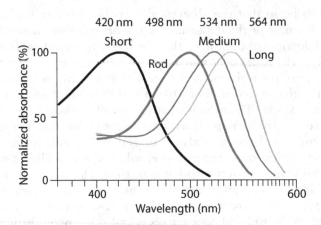

Figure 3.12 Spectral sensitivities of the Rods and Short-, Middle- and Long-wavelength cone photoreceptors.(AdaptedfromimageavailableviaOpenStaxCollege[CC-BY-3.0]fromhttps://commons.wikimedia.
org/wiki/File%3A1416_Color_Sensitivity.jpg.)

Middle- or Long-wavelength photoreceptors. The peak sensitivity of the Short-, Middle-,
and Long-wavelength photoreceptors is 420, 434, and 564 nm, respectively (Sharpe et al.
1999). How the three cones contribute to color vision is discussed below.

3.4.4 Human visual capabilities

3.4.4.1 Visual acuity

Visual acuity is a measure of the ability of the eye to resolve fine details. Acuity can be
assessed in a variety of ways using different types of stimuli. Because of the different types
of tests, it is common to express acuity in terms of the minimum separable acuity or perhaps more intelligibly the size of the smallest feature or gap between parts of a target that
the observer can reliably report. For instance, Figure 3.13 shows example of the Snellen
acuity chart (left) and the Landolt "C" acuity stimuli (right). In the case of the Landolt C
test, the task of the observer is to report which direction the gap in the letter "C" faces
(up, down, left, or right). The features in the Landolt C labeled "a" are scaled so that they
are 1/5 the size of the width labeled "b." The smallest gap that an observer can reliably
discriminate represents the minimum angle of resolution and is expressed as the reciprocal
of visual angle subtended by the features that observers must discriminate (i.e., the features
labeled "a"). The equation presented in Section 3.1.1.3.1 can be used to calculate the visual
angle subtended by the stimulus.

Observers with better visual acuity can resolve Snellen letters subtending a smaller visual
angle than a person with poorer vision. Ophthalmologists and optometrists typically measure visual acuity using the widely used Snellen acuity test. The Snellen acuity test is only
one of many ways to measure an observer's ability to resolve spatial patterns, in this case letters. The Snellen acuity test measures recognition acuity or the ability to identify individual
characters or to report their orientation. Performance is typically reported as a ratio that
indicates your performance relative to a normal observer. Most observers would be happy to
have a visual acuity of 20/15, which means that you can resolve a pattern at 20 ft (6 m) that
a normal observer can resolve at 15 ft (4.6 m), and perhaps disappointed with a visual acuity
of 20/40 means that you can resolve a pattern at 20 ft that normal observers can resolve from
40 ft (12.2 m).

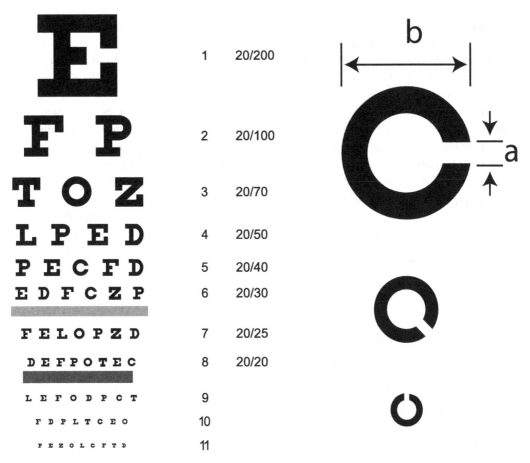

Figure 3.13 Examples of the visual stimuli used to measure an observer's ability to resolve fine details. The examples include the Snellen acuity chart (left) and the Landolt "C" acuity stimuli (right). (Snellen chart image available via a CC-SA-3.0 from https://en.wikipedia.org/wiki/Snellen_chart; Landolt "C" chart image by DarkEvil; [Public domain], via Wikimedia Commons at https://en.wikipedia.org/wiki/Landolt_C.)

BOX 3.2: PREDICTING CRASH RISK FROM MEASURES OF SNELLEN ACUITY

It would seem reasonable to require that drivers have good visual acuity. In fact, every state department of motor vehicles in the United States has certain minimum visual acuity standards for obtaining an unrestricted driver's license. If good acuity is necessary for safe driving, then one might expect that drivers with poorer visual acuity would have more accidents than drivers with better vision. Yet, it may come as a surprise that there is little experimental evidence supporting a relationship between acuity and crash history. In a classic study, Albert Burg (Burg 1967, 1968) measured basic visual functions including the acuity of drivers and correlated it to their driving records and found that visual acuity accounted for less than 1% of the variance in a drivers' accident record. In other words, acuity scores were not useful in predicting an individual's accident history (Hills 1980).

There are several possible explanations for this finding. First, this relationship is weakened by that fact that there is a significant segment of the driver population that have very good vision but who experience high accident rates—namely younger drivers. Risk-taking behaviors, drinking and driving, and limited driving experience may play a greater role in younger driver mishaps than visual acuity. Second, it also important to consider what visual acuity actually measures. Visual acuity measures an observer's best vision—the ability to resolve very small, high contrast, black letters, on a white background. Small letter targets differ considerably from the much larger objects that drivers' strike including other cars, pedestrians, road barriers, potholes, and road debris. While acuity may predict a driver's ability to read a street sign or a license plate, it is unlikely to be strongly related to a driver's failure to detect large, moving, and visually salient objects like other cars. After all, a driver's failure to detect and avoid another car is not because the other car was too small to resolve! **Critical Thinking Question:** In lieu of visual acuity, what other sensory or cognitive measures might better predict crash risk?

3.4.4.1.1 *Visual acuity and illumination level*

Personal experience offers plenty of evidence of the changes in the quality of vision as illumination is reduced. At nighttime, you might have fumbled with your keys, unable to locate the keyhole and the key that opens the car door, or attempted to navigate a darkened room wondering whether the dark form on the floor is a pet, a piece of clothing, or some other object. While the rod photoreceptors allow you to detect the presence or absence of larger objects in the dark, they do not have sufficient acuity to allow you to resolve fine details that might allow you to identify what the dark form on the floor is.

Acuity improves with luminance level reaching its peak at high luminance levels and declining steadily as the luminance is reduced. Rod or scotopic visual acuity is very poor supporting a maximum acuity of approximately 20/100, whereas best-corrected cone-mediated vision for young observers is in the range of 20/15 to 20/20. The dependence of acuity on luminance levels is directly relevant to the design of signs, warnings, and labels that we might need to read in poor environmental conditions or emergencies where lighting may be limited. Signs and warnings have to be physically larger and have greater contrast to maximize their legibility under poor viewing conditions.

3.4.4.1.2 *Acuity versus eccentricity*

Previously, it was noted that the eye is a non-uniform structure with significant variation in the distribution of rods and cones as a function of position or eccentricity on the retina. Visual acuity is another example of this general principle. For example, if we measure visual acuity by presenting letter targets at different retinal eccentricities, expressed as angular distance from the fovea, we obtain a graph like that shown in Figure 3.14. The figure illustrates how visual acuity varies as a function of retinal location. Visual acuity is greatest in the central fovea and drops rapidly; so when a target is presented 5–6 degrees from the fovea, acuity is reduced by half. Because of the much poorer acuity of the peripheral vision, you must fixate a target, thereby placing the image on the central fovea, to see details that are not visible using peripheral vision.

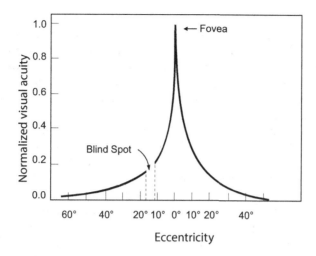

Figure 3.14 Normalized visual acuity as a function of eccentricity. (Original artwork by Vanessa Ezekowitz. Image available via CC BY-SA 3.0 https://commons.wikimedia.org/wiki/File%3AAcuityHumanEye.svg.)

3.4.4.1.3 Acuity and contrast

One way to compensate for the diminished acuity of peripheral vision is to make peripheral stimuli larger. By increasing the size and spacing between letters, we can increase their legibility using peripheral vision (Bouma 1970). Because of the poorer acuity of peripheral vision, we often have to coordinate movement of the eyes, head, and body to place an image on the fovea to scrutinize the details of the object using our best vision.

Acuity can also be further compromised by reductions in contrast which refers to differences in the intensity of light reflected from the letter versus the background against which it is presented. Examples of letters with different contrast are shown in Figure 3.15. It is readily apparent that it is more difficult to discern lower contrast letters. The contrast (C) of a visual pattern like a letter can be calculated using the formulas shown in Section 3.3.1.6.2.

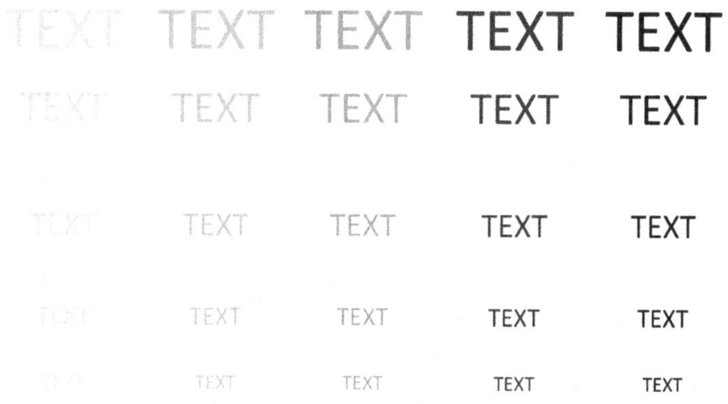

Figure 3.15 Changes in the contrast of a letter relative to its back ground as a function of letter size. (©Alex Chaparro.)

3.4.4.2 Color vision

Color offers a powerful means of coding information when implemented carefully. Color can aid in the segregation of a target from a background allowing an efficient visual search when you are attempting to find a target among non-targets. Imagine looking for a textbook on a shelf and using your knowledge that the books cover is yellow to restrict your search to yellowish book spines. Color can also aid in grouping, allowing the observer to associate related like-colored items together. Color also serves an important signaling function including revealing information about differences in the material properties of an object such as when a fruit is ripe. Color allows you to readily identify the members of a crowd that support the same team and can signal different functional states like normal versus abnormal conditions (white or green versus red).

However, the use of color coding can pose a problem for approximately 8%–9% of the male population that are color limited (Padgham and Saunders 1975). Color limitations occur infrequently in women as can be seen in Table 3.4. Table 3.5 lists the different types of color limitations and percentage of males and females in the population of persons of Western European descent who have the deficiency. Dichromatism is caused by the absence of either the Long-, Middle-, or Short-wavelength photoreceptors resulting in conditions called Protanopia, Deuteranopia, and Tritanopia, respectively. Protanopia and Deuteranopia are the two most common types of abnormal color vision, causing confusions between reds, greens, and yellows; however, discrimination of blues and yellows is preserved. Anomalous trichromacy is another form of color deficiency that is less common than dichromacy. Anomalous trichromats have three cones, but the spectral sensitivity of one of the cones is different than that found in normal observers.

Dichromatic observers report difficulty with a number of real-world tasks including recognizing when bathroom stalls are vacant (green) or occupied (red), interpreting the LED lights used to indicate normal or abnormal states on many electrical devices, deciphering colored maps and graphics in journals or news articles (Flück 2010), and discriminating between red and green signal lights. In the case of signal lights, there is redundant position

Table 3.4 Percentage frequency of occurrence of color vision defects in persons of Western European descent

Type	Male	Female	All
Monochromats**			
Rod	0.003	0.002	0.0025
Cone	Small	Small	Small
Dichromats***			
Protanope (L)	1.2	0.02	0.61
Deuteranope (M)	1.5	0.01	0.76
Tritanope (S)	0.002	0.001	0.0015
Anomalous trichromats***			
Protanomalous (L)	0.09	0.02	0.46
Deuteranomalous (M)	4.5	0.38	2.5
Tritanomalous (S)	Small	Small	Small
Total	**8.1**	**0.43**	**4.3**

Based on Table 3.3.1 from Padgham and Saunders (1975).

*L,M,S—Long-, middle-, and short-wavelength cones or commonly referred to as red, green, and blue cones, respectively.
**Monochromats have either only rods or only cones.
***Dichromats are missing a single type of cone.
****Anomalous trichromats have three cones, but the spectral sensitivity of one of them is different than that found in normal observers.

information—the topmost lamp indicates STOP and the bottom lamp means GO—which can be used in lieu of color to identify the proper action.

The effects of color limitations can be minimized through proper design. The shape, color, and text found on Stop and Yield signs offer multiple sources of redundant information that drivers can use to identify each sign under poor environmental conditions that might prevent reading the text on the sign or detection of the color of the sign from a distance. The use of redundant coding and selecting colors from a palette that color-limited observers can discriminate can ameliorate the difficulties experienced by color-limited observers.

3.4.5 Sum-up of vision

Basic visual functions like acuity, sensitivity to contrast, and color perception show considerable variation depending on illumination level, retinal location, and observer. These and other low-level sensory abilities document human sensory limitations that the design of displays must consider to ensure that the detection, recognition, discrimination, and identification tasks operators are expected to perform are actually within their capabilities.

3.4.6 Size and distance perception

The ability to move through our environment effortlessly and to interact with the objects that share that space depends on our perception of depth. Depth information also supports motor behavior allowing you to grab an object or place one in a particular location. These tasks require a number of different evaluative judgments of depth including judgments of egocentric distance and relative depth. Depth perception refers to your perception of the three-dimensional shape of an object including its depth, height, and width as well as its distance from you. When you park your car, you judge the distance of other cars or objects like gates relative to your car to ensure that you have sufficient clearance. This type of judgment is called **egocentric distance** since it is specified with respect to you. Alternatively, you might simply evaluate which of two objects is farther away, representing a **relative depth** judgment.

Studies suggest that observers use a diverse set of cues to judge the position of objects and themselves in space. Some cues are available in static views, while others are generated by movement through the environment, and still others are derived from kinesthetic cues from eye muscles when we move our eyes or through the process of accommodation.

3.4.6.1 Static cues to depth

Artists use a set of static cues to produce a sense of depth on the flat two-dimensional surface of a canvas. Many of these cues are illustrated in Figure 3.16. Depth can be conveyed by occlusion and size where nearer objects can occlude and be larger than those objects lying behind. A repeating pattern like the cobble stones produces a textured gradient regressing toward the horizon. The texture pattern becomes denser and less distinct with increasing distance. Similarly, parallel lines titled away from the viewer appear to converge in the distance as can be seen in the architecture features of the distant building.

Atmospheric perspective refers to the observation that distant objects have reduced contrast and may appear slightly bluish in color due to the scattering of light by dust particles suspended in the air. The position of scene features relative to the horizon can also provide cues to distance. Notice how pedestrians that are farther away appear lower in the painting, closer to the horizon, than pedestrians who are nearer.

Figure 3.16 The painting by Gustave Caillebotte titled Paris Street: A Rainy Day (1877) is an excellent illustration of the use of static depth cues to convey a convincing sense of depth. (Available via Wikimedia Commons [Public domain] at https://commons.wikimedia.org/wiki/File%3AGustave_ Caillebotte_-_Paris_Street%3B_Rainy_Day_-_Google_Art_Project.jpg.)

3.4.6.2 Dynamic cues to depth

Unlike static cues, kinetic cues to depth are generated by the movement of the observer or an object relative to the observer. Movement is a rich source of information concerning the spatial layout of the environment, objects in the scene, and our movement relative to the scene (Gibson 1979). If you close one eye and look at the room in front of you, you will notice that it is more difficult to judge the relative layout of the environment than with both eyes. The layout of the environment, however, can be revealed by moving your head laterally from side to side. This cue to depth is called **motion parallax** whereby objects closer to you move more than more distant objects. The movement is like that produced by traveling along a road and looking out upon the surrounding countryside. You might have noticed that objects in the distance like a tree appear to move hardly at all, whereas nearby objects whizz by the car window. The differences in the relative motion of the features in the environment offer information about their distance relative to the observer. Also, the direction of motion of the features in the scene will vary. Objects in the scene that lie beyond the point of fixation move in the same direction as the observer, whereas objects lying nearer than the point of fixation move in the opposite direction.

3.4.6.3 Kinesthetic cues to depth

As an object approaches, the accommodative state of eye's lens must change to maintain a clear image on the retina and the eyes rotate toward each other. Kinesthetic cues are generated by receptors in the muscle that respond to tension produced by the contraction of the muscles that rotate the eyes toward each other and muscles attached to edge of the lens that change its shape. Muscle tension offers cues about distance because the magnitude of the tension is related to the distance of the object because there is more angling with closer objects. However, these cues are only available over a relatively short range of distances estimated to be roughly 20 ft (6 m) or less because there is little change in accommodation or convergence of the eyes for distances beyond this range.

3.4.6.4 Binocular cues to depth

Stereopsis refers to the perception of relative depth that is produced by the slightly different views of the world offered by our two laterally displaced eyes. You can see the differences in these views by alternately closing one eye at a time while viewing your index finger at arm's length. You will notice that there is a slight shift in the position of your finger relative to the background in the scene.

Stereoscopic cues can be particularly useful in breaking the effects of camouflage when viewing pictures of a scene. Even when the outline of an object is cleverly hidden by camouflage, it can be easily revealed by viewing pairs of pictures called stereo images taken of the same scene from two slightly different positions. Stereo images have also found application in chemistry where scientists use software to create stereo views of complex molecules allowing them to visualize the complex 3D shape of chemical compounds.

3.4.6.5 Effects of illumination and distance on depth perception

It is well known that the relative effectiveness of depth cues differs and depends on distance. For instance, convergence and accommodative cues are only effective over a distance of 6 ft (2 m) or less, whereas other cues including occlusion and relative size are effective over the range of vision. This near region extending out to about 6 ft (2 m) is sometimes referred to personal space and corresponds to the space immediately surrounding an individual, a region where occlusion, retinal disparity, relative size, convergence, and accommodation are effective depth cues. Action space corresponds to a region delimited to about 100 ft (30 m) within which we can move quickly or talk to someone, and where occlusion, height in the visual field, binocular disparity, and motion perspective and relative size are effective cues to depth. The region beyond 100 ft (30 m) is called vista space and is where pictorial cues (occlusion, height in the visual field, relative size, relative density, and aerial perspective) are the dominant cues to depth.

It is instructive to consider how the effectiveness of these cues to depth might be altered by reductions in illumination level such as you might encounter at nighttime when driving. For instance, a drivers' vision at nighttime is restricted by the effective range of a vehicles headlights as well as by changes in the sensitivity of visual processes that mediate detection of the depth cues. The effective range of headlights limits a drivers' view of the driving scene to the lower range of vista space (approximately 300 ft (91 m) or less), which also limits the driver's time to detect and respond to a potential hazard.

3.5 TACTILE DISPLAYS

Tactile information or displays are often used to supplement either visual or auditory information and are rarely the primary source of information. One exception is Braille where textual information is encoded as a series of raised bumps on a sheet of paper or plastic. Typically, tactile cues serve an alerting function as in the case of the vibration of a cellular phone or the rumble strip on the side of a roadway.

Tactile cues are provided by several different receptor types found in the skin and joints that respond when stimuli are pressed against the surface of the skin and when the body or limbs move. Receptors found in the muscles and joints offer kinesthetic and proprioceptive cues. **Kinesthetic** cues provide information that allows you to sense that your body or limbs are moving or being moved and **proprioceptive** cues that allow perception of the static position of your body and limbs in space. Proprioceptive cues, for instance, allow you to

Table 3.5 Tactile receptors: sensory function, adaptation, and stimuli effective in eliciting a response

	Sensory function	Adaptation	Stimuli effective in eliciting a response
Merkel	Texture perception Pattern/form detection	Slow	Sustained pressure, very low frequency (< 5 Hz) Spatial deformation
Meissner	Low-frequency vibration detection	Fast	Temporal changes in skin Deformation (~5–700 Hz)
Unknown	Finger position, stable grasp	Slow	Sustained downward pressure, lateral skin stretch, skin slop (low sensitivity to vibration across frequencies)
Pacinian	High-frequency vibration detection	Fast	Temporal changes in skin deformation (~50–700 Hz)

Based on Tables 12.1 and 12.2 from Wolfe, Kluender et al. (2006).

recognize the position of your leg and identify whether your foot is on the accelerator or brake. Our ability to sense our movement and positions of limbs, as well as temperature, pressure, and electrical stimulation on the skin, is mediated by a diverse set of receptor types including free nerve ends and nerves with structures that encapsulate the end of the nerve fibers. Three types of encapsulated nerve fibers have been documented including Merkel disks and Meissner corpuscles that are located close to the surface of the skin and Pacinian corpuscles found deeper in the skin (Yantis 2014). It is the set of three receptor types and a fourth unidentified receptor type that are responsible for the encoding of tactile sensations including vibration, movement, shape, and texture. Table 3.5 lists the four types of receptors, their sensory function, and the stimuli effective in eliciting a response and how rapidly they adapt to stimulation. **Adaptation** refers to how quickly a receptor stops responding even if the stimulus is maintained. Adaptation accounts in part for why you no longer feel the ring on your hand, the shoes on your feet, or your clothes.

3.5.1 Tactile sensitivity

The ability to detect pressure on the skin or what is called passive touch or cutaneous sensitivity varies considerably across the body. Tactile is greatest in the fingertips, thumb, and areas of the face including the lips, cheeks, and nose. A comparison of data for males and females reveals that females are more sensitive to passive pressure than males. In addition to passive pressure, the participants' ability to distinguish when the pressure was produced by a single point or two points has also been measured. The ability to discriminate a single pressure point from two is best in the fingertips, nose, lips, and cheeks and poorest in the upper arm, calf, thighs, and back.

While the data on passive touch are useful, it is important to stress that touch typically involves active exploratory movements and that our discriminative abilities are enhanced through active exploration of an object. Craftsmen actively explore machined surfaces tactilely by moving their fingers across the surface to detect imperfections they might otherwise miss. We are very good at recognizing objects based solely on tactile cues. In fact, a test of blindfolded participants presented over a hundred different items found that they could quickly and accurately identify most of the objects when allowed to use their hands to explore them (Klatzky et al. 1985). This finding offers experimental support for the practice in aviation of using different shaped controls that allow pilots to identify the controls based on shape and textural cues, which is especially useful during phases of flight like landing and takeoff that are visually demanding.

The active form of tactile inspection involves the coordination and integration of feedback from the touch receptors and kinesthetic information of the movement and position (i.e., proprioception) of the limbs and is called **haptics**. Haptic feedback is being increasingly incorporated in many peripheral devices ranging from cellular phones to game controllers that offer vibratory feedback when selecting keys or simulated impact forces when playing a computer game. An important area of human factors research concerns the design of haptic feedback systems that would allow human teleoperators to sense resistive forces as they manipulate a tool or to feel the texture of an object as they manipulate it remotely. Some devices require users to wear an exoskeleton haptic system used to provide feedback to operators interacting with simulated objects populating a virtual world.

BOX 3.3:   THE TACTILE SENSE AS A SENSORY SUBSTITUTE FOR VISION: USING THE TONGUE TO SEE!

The primacy of visual displays including the widespread use of graphs, pictures, or maps presents an obvious problem for visually impaired or blind users. Researchers have investigated using the skin or tactile system as a *sensory substitute* for vision (Bach-y-Rita and Kercel 2003). The viability of this approach is suggested by the use of Braille where words are represented as a series of raised bumps on a sheet of paper or plastic that is inspected by moving their fingers across the bumps. But how do you represent complex images including scenes or simple line drawings of shapes, faces, or maps?

In the 1960s, scientists (Bach-y-Rita et al. 1969) developed a tactile display system that used a matrix of vibrating pins to display simple images against participant's backs. The system, called the Tactile-Vision Substitution System (TVSS), converted a video image into vibration of 400 pins arranged into a 20 by 20 array that rested against the participants' back. After some practice, blind participants learned to identify and discriminate between objects and make a variety of perceptual judgments including perspective and depth judgments, but they experienced some difficulty discriminating among stimuli with detailed internal patterns (White et al. 1970).

Researchers continue to explore ways to conveying visual information using the tactile sense. They have investigated the feasibility of using the tongue to present visual information using a matrix of electrodes that apply small controlled electric currents to the tongue (Bach-y-Rita et al. 1998). One advantage of using the tongue is that it is densely innervated by mechano-receptors having both higher sensitivity and resolution than the fingertips. To evaluate the effectiveness of the system, the investigators had subjects scan a camera over different sized Snellen E letter targets like those shown in Figure 3.14. Participants were instructed to report the orientation of the letters. A computer transduced the image acquired by the camera in a pattern of stimulation on the tongue. Using the tongue as input and without training, blind users could resolve small tactile targets estimated to be comparable to a "visual" acuity of 20/860. Participants selected for training showed continued improvements in acuity achieving acuity equivalent to 20/430.

The tactile sense can also be used to augment other sensory systems. For instance, stimulation of the tongue might hold promise as a means of improving postural control in patients with vestibular disorders (Sampaio et al. 2001; Vuillerme et al. 2008). Movement and position of the body's center of gravity can be conveyed by a small target on the tongue display allowing patients to make corrective movements to avoid falling. A commercial version of this device called the BrainPort™ is undergoing clinical trials to evaluate its safety and efficacy. **Critical Thinking Questions:** What practical considerations might limit the use of this sensory aid? Where else might tactile information be presented other than the tongue?

3.6 OLFACTORY DISPLAYS

The last type of displays we will consider are olfactory displays. There is considerably less research on the applied uses of olfactory stimuli though it is well known that olfactory stimuli can elicit compellingly detailed memories (Fields 2012) and hold promise as a means of increasing the degree of immersion an operator experiences when interacting with virtual environments. Disney, for instance, uses odorants in some of theme park rides to increase the immersion of park visitors. One of the early attempts to create a truly immersive experience was the Sensorama (Helig 2011). The Sensorama was an arcade game which sought to create the experience of riding a motorcycle through a downtown city scene. The arcade rider was seated, gripping the handlebars while placing their head in a hood where they viewed a 3D scene that was accompanied by multiple sensory cues including odors, blowing air, vibrations, and directional sound cues. The Sensorama sought to simulate the rich sensory experience of riding a motorcycle through the noisy, odorous, and uneven streets of Brooklyn New York.

To date, olfactory displays have found use primarily as a warning cue as in the case of the sulfur smelling odorant **Mercaptan**, which gives natural gas its rotten-egg smell. Similarly, oil of wintergreen is the recommended odorizer additive for Carbon Dioxide (CO_2), which is used in fire suppression systems on ships. Carbon Dioxide is used to flood enclosed spaces like cargo holds and engine rooms depriving fires of oxygen. The distinctive smelling additive warns of the presence of carbon dioxide. The widespread use of olfactory displays is hindered by a number of practical issues including the fact that we quickly become desensitized to odors, that passive diffusion of odorants limits the ability to direct odors to a particular location or person, and that the active dispersal of odorants using puffs of air or fans can be distracting and annoying.

3.7 APPLICATIONS TO SPECIAL POPULATIONS

Many consumer products including vehicles, cellular phones, computers, and software applications target a diverse population of users, which can pose a significant challenge given the wide range of capabilities of potential users. This is a very different situation from the early era of human factors where the applications were primarily military and the population of users was homogeneous, consisting of mostly of physically fit, young male conscripts with normal hearing and vision.

The range of sensory capabilities in the population at large is, however, considerably larger due to normal variation, age-related changes, the increased prevalence of diseases that affect sensory function in older adults, and the accumulated effects of exposure to injurious environmental conditions like loud noise. While individuals with physical or sensory limitations may be excluded from military service, they nevertheless constitute part of the normal variation found in the population whose needs can be accommodated through design.

3.7.1 Older drivers and highways signs

Consider the case in point of older drivers who report more difficulty driving at night and who are more likely to avoid nighttime driving (McGregor & Chaparro 2005). The reported nighttime driving difficulties may derive from several factors including their greater susceptibility to the effects of glare produced by the headlights of on-coming vehicles, reduced nighttime acuity and contrast sensitivity, which limits a driver's ability to discern faint details in the visual scene and to read signs from a distance. The driver's visual abilities may also be compromised by cataracts, which scatter light further reducing the luminance contrast of the visual image and the fact that it is effectively darker at nighttime for older adults because of reductions in the maximum diameter of the pupil under low illumination. A smaller pupil means less light enters the eye, which reduces observer's visual abilities.

The case of the Clearview highway font offers a compelling example of how design and human factors testing and evaluation principles can improve the display of information (see Box 3.1). The increasing emphasis on accommodating the abilities and limitations of older adults is partly in response to the recognition that as the baby boomer population ages, adults over age 65 represent the fastest growth segment of the American population.

To evaluate the new font, participants were asked to perform two tasks—a word legibility and word recognition task—as they sat in the passenger seat of car that approached the signs. Participants were shown signs with lettering rendered using the Clearview font and several other fonts including the approved mixed-case Standard Highway Series E (M) font. The word legibility task required participants to read correctly a single word on a road sign, whereas the word recognition task required participants to report whether a target word appeared at the top, middle, or bottom position of a sign. The reading distance was recorded and served as the main dependent performance measure. The participants in the study were older adults between 65 and 83 years of age, and the testing was conducted under both day- and nighttime viewing conditions.

The results of the study, as shown in Figure 3.17, demonstrate that while the Clearview and Series E (M) could be read from comparable distances during the daytime, at night-time, the Clearview font allowed participants to read the signs from a significantly greater distance which on a highway would translate into additional time for a driver to read and respond to the information on the sign. The new font was given interim approval by the Federal Highway Administration in 2004 and has been adopted by at least 20 states to replace posted signs. Recently, the city of New York began replacing some 250,000 streets

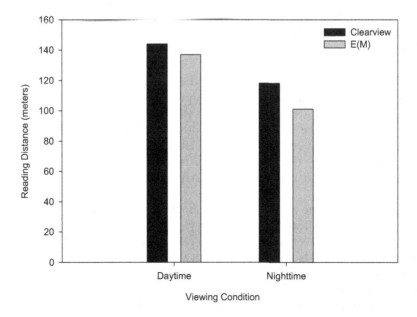

Figure 3.17 Reading distance for signs depicting Clearview and mixed-case Standard Highway Font. Data are shown for both daytime and nighttime viewing conditions. Data are from Garvey et al. (1998).

signs that have all capital letters, with new signs that use both lowercase and uppercase text and the Clearview font (See Box 3.1 for description of the new Clearview font). Changing the signs is expected to be completed in 2018!

3.7.2 Modern technology and color deficiency

It has been argued that the availability of low-cost color monitors and color printers now presents color-limited individuals with greater challenges than any time in the past (Cole 2004). Color monitors have replaced the monochrome monitors on most desks and those used to present passenger information in train stations and airports. Color displays are now found in aircraft cockpits, in vehicle navigation and entertainment systems, cellular phones, music players, and in the hospitals where they display patient vital signs. Color is used in many of these applications to code and organize information, which can pose a problem for color-limited individuals who may not be able to discriminate certain hues.

Color-limited observers report having greater difficulty noticing traffic lights and colored traffic signs than observers with normal vision (O'Brien et al. 2002), and they derive less benefit from color when it is used in complex aviation displays (Cole & Macdonald 1988) or when searching for a target in a cluttered visual display where color serves as a redundant cue (Cole et al. 2004). Medical practitioners with abnormal color vision cite difficulties noticing changes in skin color caused by rashes or jaundice, as well as difficulties with color coded charts, sides, and prints (Spalding 1997, 1999).

While abnormal color vision can be a hindrance, individuals with abnormal color vision are successfully employed as doctors, pilots, and commercial drivers. Their success can be attributed partly to the fact that the availability of redundant cues allows an observer with abnormal color vision to acquire the relevant information. The driver with abnormal color vision can use the brightness and position of a signal light to judge whether they should proceed into the intersection. The challenges faced by color-limited observers have gained some awareness, as illustrated by the availability of software programs that allow Web page designers to simulate how the page may appear to a dichromat. Also, color palettes are available that illustrate how different colors will appear to a dichromatic observer thereby allowing the selection of colors that are discriminable by an observer with abnormal color vision.

Barry Cole, a researcher who has published extensively on the effects of abnormal color vision, relates an example in one of his papers involving the design of sonar system that used yellow, red, and green color to code whether a target was stationary, moving away, or toward a reference location (Cole 2004). A significant number of the sailors with abnormal color vision could not distinguish among the colors used in the display. This incident demonstrates the importance of identifying your target user population and the characteristics that may be particularly pertinent to the design that is being developed.

The failure to consider the needs of special populations remains all too common in the design of new tools and applications. Many web pages are not assessable to blind users who must rely on text-to-speech software which can read the text on the page aloud. However, many pages lack the embedded tags that the software can use to identify pictures or other page content. Similarly, popular E-readers like the Kindle are increasingly being adopted in classrooms in lieu of textbooks, but they may not support speech-to-text and may use menus that are not designed with blind users in mind (Parry 2010).

3.8 SUMMARY

This chapter reviewed the design of visual, auditory, tactile, and olfactory displays, as well as the principles that guide the selection of a display that is appropriate for a given application. Details of basic visual abilities including acuity, color vision, contrast sensitivity, and depth perception were discussed, as well as their relevance to the design and selection of displays that are appropriate for different applications and populations of operators. The unique characteristics of the tactile and olfactory sensory channels were introduced, as well as their potential for reducing the demands on visual and auditory channels, and as a means of providing redundant coding of information. The unique challenges faced by color-limited (Color blindness) users were discussed as an example of a special population of users that many designs fail to accommodate.

LIST OF KEY TERMS

Accommodation
Acuity
Adaptation
Binocular disparity
Blind spot
Command list
Cones
Conspicuity
Contrast sensitivity function
Clearview
Cutaneous sensitivity
Dark adaptation curve
Deuteranope
Depth perception
Dichromacy
Dynamic displays
Haptics
Kinesthetic
Legibility
Mercaptan
Mesopic
Monochromacy

Passive touch
Photoreceptors
Photopic
Photopigments
Photochromatic
Proprioception
Protanope
Refraction
Relative judgment
Rods
Scotopic
Situational Awareness
Snellen Acuity
Spectral Sensitivity
Static displays
Stroke width
Tactile displays
Dichromacy
Visibility
Visual Acuity
Visual Angle

SUGGEST READINGS

Smallman, H. S., & St. John, M. (2005). Naïve realism: misplaced faith in realistic displays. *Ergonomics in Design, 13*, 14–19.
  This article challenges the popular assumption that displays that closely mimic the appearance of the real world are necessarily better than displays that use more abstract representations.
Cole, B. L. (2004). The handicap of abnormal colour vision. *Clinical and Experimental Optometry, 87*, 288–292.
  This review offers a comprehensive discussion of abnormal color vision and its effects on real-world tasks.
Sacks, O. (2010). Stereo Sue. *The Mind's Eye*. New York, NY: Alfred A. Knopf: 111–143.

This chapter offers a detailed discussion of the discovery of stereovision, the effects of stereo blindness, and the role of stereovision in normal everyday conditions.

Tufte, E. R. (1983). *The Visual Display of Quantitative Information*. Cheshire, CT: Graphics Press.
This is a very readable and very influential book concerning the design of effective graphic illustrations and the principles underlying their design.

CHAPTER EXERCISES

1. Shown in Figure 3.8 is a picture of the instrument display from a BMW. Identify the different types of displays (digital, analog, moving indicator fixed scale or fixed indicator moving indicator, etc.).
2. Calculate the visual angle subtended by a 4-in tall letter on a sign viewed from 20 ft. Note that the calculation of visual angle requires that the values of distance and stimulus size be expressed in the same units (ft., inches, mm, etc.).
3. Calculate the contrast for a stimulus with a luminance maximum and minimum of 110 and 100 units, respectively using the three equations shown on page X.
4. What are some of the principles that guide the location and arrangement of displays?
5. What kinds of tasks are better suited for use of a digital display?
6. Which wavelengths of light allow an observer to see with the cone photoreceptors while minimizing adaptation of the rod visual system?
7. What percentage of the population has abnormal color vision?
8. Which depth cue would a monocular observer lack?
9. What are some of the limitations associated with the use of olfactory displays?

REFERENCES

Anstis, S. (2002). The Purkinje rod-cone shift as a function of luminance and retinal eccentricity. *Vision Research*, 42(22), 2485–2491.

Bach-y-Rita, P., Collins, C. C., Saunders, F. A., White, B., & Scadden, L. (1969). Vision substitution by tactile image projection. *Nature*, 221, 963–964.

Bach-y-Rita, P., Kaczmarek, K. A., Tyler, M. E., & Jorge Garcia-Lara, J. (1998). Form perception with a 49-point electrotactile stimulus array on the tongue: a technical note. *Journal of Rehabilitation Research and Development*, 35, 427–430.

Bach-y-Rita, P. W., & Kercel, S. (2003). Sensory substitution and the human-machine interface. *Trends in Cognitive Sciences*, 7(12), 541–546.

BBC-News. (2002). Safety fears at air traffic centre. Retrieved April 18, 2002, from https://news.bbc.co.uk/2/hi/uk_news/1936464.stm.

Blake, R., & Sekular, R. (2006). *Perception*. New York, McGraw Hill.

Bouma, H. (1970). Interaction effects in parafoveal letter recognition. *Nature*, 226, 177–178.

Burg, A. (1967). *The Relationship between Vision Tests Scores and Driving Record: General Findings*. Los Angeles, CA: Department of Engineering, University of California.

Burg, A. (1968). *Vision Test Scores and Driving Record: Additional Findings*. Los Angeles, CA: Department of Engineering, University of California.

Cheng, K. (2005). *Designing Type*. New Haven, CT: Yale University Press.

Cole, B. L. (2004). The handicap of abnormal colour vision. *Clinical and Experimental Optometry*, 87, 288–292.

Cole, B. L., & Macdonald, W. A. (1988). Defective colour vision can impede information acquisition from redundantly colour coded video displays. *Ophthalmic and Physiological Optics*, 8, 198–210.

Cole, B. L., Maddocks, J. D., & Sharpe, K. (2004). Visual search and the conspicuity of coloured targets for colour vision normal and colour vision deficient observers. *Clinical Experimental Optometry*, 87, 294–304.

Endsley, M. R. (1995). Toward a theory of situation awareness in dynamic systems. *Human Factors*, 37(1), 32–64.

Fields, H. (2012). Fragrant flashbacks: smells rouse early memories. *Observer: Association for Psychological Science*, 25, 4.

Fiset, D., Blais, C., Ethier-Majcher, C., Arguin, M., Bub, D. N., & Gosselin, F. (2008). Features for identification of uppercase and lowercase letter. *Psychological Science*, 19(11), 1161–1168.

Flück, D. (2010). Living with color blindness. Retrieved August 20, 2010, from www.colblindor.com.

Forbes, T. W., Moskowitz, K., Morgan, G., & Loutzenheiser, D. W. (1950). A comparison of lower case and capital letters for highway signs. *Proceedings of the Highway Research Board*, 30, 355–373.

Garvery, P. M., Pietrucha, M. T., & Meeker, D. (1997). Effects of font and capitalization on legibility of guide signs. *Transportation Research Record*, 1605, 73–79.

Garvery, P. M., Pietrucha, M. T., & Meeker, D. (1998). Clearer road signs ahead. *Ergonomics in Design*, 6(3), 7–11.

Gibson, J. J. (1979). *The Ecological Approach to Visual Perception*. Boston: Houghton Mifflin.

Goldstein, E. B. (2010). *Sensation and Perception*. New York: Wadsworth Cengage Learning.

Helig, M. (2011). Sensorama. Retrieved January 12, 2011, from www.telepresence.org/sensorama/index.html.

Hills, B. L. (1980). Vision, visibility, and perception in driving. *Perception*, 9, 183–216.

Hoecks, B., & Levelt, W. (1993). Pupillary dilation as a measure of attention: a quantitative system analysis. *Behavior Research Methods, Instruments, & Computers*, 25, 16–26.

Klatzky, R. L., Lederman, S. J., & Metzger, V. A. (1985). Identifying objects by touch: an "expert system". *Perception & Psychophysics*, 37(4), 299–302.

Legge, G. E., Rubin, G. S., & Luebker, A. (1987). Psychophysics of reading. V. The role of contrast in normal vision. *Vision Research*, 27, 1165–1177.

McGregor, L. N., & Chaparro, A. (2005). Visual difficulties reported by low-vision and nonimpaired older adult drivers. *Human Factors*, 47(3), 469–478.

O'Brien, K. A., Cole, B. L., Maddocks, J. D., & Forbes, A. B. (2002). Color and defective color vision as factors in the conspicuity of signs and signals. *Human Factors*, 44, 665–675.

Padgham, C. A., & Saunders, J. E. (1975). *The Perception of Light and Colour*. New York: Academic Press.

Parry, M. (2010). Colleges lock out blind students online. *The Chronicle of Higher Education*, A1–A8.

Pelli, D. G., Burns, C. W., Farell, B., & Moore-Page, D. C. (2006). Feature detection and letter identification. *Vision Research*, 46, 4646–4674.

Pokorny, J., Lutze, M., Cao, D., & Zele, A. J. (2006). The color of night: surface color perception under dim illuminations. *Visual Neuroscience*, 23, 521–530.

Rodieck, R. W. (1998). *The First Steps in Seeing*. Sunderland, MA: Sinauer.

Sampaio, E., Maris, S., & Bach-y-Rita, P. (2001). Brain plasticity: 'visual' acuity of blind persons via the tongue. *Brain Research*, 908, 204–207.

Sanders, M. S., & McCormick, J. (1993). *Human Factors in Engineering and Design*. New York: McGraw-Hill.

Sharpe, L. T., Stockman, A., Jägle, H., & Nathans, J., Eds. (1999). *Opsin Genes, Cone Photopigments, Color Vision, and Color Blindness*. Color Vision: From Genes to Perception. New York: Cambridge University Press.

Spalding, J. A. B., Ed. (1997). *Doctors with Inherited Colour Vision Deficiency: Their Difficulties with Clinical Work*. Colour Vision Deficiencies XIII. Dordrecht, the Netherlands: Kluwer Academic.

Spalding, J. A. B. (1999). Medical students and congenital colour vision deficiency: unnoticed problems and the case for screening. *Occupational Medicine*, 49, 247–252.

Tufte, E. R. (1983). *The Visual Display of Quantitative Information*. Cheshire, CT: Graphic Press.

Tufte, E. R. (1990). *Envisioning Information*. Cheshire, CT: Graphic Press.

Võ, M. L. H., Jacobs, A. M., Kuchinke, L., Hofmann, M., Conrad, M., Schacht, A., & Hutzler, F. (2008). The coupling of emotion and cognition in the eye: introducing the pupil old/new effect. *Psychophysiology*, *45*, 130–140.

Vuillerme, N., Pinsault, N., Fleury, A., Chenu, O., Demongeot, J., Payan, Y., & Pavan, P. (2008). Effectiveness of an electro-tactile vestibular substitution system in improving upright postural control in unilateral vestibular-defective patients. *Gait & Posture*, *28*, 711–715.

Weinstein, S. (1968). *Intensive and Extensive Aspects of Tactile Sensitivity as a Function of Body Part, Sex, and Laterality. The Skin Senses*. D. R. Kenshalo. Springfield, IL: Charles C. Thomas, 195–222.

White, B. W., Saunders, F. A., Scadden, L., Bach-Y-Rita, P., & Collins, C. C. (1970). Seeing with the skin. *Perception and Psychophysics*, *7*, 23–27.

Wickens, C. D., & Carswell, C. M. (1995). The proximity compatibility principle: its psychological foundation and its relevance to display design. *Human Factors*, *37*(3), 473–494.

Wickens, C. D., Lee, J. D., Liu, Y. D., & Gordon-Becker, S. (2004). *An Introduction to Human Factors Engineering*. Upper Saddle River, NJ: Pearson Education Inc.

Wolfe, J. M., Kluender, K. R., Levi, D. M., Bartoshuk, L. M., Herz, R. S., & Merfeld, D. (2006). *Sensation & Perception*. Sunderland, MA: Sinauer Associates Inc.

Woodson, W. E. (1981). *Human Factors Design Handbook*. New York: McGraw-Hill.

Yaffa, J. (2007). The road to clarity. *New York Times Magazine*.

Yantis, S. (2014). *Sensation and Perception*. New York, NY: Worth Publishers.

Chapter 4

Audition and vestibular function

Chapter Vignette

After her late afternoon class, Sue and a friend leave campus and navigate the sidewalks and intersections surrounding the university crowded with students and office workers eager to get home, as well as all the traffic. Listening to the environment around her, she can readily distinguish the sounds of cars, trucks, construction equipment, and the voices of other pedestrians that form the din of background sounds. Each of these sounds contributes to the complex variations in air pressure that press on her eardrums, and which are the basis of hearing. She walks seemingly effortlessly while engaged in conversation with her classmate. Sue looks intently at her friend's face as the surrounding noise gets louder and her friend speaks louder to be heard over the noise at the street crossing. Distracted by the conversation and the effort of attending to her friend's story, Sue begins to step into the intersection but is alerted by a car horn and turns to her left to see an approaching car. Having safely navigated the intersection without falling, she walks along the busy sidewalk passing businesses, and patrons seated at the sidewalk café when Sue hears someone call her name from one of the cafés. Even in the presence of background noise, the sound of her name immediately drew her attention, and she shifts her gaze to the approximate location of the speaker.

This vignette illustrates the multitude of ways in which the auditory and vestibular systems contribute not only to our ability to make sense of the cacophony of sound arriving at our ears, but also the less recognized ways in which the vestibular system allows us to navigate the environment while maintaining our upright stance.

4.1 CHAPTER OBJECTIVES

After reading this chapter, you should be able to:

- Describe the major structural components of the auditory and vestibular systems.
- Describe the role of the vestibular system in maintaining balance.
- Describe the contribution of the vestibular system to motion sickness.
- Describe how sensitivity to sound varies as a function of intensity and frequency.
- Describe the relationship between the physical qualities of sound and perceptual qualities including pitch, timbre, and loudness.
- Describe the effects of masking on sound detection thresholds.
- Describe the use and design of auditory alarms.

DOI: 10.1201/9781003515463-4

- Describe the measures of speech intelligibility.
- Describe how hearing changes with age.

4.2 WHAT IS SOUND?

Sound is an important source of information that can convey unique or redundant information shared by other senses. Close your eyes and attend to the sounds around you; notice the rich acoustic environment filled with voices that reveal an individual's identity, emotional state, movements, location, and distance in space. Other sounds can convey information about the operation of devices including computers and air conditioning. In many cases, evidence of our physical actions is rendered in both visual and acoustic domains—as when you press a button to call an elevator and expect to hear the button depress and the button to light up, signaling your what? It is not surprising that we continue to press the elevator button or the delete key on a keyboard when we fail to receive visual and acoustic evidence that our actions were registered.

Sound is produced by vibrations of an object and its surfaces. The back-and-forth movement of the surface disturbs adjacent air molecules producing regions of higher and lower pressure that propagate away from the vibrating surface.

To understand sound, it is useful to consider an audio speaker diaphragm producing a simple pure tone that is characterized by its frequency and amplitude. The frequency of vibration is expressed in Hertz (Hz) or the number of cycles of movement of the diaphragm per second. One cycle of movement consists of one back-and-forth movement of the diaphragm indicated by the light and darker gray regions, as shown in Figure 4.1. The sine waves' amplitude corresponds to the height of the sine wave, or the distance between the peak and the trough. The frequency of the vibration is related to the human sensation of pitch, whereas its amplitude or intensity is associated with our perception of loudness. Over the range of human hearing that extends from 20 to 20,000 Hz, higher frequency sounds have a greater perceived pitch than lower frequency sounds; however, a sounds' pitch is also influenced by its intensity such that the pitch of high frequency tones (i.e., >3,000 Hz) shifts upward toward higher pitch when their intensity is increased. The opposite is true for lower

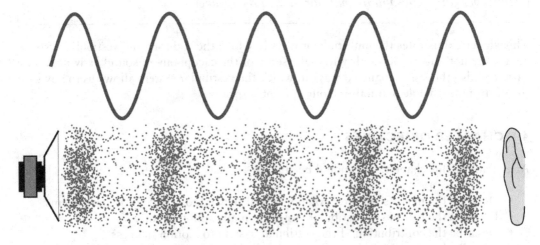

Figure 4.1 The surface or diaphragm of an audio speaker vibrates back and forth at a fixed frequency when producing a pure tone. The forward and backward movement of the speakers diaphragm creates regions of high (darker gray region) and low pressure (lighter gray region), respectively, that propagate outward. A louder tone increases the amplitude of the diaphragm's movement which translates into a louder sound. (Image from https://commons.wikimedia.org/wiki/File:CPT-sound-physical-manifestation.svg.)

frequency tones (i.e., <1,000 Hz), as the intensity of low frequency tones is increased, their pitch is perceived to shift toward lower frequency.

4.3 MEASURING SOUND INTENSITY

Sound's intensity is expressed in units of the amount of force per unit of area (N/m^2), where the unit of force is the Newton (N) and the unit of area is meters squared (m^2). The dynamic range of hearing extends from the faintest sound the human ear can detect to the loudest sound we can hear without damaging hearing representing a ratio of 1,000,000,000,000 to 1 or 1 trillion to 1 (Moore, 2001). That is a HUGE range! Because of this large range of audible sound intensities, a more concise manner for expressing sound intensity is needed. The **decibel scale** (Equation 4.1) serves this purpose by defining Sound Pressure Level (SPL) as the logarithm of the ratio between the sound pressure level of the sound of interest (P) and a reference sound (P_r). In other words, the decibel scale expresses how many times more intense a sound of interest (P) is than a reference sound (P_r). Using Equation 4.1, the $\log_{10}$ of the ratio of P and P_r is multiplied by 20:

$$SPL(dB) = 20\ Log_{10}\left(\frac{\text{Sound of interest}\ (P)}{\text{Reference sound}\ (P_r)}\right)$$

where P and P_r are the sound pressures produced by the sound you are measuring (i.e., sound of interest) and a reference sound having a pressure of $0.00002\ N/m^2$, respectively. The common parlance is to specify sound intensities as dB_{SPL} when the reference pressure corresponds to the sound pressure produced by a just detectable 1,000 Hz tone. The decibel value of sound produced by common everyday objects is shown in Figure 4.2.

Typical Sound Levels (dBA)

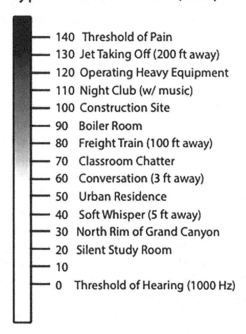

- 140 Threshold of Pain
- 130 Jet Taking Off (200 ft away)
- 120 Operating Heavy Equipment
- 110 Night Club (w/ music)
- 100 Construction Site
- 90 Boiler Room
- 80 Freight Train (100 ft away)
- 70 Classroom Chatter
- 60 Conversation (3 ft away)
- 50 Urban Residence
- 40 Soft Whisper (5 ft away)
- 30 North Rim of Grand Canyon
- 20 Silent Study Room
- 10
- 0 Threshold of Hearing (1000 Hz)

Figure 4.2 The sound pressure levels produced by common everyday sound sources. (Adapted from https://www.osha.gov/dts/osta/otm/new_noise/index.html#decibles.)

4.4 COMPLEX SOUNDS

Unlike the sound produced by a tuning fork that can be described mathematically as a simple sine wave, most natural sounds like voices or musical instruments are more complex consisting of multiple sound frequencies of different amplitudes. The French Mathematician J. B. J. Fourier demonstrated that any complex waveform can be decomposed into its constituent simpler sine waves that when added together will reproduce the original waveform.

BOX 4.1:   BETTER FOR THE ENVIRONMENT BUT MORE DANGEROUS FOR PEDESTRIANS? ANNOUNCING THE ARRIVAL OF THE ELECTRIC CAR

Electric or hybrid gas/electric cars have increased in popularity due in part to increases in gas prices and concerns about the environment. While hybrid electric vehicles (HEVs) are a more fuel-efficient means of conveyance, they have also raised concerns among transportation safety experts and advocates for the blind because they are less detectable than internal combustion engine (ICE) at low speeds where the vehicle is more likely to be powered by the electric motor, posing a risk to pedestrians. The engine noise is greatly reduced at low speeds along with other auditory cues including wind and tire noise that blind pedestrians listen for to detect the presence, distance, direction of movement, and rate of acceleration of vehicles. These acoustic cues also support orienting, allowing the blind to maintain their alignment relative to traffic on crosswalks.

Data from 12 states offer support for the concerns expressed by safety experts. They reported a significantly higher rate of pedestrian-vehicle crashes for HEV relative to ICE vehicles at low speed limits and under good visual conditions (daytime and clear weather; Hanna, 2009). Interestingly, HEVs were disproportionally involved in mishaps under conditions where a vehicle was slowing or stopping, reversing, or exiting or entering a parking space. It is precisely under these conditions—a vehicle moving slowly—that the maximum differences (7–10 dB(A)) in sound generated by the HEV and ICE vehicles are observed. HEVs were twice as likely to be involved in an incident under these conditions than a car powered by an ICE.

Tests conducted with blind pedestrians indicate that they could detect the ICE powered cars at greater distances than HEVs under both low and high ambient sound levels when the vehicles were backing up (5 mph) or approaching at a low constant speed (6 mph), but not when slowing from 20 to 10 mph. Slowing HEVs were more easily detected by blind pedestrians because they produced a distinctive 5,000 Hz hum when braking (converting mechanical kinetic energy into electricity). Vehicle manufacturers have proposed adding synthetic sounds for quieter cars that adjust the sound intensity relative to the ambient sound level to ensure they are detectable. The sounds would enhance the safety of blind pedestrians, joggers, and older adults. The flexibility of synthetic sound generating tools could allow drivers to customize the sounds their cars generate. A government report expressed concern regarding "a potential conflict between the need for standardization of vehicle sounds to enhance recognition versus possible end-user preference for unique, personalized sounds" (Garay-Vega et al., 2010, p. 69). **Critical Thinking Questions:** Do you agree with the concern expressed by the government regarding the use of personalized car ring tones? Can you identify any alternatives to using sound in this context?

4.5 THE HUMAN EAR

Figure 4.3 depicts a schematic of the human ear. The major structural components of the ear are organized into three regions: the outer, middle, and inner ear. The structure of the outer ear consists of the **pinna** (the external ear structure) and **auditory canal** that direct sound toward the **tympanic membrane** (eardrum), which vibrates back and forth in response to the changes in air pressure produced by a sound. The ear canal helps protect the eardrum and also has resonant properties that increase the sound pressure level of frequencies in the range of 2,000–5,000 Hz by as much as 12 dB. The movements of the tympanic membrane are conveyed to the inner ear by the **ossicles** of the middle ear. The ossicles are three bones called the **malleus, incus,** and **stapes,** which link the tympanic membrane and the oval window of the inner ear. Attached to the ossicles are two muscles called the **tensor tympani** and the **stapedius** that tighten in response to loud noises offering some protection from loud extended sounds. However, this tightening of the muscles, called the **aural reflex** (Reger, 1960), is relatively slow (approximately 35–140 msec) and offers little protective benefit from loud impulsive sounds like that produced by a gun.

As the ossicles move in concert with the movements of the tympanic membrane, they act as a lever on the oval window. In particular, the lever action of the Stapes on the oval window amplifies the movements of the tympanic membrane by roughly 22-fold. The greater force of the stapes acting on the oval window is necessary to overcome the inertia of the liquid found inside the cochlea.

The **cochlea** consists of a coiled tube forming a snell-shaped structure that houses receptor mechanisms that encode sounds. The tube is segmented into three compartments that run its full length and the central compartment contains the basilar membrane. The basilar

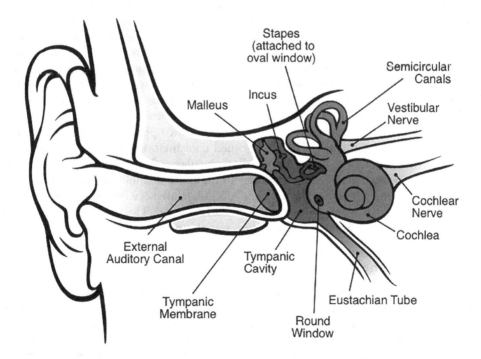

Figure 4.3 Schematic of the human ear. (Image from https://en.wikipedia.org/wiki/Auditory_system#/media/File:Anatomy_of_the_Human_Ear_en.svg.)

membrane is deformed when the liquid in the inner ear is displaced producing a wave-like movement of the liquid induced by action of the stapes on the oval window. Hair cells found along the length of the basilar membrane are bent by the resulting deformation of the basilar membrane, and it is the bending of the hair cells that initiate the generation of electrical signals that travel along the auditory nerve to higher brain centers.

4.6 THE VESTIBULAR SYSTEM

Also located in the inner ear are several other structures including three semicircular canals (anterior, posterior, and horizontal; see Figure 4.3) and otolith organs that comprise the vestibular system. The vestibular system provides information about your body's orientation in space and about the orientation and motion of your head and/or body. The three **semicircular canals** are oriented at right angles to one another and encode rotation of the head along any of three axes X, Y, and Z. Each semicircular canal contains fluid like the cochlea whereby acceleration is detected by hair cells called stereocilia, which are displaced by sudden head movements. The displacement of the fluid in the canal bends the hair cells whose response encodes information concerning the direction and rate of rotation of the head.

The **otolithic organs** that include the **utricle** and **saccule** are located at the base of the semicircular canals. The utricle and saccule detect linear acceleration and the titling of the head. The term "otolith" means *ear stones* in reference to small calcium carbonate crystals that are suspended above the hair cells by a gelatinous layer. When the head moves or is held in a static tilted position, the heavier crystals are displaced bending the hair cells thereby encoding the movement or orientation of the head.

The information provided by the vestibular system serves several critical functions including the maintenance of balance and the sense of spatial orientation and motion. Vestibular information also serves as input to parts of the nervous system that allow our musculature to make adjustments to maintain an upright posture even when the surfaces we are standing on might be moving as in the case of a bus, train, airplane, or boat. Perhaps less readily recognized is the important role vestibular input serves in allowing us to maintain a relatively stable view of the world while moving. Imagine sitting in a bus and staring out the window at a billboard or another feature in the environment. As the bus moves, the vestibular system helps you make automatic corrective eye movements in the opposite direction of motion to keep your eyes on the billboard. The finely tuned coordination of vestibular information and eye-movement control is called the **vestibulo-ocular reflex** (VOR). The coordination of head and eye movements allows the viewer to maintain a stable and clear image of the billboard on the retina while moving. Corrective VOR-related eye movements are generated in the dark and even when the observer has one's eyes closed. This result indicates that the corrective eye movements are not dependent on visual input.

4.6.1 Motion sickness

In most cases, purposeful movements under our control like walking or driving rarely cause us any discomfort. However, most if not all of us have experienced discomfort as passengers on moving cars, boats, airplanes, or amusement rides. Although referred to by different names (i.e., carsickness, seasickness, or airsickness), these forms of motion sickness are hypothesized to have a common underlying cause involving discordance between vestibular and visual cues. The most widely cited account of motion sickness is called the **sensory conflict theory** (Reason & Brand, 1975), which posits that motion sickness arises when vestibular motion cues (e.g., vestibular cues indicating motion and turning of an airplane) are in conflict with the visual cues. For example, if you are sitting inside an airplane and viewing the aircraft cabin, you see an environment that is stationary and horizontal, yet the vestibular system may be signaling movement, acceleration, and rotation associated with the movements of the aircraft. The importance of vestibular information to motion sickness is suggested by the finding that individuals without a functioning vestibular system do not experience motion sickness (Kellogg et al., 1965).

An alternative theory called the "**postural instability theory**" (Riccio & Stoffregen, 1991) proposes that postural sway or instability accounts for motion sickness and not sensory conflict per se. For instance, drivers and pilots are less likely to get motion sick than passengers in the same vehicle. Unlike the passenger, the driver or pilot can make anticipatory postural adjustments to compensate for vehicular motion, whereas the postural adjustments of the passenger are delayed and are imperfect resulting in greater instability. It is precisely this instability that is hypothesized to account for motion sickness and should precede the onset of motion sickness. Subjective reports consistent with this hypothesis indicate that discomfort is proceeded by increased movement of the head and torso (Stoffregen et al., 2000) and that participants who report discomfort show larger amplitude postural adjustments than participants who report no motion sickness (Smart et al., 1998, 2002). These findings indicate that it may be possible to identify *a priori* individuals who may be more susceptible to motion sickness and to develop interventions to prevent motion sickness. Motion sickness can affect the operational readiness of aircrews and shipboard personnel as well as the effectiveness of training programs that use simulators to train pilots and soldiers. The pace of the training regimen may be modified to reduce the incidence of motion sickness or individuals may be provided medication to reduce symptoms of motion sickness.

Motion sickness is a common side effect of playing video games with incidence rates of 50%–100% in laboratory studies (Merhi et al., 2007; Stoffregen et al., 2008). Identification of game and user properties associated with a susceptibility to motion sickness can inform the design of both games and training regiments to reduce the incidence of motion sickness. See Box 4.2 for a discussion of motion sickness caused by video games.

BOX 4.2: JANE, GET ME OFF THIS CRAZY THING!

Gamers who play popular video games like Modern Warfare II or Halo sometimes experience motion sickness. In fact, modern technology has significantly expanded the range of stomach-churning situations one might encounter. These situations range from the video games mentioned above to amusement rides, virtual reality environments, large screen 3D movie theaters, and sophisticated flight simulators used for flight training. The terms simulation sickness and **cybersickness** are used to describe the various symptoms experienced by users of video games, simulators, or virtual reality systems. They range from minor discomfort (e.g., light headache), nausea, and vomiting to severe physical disturbances (e.g., extreme vertigo).

Reports indicate that the incidence of cybersickness is very high. For instance, over 80% of users of virtual reality simulations reported symptoms of motion sickness after 20 minutes of use (Cobb et al., 1999; Nichols et al., 2000; Stanney & Kennedy, 1998; Stanney et al.,1997). The incidence of motion sickness is positively associated with the technical sophistication of the display (Merhi et al., 2007).

While the terms cybersickness and simulation sickness are often used interchangeably, some researchers distinguish between these two terms (Stanney et al., 1997) based on reported symptomology and severity of the aftereffects. The susceptibility to cybersickness or other forms of motion sickness (airsickness, sea sickness, and car sickness) vary with age (Reason & Brand, 1975), with susceptibility being greatest for children between age 4 and 12years and declining significantly between age 12 and 25. There is a continual gradual decline in susceptibility thereafter with a low incidence after the age of 50. Women are reported to be more susceptible to motion sickness, but the cause remains unknown (Benson, 2002).

Symptoms of motion sickness also decline with continual or repeated exposure to the causative conditions; however, this adaptation does not generalize to other situations or inducing stimuli. A gradual increase in exposure time to the nausea inducing stimulus may be the best method of reducing cybersickness for users who are willing to invest time in using the technology (LaViola Jr., 2000). Nevertheless, a small percentage of users, approximately 3%, may never adapt, continuing to experience discomfort, despite extensive exposure (Biocca, 1992). Cybersickness presents a challenge for video game developers and hardware manufacturers (i.e., head mounted displays for virtual environments) because an individual's susceptibility to cybersickness might influence the adoption and use of the technology. **Critical Thinking Questions:** What strategies do you use to reduce motion sickness? How effective are these strategies? Note: Recent experimental findings show that in the case of seasickness, they often recommend the strategy of looking at the horizon, which does in fact reduce body sway (Mayo, 2011)

4.7 HUMAN SENSITIVITY TO SOUND

4.7.1 Coding of loudness and frequency by the cochlea

The cochlea is the snail-shaped structure (see Figure 4.3) in the inner ear which if uncoiled would be approximately 35 mm (1.37 inches) long. The response of the basilar membrane to sounds of different amplitude and frequency offers insight regarding how the ear encodes frequency and loudness. There are two "theories" that explain how we hear low, medium, and high frequency tones. Imagine the response of your ear as you listen to a 5,000 Hz tone. The tympanic membrane and oval window oscillate back and forth 5,000 times a second as the air pressure increases and decreases sinusoidally causing a cyclic displacement of the liquid in the inner ear. As illustrated in Figure 4.4, the movement of the fluid filling the inner ear induces a traveling wave (von Békésy, 1960) in the thin basilar membrane that moves from its base at the oval window to its apex. Looking at the basilar membrane while stimulating the ear with different sound frequencies, we would see a traveling wave that peaks at different points along the basilar membrane with higher sound frequencies peaking near the base and lower frequencies peaking near the apex of the basilar membrane. Thus, the ear encodes sound frequency using a **place code** where hair cells distributed along the length of the basilar membrane encode different sounds between 20 and 20,000 Hz. The place code theory offers a good explanation of the coding of high frequency (>5,000 Hz) but not low frequency tones, and the reason is that low frequency tones produce traveling waves with broad peaks, making it more difficult to identify the place on the basilar membrane associated with a specific frequency. A **volley principle** has been proposed to account for the coding of lower sound frequencies (Wever, 1949). The volley principle proposes that groups of auditory fibers may collaboratively respond to the peak of each cycle of the stimulus by

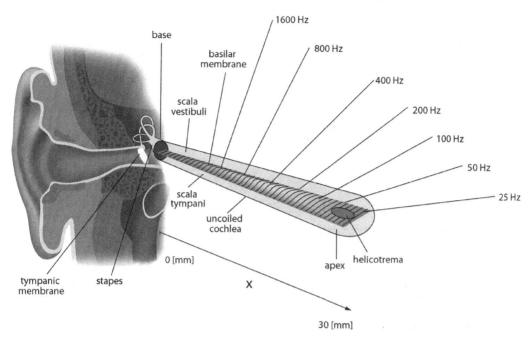

Figure 4.4 The illustration shows the uncoiled cochlea and depicts where along its length a traveling wave motion caused by different sound frequencies peak. (Image from https://commons.wikimedia.org/wiki/File:Uncoiled_cochlea_with_basilar_membrane.png.).

alternating which fibers respond to the peak of the traveling wave. The combined responses of the group of fibers—the volley—encode the stimulus' frequency. The place and volley principles are complementary—one encodes high frequencies and the other encodes low frequencies—with some overlap in the middle.

When observing the response of the basilar membrane to a loud or high amplitude tone, we would notice that the peak displacement of the basilar membrane occurs at the same point on the basilar membrane but that the displacement would be larger. More intense sounds result in larger displacements of the basilar membranes and bending of hair cells that cause neurons to fire more rapidly. Firing rate is one way that the auditory system encodes a sound's amplitude that we perceive as loudness.

4.7.2 Auditory sensitivity

Complex vibrations produce a broad range of sound frequencies, but our auditory system is only sensitive to sound frequencies between 20 and 20,000 Hz. The auditory system is also not equally sensitive to all sound frequencies.

As can be seen in Figure 4.5, we are maximally sensitive to sound frequencies between 2,000 and 5,000 Hz. The Nobel Prize winning auditory physiologist Georg Von Békésy (1957) estimated that hearing sensitivity is so acute that vibrations of the ear drum as small as 1 billionth of a centimeter or roughly one-tenth the diameter of a Hydrogen atom are detectable by listeners. Sounds outside the 2,000–5,000 Hz range require greater amplitude or intensity for a listener to detect them. For instance, we are roughly 1,000 times more sensitive to a 1,000 Hz tone than to a 100 Hz tone. Likewise, our sensitivity to high frequency sounds (>6,000 Hz) is also poorer. The upper range of hearing is limited to sounds that produce discomfort or pain and where short duration exposure to some sounds may result in permanent hearing damage.

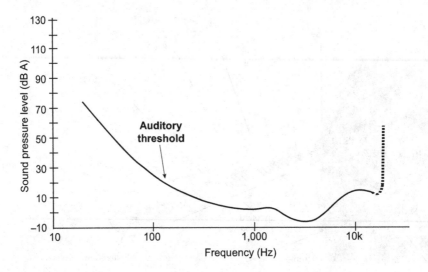

Figure 4.5 Auditory sensitivity as a function of sound frequency. (Adapted image from https://en.wikipedia. org/wiki/Psychoacoustics.)

4.7.3 Sensitivity to pitch

The perceptual correlate of sound frequency is pitch. The relationship between frequency and pitch is illustrated by the Mel scale shown in Figure 4.6 where a **mel** is a unit of pitch. The scale depicted in this figure was obtained by having participants adjust a variable stimulus so that it had either half the pitch or twice the pitch of a standard tone. The standard tone was arbitrarily defined as a 1,000 Hz tone set to 40 dB and was assigned a pitch of 1,000 mels. Importantly, the results the results reveal that the tone reveal that the tone perceived to have twice the pitch (i.e., 2,000 mels) as the standard tone is not a 2,000 Hz tone but rather is a 3,000 Hz tone. This illustrates the nonlinear relationship between frequency and perceived pitch. This finding is relevant to applications where pitch is used to code changes in the value of a variable. Consider a case where an operator is remotely monitoring parameters of an engine and pitch is used to represent engine temperature, oil pressure, or rpm's (rotations per minute). The mapping of the engine parameter (i.e., temperature)

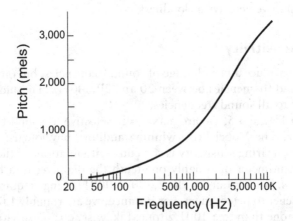

Figure 4.6 Relationship between perceived pitch and sound frequency as measured using the method of magnitude estimation. (Adapted from Stevens & Volkmann, 1940.)

to pitch must account for this nonlinear relationship. In other words, a doubling of engine temperature cannot be encoded by simply doubling the tones frequency since this will not produce a doubling of the sound's perceived pitch.

A tone's perceived pitch is affected by other stimulus properties including intensity and duration. For instance, the pitch of low frequency sounds (e.g., 300 Hz) decreases as the tones' intensity increases, but the opposite is true for higher frequency tones, i.e., their pitch increases as their intensity increases. The perceived pitch of tones with frequencies of 1,000–2,000 Hz are largely unaffected by changes in intensity.

Stimulus duration also influences the perceived pitch of a tone. Listeners report short duration tones, lasting 10 msec (.01 sec) or shorter, sound like clicks and take on their characteristic pitch only when their duration is increased to approximately 40 msec.

4.7.3.1 Pitch perception and the missing fundamental

As noted earlier, complex sounds produced by human voices and instruments like a trumpet or piano have a harmonic structure consisting of a **fundamental frequency**, say 200 Hz, which is the lowest sound frequency present in the sound and **harmonics** (i.e., 400, 600, and 800 Hz) that are tones with frequencies that are integer multiples of the fundamental frequency. The perceived pitch of the complex sound, however, is most closely associated with the fundamental frequency (i.e., 200 Hz). However, what would happen if the sound were filtered so that the fundamental frequency was eliminated, but the harmonic frequencies were retained—would our perception of pitch change? Interestingly, our perception of pitch does not change; we continue to hear the pitch associated with the missing fundamental. This finding demonstrates that the perception of pitch depends on the fundamental frequency and the pattern of harmonic frequencies. The result of the missing fundamental is advantageous given that the small inexpensive speakers found on phones cannot reproduce the fundamental frequencies of male voices that range from 85 to 155 Hz (Fitch & Holbrook, 1970). Nevertheless, when we listen to a male speaker's voice over the phone, the pitch sounds normal even though the fundamental frequency is absent. This is a result of our auditory system's ability to synthesize the missing fundamental from the pattern of the harmonics.

The harmonic frequencies are also critical to the perception of **timbre** or the quality of sounds that allows listeners to discriminate between a piano and trumpet playing the same note (say a "C" sharp). While the fundamental frequency and consequently the pitch of the sounds are identical, listeners have no difficulty discriminating between the sounds produced by each instrument. The ability to judge this quality of a sound relies on several properties including the harmonic structure; that is the composition and amplitude of the harmonic sound frequencies as well as differences in how rapidly the sounds rise and decay.

4.8 MASKING

The ideal listening conditions used to measure auditory threshold sensitivity are different from the environment described in the opening vignette where Sue seeks to detect, recognize, and discriminate auditory signals in the presence of other sounds. Understanding the effects of noise on the ability to detect other signals, or what is technically referred to as **masking**, is obviously important given that this is much more representative of the conditions we experience in everyday situations. Studies of masking investigate how detection of one sound is affected by the presence of another sound. The masking effects of a sound are studied by evaluating the thresholds for detecting a target sound presented alone under quiet conditions and when presented simultaneously with a second sound (i.e., the masking sound).

When a masking sound is low in intensity, it selectively raises thresholds for a narrow range of tones that are similar to it in frequency. However, as the intensity of the masking frequency increases, say 40 or 50 dB, its influence broadens, affecting sensitivity for a larger range of sound frequencies. Importantly, the effects of the masker are asymmetric, so that at higher intensities its effects extend more broadly to higher frequency sounds than to tones that are lower in frequency than the masking tone. These results have several implications for real-world environments where listeners have to detect sounds in the presence of ambient noise caused by other equipment. In general, noise sources that are similar or lower in frequency than a target tone may pose the greatest difficulty to detecting a target signal, and the maskers' effect becomes more severe as the noise source is more intense. While the interaction of target and masking noise is complex depending on the intensity and frequency properties of the two sounds, these data nevertheless serve to illustrate the effects of masking using relatively simpler auditory stimuli.

4.9 RELATIONSHIP BETWEEN INTENSITY AND LOUDNESS

S. S. Stevens, an experimental psychologist, developed a special experimental technique called **Magnitude Estimation** to study how the subjective perception of loudness was related to the sound's physical amplitude. He had the subjects first briefly listen to a standard tone which was a 1,000 Hz tone set to 80 dB and then listen to a comparison tone of the same frequency, but different amplitude. He then instructed participants to assign the standard tone a value of 10 units and then assign the comparison tone a value depending on how loud they perceived it to be relative to the standard tone. If the comparison tone was perceived to be half or twice as loud as the standard tone, then they would assign it a value of 5 or 20, respectively. As is apparent from Figure 4.7, loudness and amplitude are not linearly related to each other. As noted earlier, at low sound amplitudes, a doubling of a sounds' magnitude from 20 to 40 dB does not produce a doubling of a sound's perceived loudness (plotted on the y axis), whereas at greater sound amplitudes, say 80–90 dB, small increases in intensity cause a greater than doubling in the perceived loudness of a tone.

The loudness of a tone also depends on its frequency. This can be demonstrated by having participants perform a loudness-matching task wherein they vary the loudness of a sound to match the loudness of a reference sound. By repeating this procedure for different sound

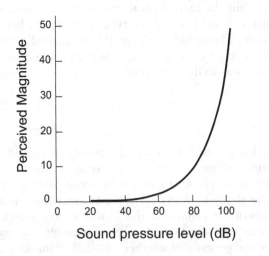

Figure 4.7 Relationship between sound intensity and perceived magnitude (expressed in mels) measured using the magnitude estimation technique developed by S.S. Stevens. (Adapted from Stevens, 1956.)

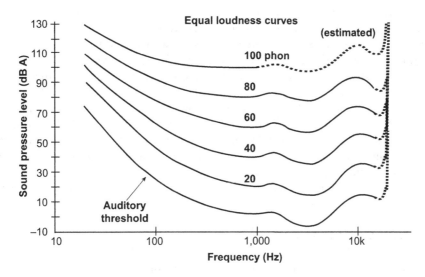

Figure 4.8 A plot of equal loudness curves as a function of frequency for different matching loudness (phon) levels.(Imageadaptedfromhttps://commons.wikimedia.org/wiki/File:Equal_loudness_no_caption.svg.)

frequencies and for different intensities of the reference sound, we obtain **equal loudness contours** like those depicted in Figure 4.8. Consider the equal loudness curve labeled 20; this curve was obtained by adjusting the intensity of other sound frequencies so that their perceived loudness matches that of a 1,000 Hz, 20 dB reference tone. The other curves were obtained in an identical manner, but with the intensity of the 1,000 Hz tone set to 40, 60, and 80 dB. Each of these curves then offers a summary of the intensity level of different sound frequencies that make them all sound equally loud as the 1,000 Hz reference tone.

The equal loudness contours, also called **Fletcher–Munson curves** (Fletcher & Munson, 1933), illustrate several important points about our perception of loudness. First, sounds in the range of 3,000–5,000 Hz require the least amplitude to match the perceived intensity of the 1,000 Hz tone, which is consistent with the fact that listeners are maximally sensitive to sounds in this range as can be seen in the bottom most line indicating hearing thresholds for different sound frequencies in Figure 4.8. Second, equal loudness contours are flatter at higher intensities indicating that low, medium, and high frequency tones are perceived to be more similar in loudness at higher intensities. This property of hearing is also reflected in sound meters that employ different frequency weighting schemes depending on sound intensity. As discussed in Section 4.8.2, under relatively quiet conditions, where sensitivity to low and high frequency sounds is considerably better, sound meters use a weighting scheme that gives more importance to energy found in medium sound frequencies. The Fletcher–Munson curves are also relevant to the perception of music because the perception of music is expected to change given the unequal contribution of low frequency tones to loudness at lower volume settings. In fact, the loudness button found on some stereo systems selectively increases the intensity of low frequency tones when the stereo is played at lower volume settings thereby compensating for their reduced contribution to loudness (Foley & Matlin, 2015).

A special unit called the phon is used by researchers to express the perceived loudness of sound. The **phon** is defined as the decibel value of a 1,000 Hz reference tone. Confused? In other words, 100 and 500 Hz tones perceived to be equally loud as a 40 dB, 1,000 Hz tone are all said to have a loudness of 40 phons. Similarly, tones perceived to be equally loud as a 60 dB, 1,000 Hz tone having a loudness of 60 phons.

A word of caution is in order with regard to the interpretation of phons. Phons indicate the subjective equality of different tones in loudness—that is sounds that have the same phon value are perceived to be the same loudness—but differences in phons, say 40 versus 50 phons, do not say anything about the *relative* loudness or how much louder one sound is than another. Instead, a unit called a **Sone** is used to express relative loudness, where one sone is defined as the loudness of a 40 dB, 1,000 Hz tone (see Figure 4.8). A sound that is twice as loud as the reference 1,000 Hz tone has a loudness of two sones, a tone that is 3 times as loud as the reference tone has a loudness of three sones, and a tone that is half as loud as the reference has loudness value of 0.5 sones. Sone values are provided by manufacturers of commercial products including bathroom and oven fans although you would be hard pressed to find consumers and contractors who know what the sone values means. For comparison, the sone values of an open office and a TV or radio are 1.5–2 and 3–4 sones, respectively.

4.10 SOUND LOCALIZATION

When we listen to our environment, we can readily identify the relative location of sounds even if we do not directly look at them. Identifying the location of a sound is important given that we normally want to look at the person addressing us, or the location of an on-coming car as described in the scenario at the start of the chapter. Our ability to localize objects using sound relies on two auditory cues, namely interaural intensity differences and interaural timing differences. **Interaural intensity differences (IID)** refer to differences in the intensity of sounds due to their position relative to the head of the listener. The listener's head creates an acoustic shadow for sound frequencies greater than 3,000 Hz such that sounds originating from positions directly beside a listener are less intense in the opposite ear.

The change in position relative to the head along the horizontal plane is called the azimuth of sound, with an azimuth of 0° corresponding to straight ahead, an azimuth of +90° corresponding to the right of the listener, and an azimuth of 180° corresponding to a position directly behind the listener. For low frequency sounds, say <500 Hz, there is little IID regardless of the sounds position due to diffraction which causes sound waves to bend around the head thereby reducing any IID between the ears. However, for higher frequency sounds (5,000–10,000 Hz), the IID increases especially for sound positions lying between 30 and 120 degrees.

Interaural time differences (ITI) arise from differences in the arrival time of sounds to each ear. Because the ears are physically separated, sounds originating from a source located off to one side will on average reach one ear about 640 msec (1 sec = 1,000 msec) earlier than the other ear. While the time differences may seem small, evidence suggests that listeners are capable of discriminating times differences as small as 10 msec (1/100 of a second), which may be sufficient to discriminate the positions of two tones separated by 2 inches from 3.05 meters (10 ft). As expected, ITI is greatest when a sound originates from a position directly to the side and is zero when the sound originates equally between the ears such as in front, above, or behind the listener. This cue is most effective for low frequency sounds (<1,500 Hz) because the ability to discriminate time differences is frequency dependent and is not very reliable for higher frequency sounds. In normal circumstances, confusion in localizing a sound location can be readily remedied by head movements that generate asymmetries in the stimulation of the ears.

While IID and ITI provide information about the azimuth of a sound source, what cues allow a listener to judge the elevation or vertical position of a sound relative to their head? It turns out that shape of the pinna may play an important role in generating cues to elevation. Sound recordings show that the pinna selectively enhances or attenuates different sound frequencies depending on its unique folds and shape and the elevation of the sound source. Listeners are attuned to these characteristics and use these cues to identify the elevation of a sound source.

4.11 PRECEDENCE EFFECT

The experiments described above concerning sound localization were carefully designed to minimize any potential confounding effects of reverberations or echoes produced by walls or other surfaces in the environment. However, sound propagates in all directions, with some traveling directly to ears and arriving earlier than sound being reflected off other surfaces. So how does the auditory system resolve the conflicting localization cues caused by leading (sound traveling a direct path to the ear) and lagging (sound reflected off other surfaces) sounds? Does the perceived position of a sound change as each sound arrives at the ear? You can test this at home by sitting between a set of speakers and shifting your sitting position from the middle toward one or another speaker. What you will notice is that as you shift your position to the left or right, the origin of the sound shifts toward the closer speaker. Experimental evidence suggests that localization is based on cues available in the sounds arriving first (i.e., hence the term precedence effect) and that the localization information available in lagging sounds may be actively suppressed at some site in the nervous system (Litovsky et al., 1999).

BOX 4.3: SPATIAL AUDIO

In important tasks such as air traffic control, the controller listens to multiple, mostly male talkers, over a monaural audio channel lacking any cues that allow the localization of the speakers in space. It is well known that speech intelligibility under these circumstances declines by about 40% with each additional talker (Brungart & Simpson, 2005). Techniques have been developed for processing sounds such that when presented over headphones they appear to originate from different points in space around the listener. The software processes the audio signal and constructs an audio stimulus that simulates the primary cues (i.e., interaural time differences and interaural levels differences) used to localize the position of a sound source in the horizontal plane (i.e., azimuth).

Spatial audio, as this technique is called, significantly improves speech intelligibility in multi-talker listening situations. Research shows the results of an experiment comparing speech intelligibility performance for a spatial audio condition and a dichotic listening condition (technical jargon meaning identical inputs to both ears). Listeners wore headsets and were instructed to shadow one speaker, and the number of competing speakers was varied between trials. In the spatial audio condition, the competing speakers were assigned one of seven different spatial positions. In the dichotic listening condition, the overlapping speakers are all perceived to originate along a line between the ears. These data shown in Figure 4.9 suggest that performance with spatial audio exceeds performance under the dichotic condition regardless of the number of talkers and that the performance gains exceed 20% in some multi-talker conditions. These data illustrate the effectiveness of spatial audio cues in allowing a listener to segregate and attend to a target message amongst a cacophony of competing voices. **Critical Thinking Questions:** Can you think of any others auditory cues that listeners can use to discriminate between different speakers?

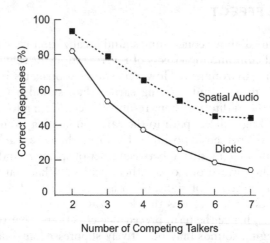

Figure 4.9 Results of experiment comparing speech intelligibility with and without spatial audio as a function of the number of competing talkers. The illustrations on the left depict the spatial location of the seven voices heard over the headset. In the absence of spatial audio, all the voices are perceived to originate in the center of the listeners head. (Adapted from Ericson et al., 2004.)

4.12 SUMMARY OF AUDITORY SENSORY THRESHOLDS

The basic auditory discriminative abilities described above offer a view of the complex and rich perceptual abilities that users possess. These abilities represent the building blocks for the design of effective auditory displays and are the basis of other complex perceptual abilities; perhaps none more important than the remarkable ability to understand speech under less-than-optimal listening conditions. These abilities are described in the following sections.

4.13 AUDITORY DISPLAYS

Auditory displays are ubiquitous. The term *auditory display* encompasses a wide range of uses of sound to convey information to listeners. The use of sounds includes alarms, warnings, status indications, instructions (Pull up!), and sonification. They are widely used in medical and aviation settings as well as in industrial control rooms. Their principal purpose is to *notify* or *alert* the listener that certain events have occurred (e.g., an email has arrived, the elevator has arrived at the designated floor), or that a certain amount of time has elapsed (e.g., cooking timer) or has been reached (e.g., 6 AM clock alarm) without also disrupting other ongoing tasks. **Cautions** and **warnings** are considered a special class of auditory displays since they are used to indicate an adverse event that is typically associated with a particular sound such as a tornado warning or fire alarm. The information communicated by alarms or notification sounds is rather limited since they only indicate a change in state— the tornado warning indicates that a tornado has been seen or is imminent—but it does not indicate where the tornado is located, which direction it is moving, or its severity. Speech may also be used as an auditory display (McGookin & Brewster, 2004). Synthesized speech or concatenated segments of recorded words are generated by some system as instructions for system users. For instance, vehicle navigation systems offer the driver synthesized verbal instructions alerting them of lane changes or upcoming turns.

Auditory displays can be organized into four general classes including personal devices, transportation, military, and control room applications (Stanton & Edworthy, 1999). Personal devices include car alarms, alarm clocks, and cellular phone chimes intended for a single user and that are not part of a larger integrated system. Transportation alarms include those designed for different modes of transportation including cars, trucks, buses, or airplanes that typically have more than one alarm and more than one operator. Military applications are primarily distinguished from the other two categories by the fact that they require specialized training and may be used to signify external threats in addition to internal system failures or states. Finally, central control room applications include alarms designed to support supervisory control where operators may monitor a complex system like a power plant or multiple patient vital signs from a central location or room.

One advantage of auditory displays is that they are omnidirectional, thus the operator does not have to be in a specific position to become aware of an alarm. When sounded, single auditory stimuli tend to elicit faster operator responses (Brebner & Welford, 1980) and their sound qualities can be designed or modified in real time to indicate urgency or priority. In real-world settings, an operator may be confronted by visual, auditory, and haptic information. Under such conditions, experimental findings indicate that visual information may dominate over auditory and haptic information (Hecht & Reiner, 2009).

The choice of auditory versus visual display of information is a complex one influenced by many factors. Sanders and McCormick (1993) offered a number of useful guidelines to consider when choosing between auditory or visual displays that are shown in Table 4.1. Unlike visual displays, auditory information is by its very nature transitory, and the listener has a short amount of time to attend and process the information or risk losing it. The processing of this information also places demands on limited capacity memory systems like verbal working memory that severely restrict a listener's ability to encode and maintain complex verbal messages such as instructions, descriptions, or directions. While the permanence of visual displays is a distinct advantage, they also restrict the mobility of the operator, and this sensory channel is often already burdened by other competing visual information sources that may impair detection of system warnings.

4.13.1 Design of warning and alarm signals

The chief challenge in designing or selecting an auditory alarm is ensuring that it is detectable in the environment where it is to be used. As discussed earlier, masking by ambient noise with components that are similar in frequency to the alarm pose the greatest risk to detection. Therefore, it is important to know as much as possible about the acoustic environment including noise intensity across the spectrum in the environment where the alarm will be used. Alarms that are 15–25 dB above the masked thresholds are unlikely to be

Table 4.1 Recommended uses of auditory and visual displays

Auditory presentation recommended when:	The message deals with events in time.
	The message calls for immediate action.
	The visual system is overburdened.
	The receiving location is too bright or dark adaption integrity is necessary.
	The person's job requires moving about continually.
Visual presentation recommended when	The message is complex.
	The message is long.
	The message will be referred to later.
	The message deals with locations in space.
	The message does not call for immediate action.
	The auditory system is overburdened.
	The received location is too noisy.
	The person job allows them to remain in one position.

Adapted from Sanders and McCormick (1993).

missed (Patterson, 1990). Although more intense tones could be used, they are more likely to startle the operator and interfere with any ongoing task during high workload conditions.

Early attempts at designing alarms relied primarily on frequency, intensity, and pulse rate (the fluctuation of the intensity) to produce prototypical alarms associated with school bells, foghorns, and klaxon and police sirens. The nuanced approach favored by Patterson (1990) emphasizes the purposeful construction of alarms designed to draw attention and convey urgency and meaning, while allowing communication between operator's and reducing interference with ongoing activities. Using bursts (100–300 msec) of sounds selected to optimize detectability and varying timing and loudness can convey urgency to the listener. The sound waveform envelope can also be sculpted to eliminate rapid sound onsets that can startle the operator.

4.13.2 Auditory icons, earcons, and sonification

Many cellular phone users rely on individualized ring tones to alert them of incoming calls from certain callers. The user can identify the caller without looking at the phone, which may be useful in situations where they are occupied with other tasks and their hands are not free to handle the phone. The individualized sounds may consist of music clips (e.g., Wild thing), natural sounds (braking glass), or more abstract computer-generated tones.

The use of sounds in this manner constitutes a novel form of auditory display called **auditory icons** and **earcons**. Auditory icons are an extension of the concept of visual icons where graphical symbols are used to represent a target action or process (see Chapter 3). Visual icons are effective because they can convey a lot of information very concisely and their intended meaning can be readily recognized. In the case of auditory icons, sound is used to represent the attributes of the object or action that is being carried out by the user.

In some cases, it may be difficult to find natural sounds to associate with more complex or abstract actions. For these situations, an option might be to develop an "earcon," which is a synthetic tone specifically designed to symbolize a state or action. Whereas an auditory icon relies on an intuitive link between a sound and the data or state, in the case of an earcon, the relationship must be explicitly learned. Consider the beeping sound now used to indicate that a vehicle is backing up. It is unlikely that without the explicit association (hearing the sound and seeing a vehicle backing up) that a listener would know what the sound signifies. Earcons have found wide application in computers where they are used as feedback when users scroll through menus options or when they select an option on a menu.

Sonification represents the last class of auditory displays. **Sonification** involves mapping numerical relations on to an acoustic domain so that users can interpret, understand, or communicate these relationships. Perhaps the most successful and widely known examples of sonification are Sonar and the Geiger counter. The rate and frequency of the audible clicks produced by the Geiger counter are directly related to the radiation level in the immediate area surrounding the counter and has been shown to be a superior way of monitoring radiation levels to a visual display or a visual plus audio display (Tzelgove et al., 1987).

Recent applications of sonification include auditory graphs where sound is used to represent quantitative data (Flowers & Hauer, 1993). The modulation of sound frequency or pitch is used to represent changes in the Y-axis values of data, whereas time represents the x-axis value of the graph. Changes in pitch over time can be a very effective means of describing auditorily the shapes of different curves on a graph. Because of the auditory system's acute sensitivity to temporal changes in sound, it may offer a more effective means of detecting temporal or periodic patterns in complex data that might otherwise be difficult to discern visually (Nees & Walker, 2009).

BOX 4.4:   LISTEN—CAN YOU SEE THAT?

Vision is the primary sensory system used to present information to human operators. This sensory channel bears a considerable burden creating a potential bottleneck in information processing. The auditory channel has been explored as a way to detect patterns in data that visually might appear a lot like noise. The auditory channel is particularly well suited to the detection of changes in a pattern across time. Doctors use stethoscopes to monitor breathing and heart beats to detect changes that might be indicative of serious medical conditions. Similarly, sonar operators on submarines listen for sounds that indicate the presence of an enemy submarine.

The discriminative power of the ear was recognized by scientists who sought to enlist it in the analysis of data. One of the early applications (Speeth, 1961) was the conversion of seismographs, which are recordings of **seismic waves** produced by earthquakes or explosions into sounds in an attempt to facilitate discriminating between records generated by underground nuclear explosions and naturally occurring earthquakes. By speeding up the seismograph recordings, the slower ground vibrations are shifted into frequencies that the human ear can detect, which represents a form of auditory display called **audification**. This allows the listener to hear natural phenomena like earthquakes, the solar wind generated by the sun, or high frequency radio waves that bathe the planet. In the case of seismographs, listeners become adept at discriminating between nuclear explosions and natural earthquakes, with accuracy levels approaching 90% (Hayward, 1994). Audification has been explored as a non-invasive technique for the diagnosis of knee-joint problems (Krishnan et al., 2001) and heart problems by making faint sounds audible so doctors can detect heart (Zhang et al., 1998) and joint abnormalities.

Unlike audification, sonification represents a more deliberate form of auditory display where the raw data are processed to extract features that are purposefully mapped onto sounds whose properties (i.e., pitch, amplitude, and tempo) change as the values of the variables change. An illustrative example involves the use of sonification to explore wave action information collected by buoys anchored to the sea bottom (Sturm, 2005). The inaudible range of ocean waves (i.e., 0.025–0.58 Hz) was made audible by multiplying the frequencies by 10,000, thereby placing them within the range of human hearing (20–20,000 Hz). Selective filtering of the spectrum can emphasize portions of the spectrum that are then mapped onto sound properties like pitch. By associating changes in wave frequency with pitch, listeners can discern seasonal changes in sea state as well as detect the presence and distance of storms, and the action of the wind on the sea.

The sonification of data holds promise as a means to explore complex data sets to identify patterns that are not readily recognizable by the eye. By filtering and amplifying components of the data, scientists can detect structures or patterns that they might otherwise miss. There have been a number of notable examples where sonification led to novel and important scientific insights in space sciences and physics (see paper by Barrass & Kramer, 1999). **Critical Thinking Questions:** How do you use sound to learn about your environment? For instance, can you identify who is walking down a hallway or who is at your door from the characteristic sound of their car or their walking? How do changes in the sound produced by your car reveal information about the road surface or state of your engine?

4.14 NOISE

Noise is defined as any unwanted sound or sound that is not task related and it need not be loud. Stop for a minute and listen to your surroundings. As you read this text, individuals talking nearby or the music you are listening to are noise. If you are typing notes as you read, then the sound of the keys might not be bothersome to you and are not noise, but nevertheless it is noise to persons seated nearby and may be irritating or distracting to them. The effects of noise are complex, as noise can be both a distraction—as in the case of two people talking in an otherwise quiet library—and a help to selective attention as in the case of the background din of your favorite coffee shop where you go to study.

Another common effect of loud sound is hearing loss. Loud (>140 dB), brief sound sources can produce permanent hearing loss. A loud **impulsive sound** source like detonating fire-crackers (170 dB SPL) or a gun fired (160–170 dB SPL) near the head can cause significant hearing loss characterized by a permanent increase in hearing thresholds or what is techni-cally called a **permanent threshold shift**—an immediate and enduring loss in sensitivity to some sound frequencies. Hearing damage can also result from repeated or sustained expo-sure to loud sounds at many work sites (e.g., construction, airports, and farming) or during recreation (sporting events, music concerts, or using earbuds or headphones). The hearing loss experienced by older listeners may represent the combined effects of age-related changes in hearing known as **presbycusis**, work related, or occupational hearing loss, as well as exposure to non-work-related noises such as listening to loud music called **sociocusis**.

While not all sound exposure will result in a permanent hearing loss, exposure to everyday sounds like a hairdryer (80 dB), vacuum (70 dB), and garbage disposal (80 dB) is sufficiently loud to cause a temporary elevation in auditory detection thresholds. This phenomenon, appro-priately called a **temporary threshold shift**, is assessed by measuring detection thresholds for a tone following a 2-minute exposure to a target stimulus (Jones & Broadbent, 1991). In general, sound levels equal to or greater than 60–65 dB(A) can produce temporary thresholds shifts that last from a few minutes to several hours or longer. The duration and magnitude of the threshold shift depend on the duration and intensity of the noise. Recovery from extended exposure to a loud sound—say a rock concert, which averages 103 dB (A) —can raise thresholds by 50 dB (representing a ~315-fold increase in threshold) and take as long as 16 hours for full recovery of sensitivity (Clark & Bohne, 1986; Davies & Jones, 1982). The amount of recovery may lessen with repeated exposure to the offending sound resulting in a permanent impairment.

4.15 SPEECH COMMUNICATION

Auditory alarms are not the only way of conveying information to an operator. Another major source of information is speech. Operators listen to voice communication communi-cated face-to-face or via another device—a cellular phone or public announcing system. The origin of the message may be another human or a computerized system that uses computer algorithms to produce synthesized voices as in the case of GPS navigation systems that pres-ent either alerting or instructional messages to human operators.

4.15.1 Evaluating speech quality

Several different methods can be used to evaluate the effectiveness of electronic communica-tion systems including phones, public announcement systems, as well as the suitability of different architectural environments such as classrooms or an airport terminal for effective voice communication. Depending on the application, other considerations such as voice or music quality may be paramount. Below we introduce two of these techniques.

4.15.1.1 Speech intelligibility

The simplest of these techniques is the measure of **speech intelligibility** that evaluates how well speech is understood. Intelligibility can be assessed by presenting a set of test stimuli and having the listener write down or repeat what was heard. The test stimuli may consist of consonant-vowel or vowel-consonant nonsense syllables, monosyllabic words, or individual sentences that allow the investigator to assess the comprehension of message, sentence, words, or syllables (Committee on Hearing, Bioacoustics and Biomechanics [CHABA], 1988).

The test stimuli may be presented via an audio recording or a practiced live speaker. In the case of sentences, performance is calculated based on the percentage of words in the sentence or target words in a set of sentences correctly reported by the participant. Examples of standardized test sentences from the Central Institute for the Deaf (CID; Davis & Silverman, 1978) and SPIN test are shown in Table 4.2. The acronym "SPIN" stands for "SPeech In Noise" (Kalikow et al., 1977), and this test requires the listener to repeat the last word of each sentence where the last word is either predictable or unpredictable from the context. The performance is based on the percentage of the keywords correctly reported by the listener in the presence of noise.

While tests of intelligibility can be very useful, they are also time-consuming and can be costly, requiring laboratory testing conditions, trained personnel, and the selection of representative participants for testing. To facilitate the evaluation of the effects of background noise on communication, researchers have developed a number of alternative methods to predict speech intelligibility by characterizing the properties of the noise. One of these methods discussed below is the **articulation index**.

4.15.1.2 Articulation index

The articulation index (AI) was developed specifically to reduce the time and cost of laboratory testing and to quantitatively predict speech intelligibility from the qualities of speech and noise of the communication system (French & Steinberg, 1947). The AI method (now called the **Speech Intelligibility Index**, SII) involves independently measuring

Table 4.2 Examples of the test sentences from the CID and SPIN tests. The underlined words indicate the keywords for the SPIN sentences use to measure performance

CID sentences (Low Predictability)
  1. It's time to go.
  2. If you don't want these old magazines, throw them out.
  3. Do you want to wash up?
  4. It's a real dark night so watch your driving.
  5. I'll carry the package for you.
  6. Did you forget to shut off the water?
  7. Fishing in a mountain stream is my idea of a good time.
  8. Fathers spend more time with their children than they used to.
  9. Be careful not to break your glasses.
10. I'm sorry.

SPIN Sentences (High predictability)
  1. The watchdog gave a warning growl.
  2. She made the bed with the clean sheets.
  3. Close the window to stop the draft.
  4. My TV has a 12-inch screen.
  5. The old train was powered by steam.
  6. They might have considered the hive.
  7. The old man discussed the dive.
  8. Bob heard Paul called about the strips.
  9. I should have considered the map.
10. Miss Brown shouldn't discuss the sand.

Table 4.3 Calculation of Articulation Index. Noise amplitude (dB) measured at different frequency bands is listed in column 1, differences in the measured dB value of speech and noise are shown in column 2, and weights assigned to each band are listed in column 3. The differences (column 4) multiplied by their weights are listed in column 4 and are summed below

Column 1: Frequency band	Column 2: Difference between speech peak and noise, dB	Column 3: Relative weighting	Column 4: Multiplication of columns 2 and 3.
200	30	0.0004	0.0120
250	26	0.0010	0.0260
315	26	0.0010	0.0270
400	28	0.0014	0.0394
500	26	0.0014	0.0364
630	22	0.0020	0.0440
800	16	0.0020	0.0320
1,000	8	0.0024	0.0192
1,250	3	0.0030	0.0090
1,600	0	0.0037	0.0000
2,000	0	0.0038	0.0000
2,500	12	0.0034	0.0408
3,150	22	0.0034	0.0758
4,000	26	0.0024	0.0624
5,000	25	0.0020	0.0500
		AI=	0.04738

Adapted from Sanders and McCormick (1993).

the intensity of speech and noise at a series of different frequency bands and subtracting the difference. An example of these calculations is shown in Table 4.3. The first column specifies the different frequency bands, while column two is the difference in the intensity in dBs of the signal and noise measured at each corresponding sound band. A value of zero is assigned in cases where the noise intensity exceeds that of speech intensity, as is the case for sound frequency bands of 1,600 and 2,000 Hz. In cases where the voice intensity exceeds that of noise by 30 or more dBs, a difference of 30 is recorded in column 2.

The values in column two are multiplied by weights shown in column 3, and the product of these two numbers is shown in column 4. The sum of column 4 provides the AI (AI or SII) score. Intelligibility improves as the AI score increases and is affected by the choice of stimuli where performance is better for restricted vocabulary sets than larger ones. The higher the AI value, the higher the expected user satisfaction with the communication system. AI values between 3 and 4.999 are considered acceptable, values between 0.5 and 7 are considered good, and AI values higher than 0.7 are rated as very good to excellent. Depending on the application and criticality of information, lower levels of intelligibility may be acceptable.

Techniques for predicting speech understanding have improved and alternative methods, including the Speech Transmission Index (STI), have been developed that are more robust, and account for the effects of speaker gender, variability in noise intensity across time (Ma et al., 2009), reverberation, and distortions on speech communication.

4.15.2 Effects of environmental factors on speech intelligibility

Human voices produce sound frequencies between 400 and 4,000 Hz with amplitudes of 40–70 dB during conversational speech. As noted in the scenario at the opening of this chapter, our ability to understand speech is affected by ambient noise like that which might be found in a crowded airport, a train station, or near a construction site. Depending on its amplitude and frequency, noise can mask the voice of speakers making it more difficult to discern what they are saying. The effectiveness of a masking noise stimulus depends on the intensity and frequency spectrum of the noise. Ambient sources of noise like that found at a construction site make it difficult to understand speech because it is often loud and overlaps in frequency with that produced by the speaker. Sounds with a wide range of sound frequencies between 20 and 4,000 Hz are particularly effective maskers as are the voices or babble produced by other speakers given the overlap in frequency and amplitude.

4.15.3 Effect of vocabulary set size

Message properties can ameliorate some of the effects of noisy conditions on speech intelligibility. For example, restricting vocabulary set size improves intelligibility. Miller and colleagues (Miller et al., 1951) evaluated speech intelligibility while using vocabulary set sizes of 2, 4, 8, 16, 32, and 256 single-syllable words under different signal-to-noise ratios (SNR), where SNR is defined as the ratio of the speech intensity to noise intensity expressed in decibels (dB). The ability to recognize the words improves when the vocabulary set is smaller. The benefit of a smaller vocabulary set is largely due to the fact that it is easier to infer the identity of an unintelligible word when the vocabulary set is smaller since the number of potential alternatives is also correspondingly smaller.

BOX 4.5: SELECTIVE LISTENING AND THE "COCKTAIL PARTY PROBLEM"

The laboratory conditions often used to investigate auditory function and abilities are unlike the typical conditions we find ourselves in as we go about our daily activities. This belies an important and sometimes unappreciated fact that we can follow a specific conversation even though the acoustic environment is awash in other sounds, including other conversations, that all add together forming a complex waveform that enters the outer ear. What cues allow the listener to segregate this complex waveform into distinct auditory streams that can be localized in space and assigned to specific objects or talkers?

Research in this areas was initiated by E. Colin Cherry (1953), a British experimental psychologist, who referred to this issue as the "Cocktail Party Problem." Cherry employed binaural and dichotic listening tasks to study the ability of listeners to segregate spoken messages. In the **binaural** task (i.e., both ears), participants were presented identical recordings of two speakers talking simultaneously to both ears. In the dichotic listening task, a different message was presented to each ear. Participants were asked to shadow one of the messages, a task requiring them to repeat one of the messages as soon as possible after hearing it while ignoring the other speaker.

Cherry reported that participants experienced "great difficulty" and found it almost impossible to follow one message when presented with two concurrent messages binaurally. By contrast, in the dichotic listening condition, they "experience no difficulty in listening to either message at will and 'rejecting' the unwanted one" (Cherry, 1953, p. 977). What explains the difference in shadowing performance under the binaural and dichotic listening tasks? Research suggests that listeners use a few cues including sound intensity (Treisman, 1964), perceived spatial location, and sound properties such as fundamental frequency (male versus female voice) to selectively attend to and segregate different talkers. Listeners find it easier to attend to a target voice if it is a woman's voice among men's voices (or vice versa), if the voice appears to originate from a distinct position in space or if it is louder than the other voices. **Critical Thinking Questions:** What are the cognitive implications of trying to attend to one voice in a complex mix of voices? How cognitively effortful is it to try to attend to one voice? Does the associated attentional demand decrease the listener's ability to detect or process other scene information?

4.15.4 Role of context

One might expect that also knowing the subject matter of communication might also influence performance since it would allow the listener to restrict the set of words they expect to hear in the communication. Knowledge that someone is speaking about psychology versus sports may facilitate speech recognition. In addition, speech has a structure that allows the listener to predict with some certainty what follows. For instance, in the English language, nouns follow verbs so that in a sentence like "He jumped over the ____." we know that it is some sort of object, and given that communication occurs in a situational context (it was raining), we might infer that the missing word was "puddle."

The benefits of context can be quite large. Miller et al. (1951) assessed benefits by comparing individual's ability to correctly report keywords presented singly or in the context of sentences. Performance was evaluated at different signal-to-noise levels ranging from negative (i.e., the masking noise is louder than the target voice sound) to positive (i.e., the target voice sound is louder than the masking noise; see Figure 4.10; Miller et al., 1951). For comparison, a signal-to-noise level of between +9 and +14 dB is comparable to a quiet living room. Similarly, the SNR level at a well-attended cocktail party is estimated at between –2 and +1 dB (Plomp, 1977) and the average SNR of a subway or aircraft is approximately –2 dB (Pearsons et al., 1977). The results show that at intermediate levels of signal to noise (+6 dB), a listener's performance improves by approximately 30% when the words are part of a sentence. At worse signal-to-noise levels, the benefits of context are greatly reduced because the louder noise makes it harder to identify most of the sentence, hence reducing the available contextual cues.

4.15.5 Contribution of vision to speech perception

While speech perception has often been regarded as largely an acoustic phenomenon, listeners also use visual cues to disambiguate speech in noisy conditions much like Sue did in the vignette at the start of the chapter. The classic study of Sumby and Pollack (1954) demonstrated that visual cues are particularly important in noisy conditions where the auditory signal is masked by extraneous noise. They tested speech recognition with and without view of the talker's facial and lip movements under different speech-to-noise ratios (SNR). View of the talkers face significantly improved speech recognition when the noise made speech unintelligible. Recent findings have shown that perception of a talkers' face not only facilitates speech recognition but also improves the detection thresholds of speech embedded in noise (Grant & Seitz, 2000). This is illustrated by the graph shown in Figure 4.11 from

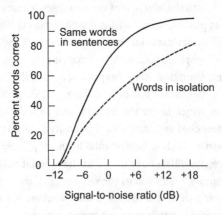

Figure 4.10 Effects of context on speech intelligibility under different signal-to-noise conditions for isolated words (top curve) and words spoken as part of a sentence (bottom curve). (Adapted from Miller et al., 1951.)

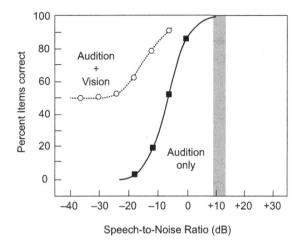

Figure 4.11 Effects of signal to noise on speech intelligibility when the listener can (audition+vision) or cannot (audition only) see the speakers face. The gray bar denotes the signal-to-noise ratio typical of quiet living room. (Adapted from Erber, 1969.)

Erber (1969) where speech intelligibility was measured at different signal-to-noise ratios (SNR) under an *audition-only* condition where the listener could not see the speakers face and under an *audition plus vision* condition where the listener could see as well as hear the speaker. Consider the data, note that in the audition-only condition as the amplitude of the noise increases (more negative values), the listener's ability to recognize the words decline precipitously so that by an SNR value of –20 the listener cannot identify anything the speaker said (0% correct). Surprisingly, when the listener can see the speaker's face performance jumps from 0% to 50%! This is indicated by the vertical dashed line labeled "a." Consequently, the listener goes from not being able to identify anything the speaker said to being able to identify at least half the words solely by also being able to see the speaker's face. As the horizontal arrow labeled "b" illustrates, seeing the speakers face has an effect equivalent to improving the SNR by +20 dB or a factor of 10.

Listeners use a variety of visual cues including the shape and movements of the lips and tongue, features of the upper face including the eye brows, and movements of the head made in synchrony with speech (Munhall et al., 2004; Rosenblum et al., 1996) to disambiguate speech in noise. These sources of information provide a variety of cues that help not only identify which speech sounds the speaker is producing but also the nature of the comment (serious, jocular), stress on segments of the message, and cues concerning segmenting the message.

Listeners with normal vision derive a benefit from viewing a speaker's face in noisy conditions like those we experience in many everyday situations like in a crowded restaurant, cocktail party, airport, and construction areas. This also emphasizes the importance of facilitating conditions like environmental lighting that can make it easier for the listener to see the facial cues.

4.16 APPLICATIONS TO SPECIAL POPULATIONS

Hearing loss is one of the most common sensory declines experienced by older adults. It is estimated that roughly 40%–45% of adults over the age of 65 have significant hearing impairment, and the prevalence is higher among men than women. As can be seen in Figure 4.12, by age 65+, almost twice as many men as women have some form of measurable

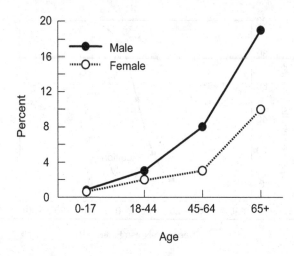

Figure 4.12 Percentage of males and females with hearing loss for different age groups. (Adapted from O'Neill, 1999.)

hearing loss. Some of the gender differences are attributable to the greater exposure of men to damaging sounds at the workplace.

Sensitivity to sound shows a gradual but continuous decline over an individual's lifespan. The rate of decline increases between age 20 and 30 for men and above age 50 for women with the rate of decline being steeper for men than women (O'Neil, 1999). Pure tone thresholds for men and women are depicted in Figure 4.13. The main difference as a function of gender is that men show a steeper loss in sensitivity for higher frequency tones with this loss being in the moderate to severe range.

Normal age-related hearing loss is called presbycusis, meaning "old ear," and is exacerbated by exposure to damaging sounds. Figure 4.14 illustrates that the effect of noise (exposure to 101 dB(A)) induced hearing loss as a function of age. Note that initially the change in sensitivity is associated with a notch or loss of sensitivity centered at 4,000 Hz that deepens and widens with continued exposure to the damaging sound so that by age 60 sensitivity to a broad range of sound frequencies including sounds typical of voices are significantly diminished.

A common complaint among older adults is that speakers mumble, talk too fast (Schneider et al, 2002), or "don't speak up" (Dye & Peak, 1983). These difficulties in understanding speech are partly a result of reduced sensitivity to high frequency sounds that is evident in Figure 4.13. High frequency sounds convey information about consonants, and it is consonants that carry important acoustic cues essential for speech intelligibility. Among older adults, much of the variability in speech understanding can be attributed to diminished high frequency sensitivity (Humes, 2007). The loss of sensitivity to high frequency sounds can be used to some advantage, as illustrated by the Welsh security company that markets a 17,000 Hz buzzer to store owners as an auditory repellent to teenagers who loiter in front of their shops (Vitello, 2006). The buzzer leverages the fact that adults with presbycusis are much less sensitive to higher frequency sounds than young listeners. In an interesting turn of events, the high frequency noise has been reemployed to create a ring tone called Mosquito Noise (www.freemosquitoringtone.org) that young cellular phone users can hear but their parents or teachers cannot!

Many everyday listening conditions are less than optimal requiring the listener to contend with background noise, the presence of multiple talkers, and reverberation. Reverberation is a common property of large rooms with high ceilings and hard surfaces like wood, cement,

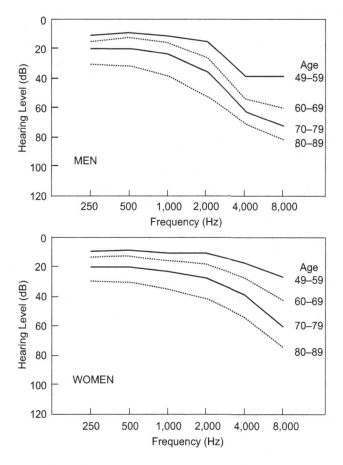

Figure 4.13 Average pure tone threshold sensitivity for men (top panel) and women (bottom panel) for different age groupings. (Adapted from Cruickshanks et al., 1998.)

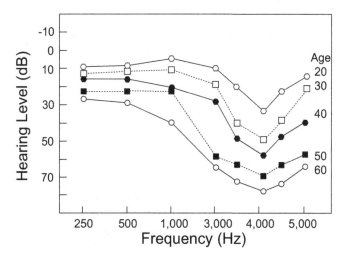

Figure 4.14 Effects of noise exposure (101 dB(A)) and aging (presbycusis) on hearing sensitivity with time. (Adapted from Jones, 1996.)

or glass that reflect sounds slowing their decay. Reverberation, defined as the time it takes a sound to decay 60 dB after it is terminated, is particularly difficult for older adults.

Older listeners, especially those over age 60, experience greater difficulties following and understanding speech when talkers speak faster, there are multiple talkers, or the acoustic environment is reverberant. Listening conditions that have nominal effects on younger adults nevertheless may represent a demanding listening environment for older adults. Evidence suggests that under these conditions that older listeners rely on (Gordon-Salant & Fizgibbons, 1997) and are better at using cues such as linguistic context (Schneider et al., 2002) to help support speech understanding.

Older adults experience greater difficulty listening for messages over loudspeakers at airports and train stations or attempting to communicate in a loud restaurant. These data illustrate the importance of improving the acoustic environment by reducing reverberation, using higher quality audio systems that do not distort or filter the voice stimulus, and minimizing other controllable sources of noise, thereby improving ratio of the intensity of the signal (voice) to noise (background noise level).

Individuals with hearing disabilities and listeners with normal hearing benefit from **closed captioning** wherein text, transcribing the audio portion of the programming, is displayed as text (Jensema et al., 1996). The display rate of the captioning for a range of programs (soap operas, documentaries, film talk shows, sports, etc.) is on average about 141 words per minute. Closed captioning eliminates the difficulties associated with not only fast talkers, poor speakers, but also poor room acoustics or noisy environments.

4.17 SUMMARY

While it may be true that vision may be our dominant sense, hopefully it should be evident from this chapter that we possess remarkable auditory abilities and that sound represents a rich and complex source of information. It can be argued that the auditory channel has been underutilized as a means to purposefully present information to users and as a means to explore complex data sets. In this chapter, we have sought to provide a description of basic auditory and vestibular abilities, but also to highlight examples of novel (e.g., spatial audio) and innovative (e.g., sonification) uses of this sensory channel to improve human performance. In addition, we reviewed how noise stimulus properties including the vocabulary set size and context affect speech intelligibility. Finally, this chapter also reviewed how aging affects hearing and speech understanding under different environmental conditions.

LIST OF KEY TERMS

Amplitude	Decibel (dB)
Articulation index (AI)	Decibel scale
Auditory canal	Dichotic
Auditory display	Earcon
Auditory icons	Equal loudness curves
Audification	Fletcher–Munson cures
Aural reflect	Hertz
Binaural	Impulsive noise
Cautions	Incus
Closed captioning	Intelligibility
Cochlea	Interaural intensity difference (ITI)

Interaural timing difference (IDI)
Loudness
Magnitude estimation
Malleus
Masking
Mel
Missing fundamental
Natural frequency
Ossicles
Otolith organs
Permanent threshold shift
Pinna
Pitch
Phon
Place code
Postural instability theory
Precedence effect
Presbycusis

Saccule
Sensory conflict theory
Sociocusis
Sonification
Semicircular canals
Sone
Stapes
Stapedius
Spatial audio
Temporary threshold shift
Tensor tympani
Timbre
Tympanic membrane
Utricle
Vestibulo-ocular reflex
Volley principle
Warnings

SUGGESTED READINGS

Nees, M. A., & Walker, B. N. (2009). Auditory interfaces and sonification. In C. Stephanidis (Ed.), *The Universal Access Handbook*, pp. 507–521. New York: CRC Press.
    The paper offers a very readable review of the use and design of auditory interfaces and sonification.
Schneider, B. A., Daneman, M., & Pichora-Fuller, M. K. (2002). From discourse comprehension to psychoacoustics. *Canadian Journal of Experimental Psychology*, 56(3), 139–152.
    The review offers an excellent discussion of the effects of aging, cognition, and auditory sensitivity on speech communication.
Wingfield, A., Tun, P. A., & McCoy, S. L. (2005). Hearing loss in older adulthood. What it is and how it interacts with cognitive performance. *Current Directions in Psychological Sciences*, 14, 144–148.
    The review offers an excellent discussion of the interactions aging, cognition, and working memory on speech communication.

CHAPTER EXERCISES

1. What are the perceptual qualities related to the physical properties of sound including intensity and frequency?
2. Using equation 1 shown below, calculate the decibel values for each of the following examples:

$$\text{SPL(dB)} = 20 \text{ Log}_{10}\left(\frac{\text{Sound of interest}, (P)}{\text{Reference sound}, (P_r)}\right)$$

1. $P = 0.0002,$ $P_r = 0.0002$ dB_____
2. $P = 0.002,$ $P_r = 0.0002$ dB_____
3. $P = 0.00015,$ $P_r = 0.0002$ dB_____
4. $P = 0.00025,$ $P_r = 0.0002$ dB_____

3. Describe the major structural components of the auditory system associated with the outer, middle, and inner ear.
4. Describe the three sites where the amplitude of sound frequencies is amplified by the ear.
5. What are the otolithic organs? What sensory abilities do they support?
6. Which sound frequencies is the auditory system most sensitive to?
7. Describe the effects of masking on auditory sensitivity?
8. How does the effect of masking change as the intensity of the masking tone is increased?
9. What are the units for subjective loudness, relative loudness, and perceived pitch?
10. What are the auditory cues that listeners use to locate the positions of sounds in the environment?
11. Describe three factors that influence speech intelligibility.
12. Why might auditory spatial cues improve the ability of listeners to segregate auditory messages?
13. Why might men experience greater hearing loss than women?

REFERENCES

Barrass, S., & Kramer, G. (1999). Using sonification. *Multimedia Systems, 7*(1), 23–31.
Benson, A. J. (2002). Motion sickness. In K. B. Pandolf & R. E. Burr (Eds.), *Medical Aspects of Harsh Environments Textbooks of Military Medicine*, Vol. 2, pp. 1059–1094. Washington, DC: Office of the Surgeon General at TMM Publications Borden Institute Walter Reed Army Medical Center. https://www.bordeninstitute.army.mil/published_volumes/harshEnv2/harshEnv2.html.
Biocca, F. (1992). Will simulation sickness slow down the diffusion of virtual environment technology. *Presence: Teleoperators and Virtual Environments, 1*(3), 334–343.
Brebner, J. T., & Welford, A. T. (1980). Introduction: an historical background sketch. In A. T. Welford (Ed.), *Reaction Times*, pp. 1–23. New York: Academic Press.
Brungart, D. S., & Simpson, B. D. (2005). Improving multitalker speech communication with advanced audio displays. *Paper presented at the New Direction for Improving Audio Effectiveness*, April 1, 2005. Neuilly-sur-Seine, France.
Cherry, E. C. (1953). Some experiments on the recognition of speech, with one and with two ears. *Journal of the Acoustic Society of America, 25*, 975–979.
Clark, W. W., & Bohne, B. A. (1986). Temporary hearing losses following attendance at a rock concert. *Journal of the Acoustical Society of America, 79*(S1), S48.
Cobb, S. V. G., Nichols, S., Ramsey, A., & Wilson, J. R. (1999). Virtual reality-induced symptoms and effects (VRISE). *Presence: Teleoperators and Virtual Environments, 8*, 169–186.
Committee on Hearing, Bioacoustics and Biomechanics (CHABA). (1988). Speech understanding and aging. *Journal of the Acoustic Society of America, 83*, 859–895.
Cruickshanks, K. J., Wiley, T. L., Tweed, T. S., Klein, B. E., Klein, R., Mares-Perlman, J. A., & Nondahl, D. M. (1998). Prevalence of hearing loss in older adults in Beaver Dam, WI: the Epidemiology of Hearing Loss Study. *American Journal of Epidemiology, 148*(9), 879–886.
Davies, D., & Jones, D. (1982). Hearing and noise. In W. Singleton (Ed.), *The Body at Work*. New York: Cambridge University Press, 365–413.
Davis, H., & Silverman, S. R. (1978). *Hearing and Deafness*. New York: Holt, Rinehart and Winston.
Dye, C. M., & Peak, M. F. (1983). Influence of amplification on the psychological functioning of older adults with neurosensory hearing loss. *Journal of the Academy of Rehabilitative Audiology, 16*, 210–220.
Erber, N. P. (1969). Interaction of audition and vision in the recognition of oral speech stimuli. *Journal of Speech and Hearing Research, 12*, 423–425.
Ericson, M. A., Brungart, D. S., & Simpson, B. D. (2004). Factors that influence intelligibility in multi-talker speech displays. *International Journal of Aviation Psychology, 14*(3), 313–334.
Fitch, J. L., & Holbrook, A. (1970). Modal fundamental frequency of young adults. *Archives of Otolaryngology, 92*, 379–382

Fletcher, H., & Munson, W. A. (1933). Loudness, its definition, measurement and calculation. *Journal of the Acoustic Society of America, 5*, 82–108

Flowers, J. H., & Hauer, T. A. (1993). "Sound" alternatives to visual graphics for exploratory data analysis. *Behavior Research Methods, Instruments, & Computers, 25*, 242–249.

Foley H., & Matlin, M. (2015). Sensation and Perception. Psychology Press.

French, N. R., & Steinberg, J. C. (1947). Factors governing the intelligibility of speech sounds. *Journal of the Acoustic Society of America, 19*, 90–119.

Garay-Vega, L., Hastings, A., Pollard, J. K., Zuschlag, M., & Stearns, M. D. (2010). *Quieter Cars and the Safety of Blind Pedestrians: Phase 1.* (DOT HS 811 304). Washington DC.

Gordon-Salant, S., & Fizgibbons, P. J. (1997). Selected cognitive factors and speech recognition performance among young and elderly listeners. *Journal of Speech Language and Hearing Research, 40*, 423–431.

Grant, K. W., & Seitz, P. F. (2000). The use of visible speech cues for improving auditory detection of spoken sentences. *Journal of the Acoustic Society of America, 108*, 1197–1208.

Hanna, R. (2009). *Refaat Hanna.* (DOT HS 811 204). Washington, DC: National Highway Traffic Safety Administration.

Hayward, C. (1994). Listening to the Earth Sing. *Paper presented at the Auditory Display: Sonification, Audification and Auditory Interfaces. SFI Studies in the Sciences of Complexity.*

Hecht, D., & Reiner, M. (2009). Sensory dominance in combinations of audio, visual and haptic stimuli. *Experimental Brain Research, 193*(2), 307–314.

Humes, L. E. (2007). The contributions of audibility and cognitive factors to the benefit provided by amplified speech to older adults. *Journal of the American Academy of Audiology, 18*(7), 590–603.

Jensema, C., McCann, R., & Ramsey, S. (1996). Closed-captioned television presentation speed and vocabulary. *American Annals of the Deaf, 141*, 284–292.

Jones, C. M. (1996). ABC of work related disorders: occupational Hearing loss and vibration induced disorders. *British Medical Journal, 313*, 223–226,

Jones, D. M., & Broadbent, D. E. (1991). Human performance and noise. In C. M. Harris (Ed.), *Handbook of Acoustical Measurements and Noise Control*, 24.21–24.24. New York: McGraw-Hill.

Kalikow, D. N., Stevens, K. N., & Elliott, L. L. (1977). Development of a test of speech intelligibility in noise using sentence materials with controlled word predictability. *Journal of the Acoustic Society of America, 61*, 1337–1351.

Kellogg, R. S., Kennedy, R. S., & Graybiel, A. (1965). Motion sickness symptomatology of labyrinthine defective and normal subjects during zero gravity maneuvers. *Aerospace Medicine, 4*, 315–318.

Krishnan, S., Rangayyan, R. M., Bell, G. D., & Frank, C. B. (2001). Auditory display of knee-joint vibration signals. *Journal of the Acoustical Society of America, 110*(6), 3292–3304.

LaViola Jr, J. J. (2000). A discussion of cybersickness in virtual environments. *SIGCHI Bulletin, 32*, 47–50.

Litovsky, R. Y., Colburn, H. S., Yost, W. A., & Guzman, S. J. (1999). The precedence effect. *Journal of the Acoustic Society of America, 106*(4), 1633–1654.

Ma, J., Hu, U., & Loiizou, P. C. (2009). Objective measures for predicting speech intelligibility in noisy conditions based on new band-importance functions. *Journal of the Acoustical Society of America, 125*(5), 3387–3405.

McGookin, D. K., & Brewster, S. A. (2004). Understanding concurrent earcons: applying auditory scene analysis principles to concurrent earcon recognition. *ACM Transactions on Applied Perception, 1*(2), 130–155.

Merhi, O., Faugloire, E., Flanagan, M., & Stoffregen, T. A. (2007). Motion sickness, console video games, and head-mounted displays. *Human Factors, 49*(5), 920–934. DOI: 10.1518/001872007X230262

Miller, G. A., Heise, G. A., & Lichten, W. (1951). The intelligibility of speech as a function of the context of the test material. *Journal of Experimental Psychology, 41*, 329–335.

Moore, B. C. J. (2001). Loudness, pitch and timbre. In E. B. Goldstein (Ed.), *Blackwell Handbook of Perception*, 408–436. Malden, MA: Blackwell Publishers.

Munhall, K. G., Jones, J. A., Callan, D. E., Kuratate, T., & Vatikiotis-Bateson, E. (2004). Visual prosody and speech intelligibility: head movement improves auditory speech perception. *Psychological Science, 15*, 133–137.

Nees, M. A., & Walker, B. N. (2009). Auditory interfaces and sonification. In C. Stephanidis (Ed.), *The Universal Access Handbook*, 507–521. New York: CRC Press.

Nichols, S., Haldane, C., & Wilson, J. R. (2000). Measurement of presence and its consequences in virtual environments. *International Journal of Human-Computer Studies, 52,* 471–491.

O'Neill, G. (1999). Hearing loss: A growing problem that affects quality of life. In *Challenges for the 21st Century: Chronic and Disabling Conditions*, Vol. 2, pp. 1–6. Washington, DC: National Academy on an Aging Society.

Patterson, R. D. (1990). Auditory warning sounds in the work environment. *Philosophical Transactions of the Royal Society of London, Series B: Biological Sciences, 327,* 485–492.

Pearsons, K. S., Bennett, R. L., & Fidell, S. (1977). *Speech Levels in Various Noise Environments.* Washington, DC: Office of Health and Ecological Effects.

Plomp, R. (1977). Acoustical aspects of cocktail parties. *Acustica, 38,* 186–191.

Reason, J. T., & Brand, J. J. (1975). *Motion Sickness.* London: Academic Press.

Reger, S. N. (1960). Effect of middle ear muscle action on certain psychological measurements. *Annals of Otology, Rhinology and Laryngology, 69,* 1179–1198.

Riccio, G. E., & Stoffregen, T. A. (1991). An ecological theory of motion sickness and postural instability. *Ecological Psychology, 3,* 195–240.

Rosenblum, L. D., Johnson, J. A., & Saldaña, H. M. (1996). Visual kinematic information for embellishing speech in noise. *Journal of Speech and Hearing Research, 39*(6), 1159–1170.

Sanders, M. S., & McCormick, E. J. (1993). *Human Factors in Engineering and Design.* New York: McGraw-Hill, Inc.

Schneider, B. A., Daneman, M., & Pichora-Fuller, M. K. (2002). From discourse comprehension to psychoacoustics. *Canadian Journal of Experimental Psychology, 56*(3), 139–152.

Smart, L. J., Pagulayan, R., & Stoffregen, T. A. (1998). Self-induced motion sickness in unperturbed stance. *Brain Research Bulletin, 47,* 449–457.

Smart, L. J., Stoffregen, T. A., & Bardy, B. G. (2002). Visually-induced motion sickness predicted by postural instability. *Human Factors, 44,* 451–465.

Speeth, S. D. (1961). Seismometer sounds. *Journal of the Acoustical Society of America, 33,* 909–916.

Stanney, K. M., & Kennedy, R. S. (1998). Aftereffects from virtual environment exposure: how long do they last? *Paper presented at the Human Factors and Ergonomics Society 42nd Annual Meeting.*

Stanney, K. M., Kennedy, R. S., & Drexler, J. M. (1997). Cybersickness is not simulator sickness. *Paper presented at the Human Factors Society and Ergonomics Society 41st Annual Meeting.*

Stanton, N. A., & Edworthy, J. (1999). Auditory warnings and displays: an overview. In N. A. Stanton & J. Edworthy (Eds.), *Human Factors in Auditory Warnings.* Brookfield, VT: Ashgate Publishing Co, 3–30.

Stevens, S. S. (1956). The direct estimation of sensory magnitudes: loudness. *The American Journal of Psychology, 69*(1), 1–25.

Stevens, S. S., & Volkmann, J. (1940). The relation of pitch to frequency: a revised scale. *The American Journal of Psychology, 53*(3), 329–353.

Stoffregen, T. A., Faugloire, E., Yoshida, K., Flanagan, M. B., & Merhi, O. (2008). Motion sickness and postural sway in console video games. *Human Factors, 50,* 322–331.

Stoffregen, T. A., Hettinger, L. J., Haas, M. W., Roe, M., & Smart, L. J. (2000). Postural instability and motion sickness in a fixed-base flight simulator. *Human Factors, 42,* 458–469.

Sturm, B. L. (2005). Pulse of an ocean: sonification of ocean buoy data. *Leonardo, 38*(2), 143–149.

Sumby, W. H., & Pollack, I. (1954). Visual contributions to speech intelligibility in noise. *Journal of the Acoustic Society of America, 26,* 212–215.

Tzelgove, J., Srebro, R., Henik, A., & Kushelevsky, A. (1987). Radiation search and detection by ear and by eye. *Human Factors, 29*(1), 87–95.

Vitello, P. (2006). A ring tone meant to fall on deaf ears. *New York Times.*

von Békésy, G. (1957). The ear. *Scientific American, 197*(2), 66–78.

von Békésy, G. (1960). *Experiments in Hearing.* New York: McGraw-Hill.

Wever, E. G. (1949). *Theory of Hearing.* New York: Dover

Zhang, X., Durand, L., Senhadji, G. L., Lee, H. C., & Coatrieux, J. L. (1998). Analysis-synthesis of the phonocardiogram based on the matching pursuit method. *IEEE Transactions on Biomedical Engineering, 45*(8), 962–971.

Chapter 5

Methods of evaluation

Chapter Vignette

Elvin is a college junior excited to start his third year of university. He heard that the university implemented a new computer system over the summer to which most of the student records and course management materials had been transitioned. Not thinking too much about how this affected him personally, Elvin waited until the last day of registration to sign up for his classes. He had done this using the university website many times before and found it to be very easy. However, this year when he got on the university website, he learned that he needed a new password to log into the course registration portion of the website. Elvin clicked on a link to get a new password but discovered that he first had to enter his old password and a new student ID number. Elvin was unaware that he had a new student ID or where to get one, and no additional help was offered on the website. In addition, he could not remember his old registration password and was unable to retrieve it from the system without his new ID number. Elvin clicked on every link on the website but always ended up on the same page saying "System error. Insufficient privilege to access this page." No phone number or contact information was provided on the site. Frustrated, Elvin drove to the university and stood in a long line of students to acquire his new student ID and enroll in classes.

Elvin's frustration with the new course enrollment system was due to many factors including a lack of awareness and understanding of the new system, a lack of communication as to how it works, and a software interface that was not intuitive for the target end users. A primary problem was that the new system was developed and implemented without involving the people who would be using it. Had they considered the users, they could have anticipated Elvin's situation and provided solutions to ensure that he could still register using the new system. This chapter examines how human factors scientists can evaluate systems and products to meet the end user's needs and provide a satisfying experience.

5.1 CHAPTER OBJECTIVES

After reading this chapter, you should be able to:

- Describe the importance of user-centered design.
- Explain methods of gathering end-user requirements and how they contribute to the design of a product.
- Explain the pros and cons of user interviews and questionnaires

DOI: 10.1201/9781003515463-5

- Be able to develop an interview and questionnaire with different question types.
- Discuss how to evaluate a design's usability.
- Understand how to conduct a usability test.
- Understand the measures used to evaluate usability, user satisfaction, and mental workload

5.2 WHAT ARE METHODS OF EVALUATION AND WHY DO WE STUDY THEM?

In the example above, Elvin had trouble with the university system most likely because it was not developed with him, or users like him, in mind. Unsurprisingly, these problems arose if the university failed to test this new system and provide students with sufficient information (e.g., online help) and training to use the new system. Elvin had been using the online university system for 2 years and assumed that the new system would be similar, or at least require similar information to use it. By using various design and evaluation methods, the university could have collected information about the users' (i.e., student, faculty, and staff) initial requirements, as well as completed user testing of the new system before it was launched. This evaluation of the new computer system would have helped the university discover if other students or staff members encountered the same difficulties that Elvin experienced. If so, these problems could have been reduced, if not eliminated. Using various evaluation methods, we can better understand the end user needs and capabilities so that designs can be created to reduce errors and problems.

5.3 METHODS OF EVALUATION

A diverse set of evaluation methods have been developed that can assist us in understanding human interactions in systems and products. If we are attempting to design a new system, we might want to interview or observe the users first to better understand their work functions and needs. Although some evaluation methods such as surveys and questionnaires are also used as research methods tools (see Chapter 2), the purpose of evaluation methods is to gather information or answer a specific question about a particular applied problem. In Elvin's case, we would be interested in knowing how students, faculty, staff, and prospective students or their parents will interact with the new computerized registration system. This can be examined using an evaluation technique called usability testing. In other evaluation methods, we also assess the actual individuals who work in the environment or will use a product. This technique can be applied to any product design—from the chair that you may be sitting in now, to the computer you use to take notes in class, to a new aircraft cockpit, to the space station control center. This illustrates the importance of user-centered design.

5.4 USER-CENTERED DESIGN

Elvin's situation is just one example of how products can be designed without the target user in mind. We have all been in similar frustrating situations—think about the last time you experienced a new version of a software program, website, smartphone app, or classroom technology. We might get angry and frustrated if it doesn't work the way we think it should. Sometimes we blame ourselves and attribute the problems we are having to our inefficiencies or incompetence. To avoid these problems and frustrations, human factors practitioners employ **user-centered design** (UCD), the practice of involving end users in the product design process from the beginning. UCD stresses the importance of first identifying user needs and requirements to understand how users work with existing systems before new systems are designed.

Human factors researchers use many methods to document user needs and requirements including observation, surveys, and interviews. **Participatory design** is a methodology that gathers different types of knowledge from individuals. **Explicit knowledge** is that which can be easily articulated. For example, if someone can explain exactly how to make Kool-Aid (combine one cup of mix, one cup of sugar, and one quart of water; stir until dissolved), we can determine the person's understanding of the task of making Kool-Aid. In contrast, **tacit knowledge** involves the information, habits, and processes that one knows, but is unable to articulate. For instance, although many of us can ride a bicycle and explain that we need to pedal to propel the bike forward, the tacit knowledge of how to balance along with pedaling and steering is difficult to verbalize exactly. The goal of participatory design is to tap into an individual's explicit and tacit knowledge of how products and systems function. Based on this information, new products or systems can be designed to fit the user's needs rather than designed to serve a function and then require users to learn a new process to use it.

There are three stages to participatory design research: Stage 1: initial exploration of work; Stage 2: discovery process; and Stage 3: prototyping. During Stage 1, designers observe the users in their work environment to better understand how they do their jobs and how they work with other users and systems. This often taps into a user's tacit knowledge. During Stage 2, designers and users meet to talk through the observations from Stage 1 and develop workflow models and other process diagrams. In this phase, both explicit and tacit knowledge are revealed as the users discuss their environment and tasks. In Stage 3, both designers and users work together to create prototype mock-ups (on paper and electronically) of new systems based on their joint understanding of the processes from Stages 1 and 2 (see Case Study 1 in Chapter 1 as an example).

To gather user requirement information for Stage 2 of the typical participatory design process, two of the most common methods include user interviews and questionnaires.

5.4.1 Interviews

Interviews are one-on-one discussions with end users that provide a flexible method of gathering a large amount of information from a few individuals. Typically, there is one interviewer and one respondent. When there is more than one respondent and possibly more than one interviewer, this is a focus group (see Section 5.4.3 on focus groups). To conduct an interview, interviewers and respondents meet face-to-face or talk via a video conferencing application, telephone, or other technologies.

There are three basic types of interviews used in participatory design: structured, semi-structured, and unstructured. **Structured interviews** involve a pre-defined set of questions that a researcher asks a respondent. Respondents are instructed to answer the question only and not to deviate to tangential topics or discussions. In **semi-structured interviews**, the researcher begins with a set of pre-defined questions but also probes the respondent's answers with additional questions. **Unstructured interviews** have no pre-defined questions. Researchers and participants simply discuss an area of interest. The researcher may guide the discussion initially but allow the respondent to talk freely on the topic at hand and related topics.

The decision of when to use a particular interview style is dependent on the purpose of the interview and the demographics of the target population, and impacts the method of analysis. If the purpose is to gather a large amount of information about a topic, one might want to use an unstructured interview. An interviewer can acquire a large quantity of information in an unstructured interview, even more than what might be gathered in a structured interview. Although the data collection is relatively easy with an unstructured interview, the analysis of the unstructured interview data is often much more difficult than data collected with a structured or semi-structured interview. The free-form nature of the unstructured interview often results in a lack of consistency in the data gathered that prohibits direct comparisons of the data across participants. Researchers use a process called **content analysis**, or thematic analysis, to identify the common underlying themes in responses.

If you have a new situation that you know little about, your purpose for the interview might be to gather as much information about the environment or situation. In this case, using an unstructured interview would be appropriate. The information learned after the content analysis can then inform semi-structured or structured interviews or even other methods.

If you decide to use a semi-structured or structured interview to start the evaluation process, and you use more categorical answers (e.g., yes/no, Likert scale of 1 to 5), it will be easier to analyze. The population you wish to interview, however, can influence the type of interview you use. For example, if the population of interest is children (e.g., for the design of toys), a semi-structured interview might be best to guide the interview with the option to seek clarification of the children's responses. As you might suspect, a combination of methods is often most appropriate in some design situations. In a domain that may be new to a designer, the researcher might decide to first conduct a series of unstructured interviews to identify the critical areas to probe further with a structured or semi-structured interview.

Regardless of the style, interviews employ a variety of question types. **Closed-ended questions** require respondents to select from a limited number of possible answers, such as Yes/No, or a rating on a pre-defined scale. **Open-ended questions** allow respondents to answer any way they want. Researchers sometimes also use **probing questions** after either a closed- or open-ended question to gather more information about the participant's response. Probing questions might be established beforehand, but they also might be created on the spot based on a respondent's answer.

As a side note, we often refer to these items as questions; however, not all items are written as a question, but rather as a request for information (see Table 5.1). We should refer to the questions or requests as items within the interview.

Table 5.1 Types of interview questions for unstructured, semi-structured, and structured interviews

Type of interview	Type of question	Example
Unstructured	Open-ended	How do you start your day/the process/...? Tell me about your job. Walk me through the process...
	Closed-ended	Is that done during the night or the day?
Semi-structured	Closed-ended	How often do you complete the Alert checklist? Every 30 minutes, every hour, every 2 hours?
	Probing question	What might be a reason for the need to frequently complete the Alert checklist?
	Open-ended	How do you process an Exception form?
	Probing question	The step where you make an exception is unclear. How do you decide whether it is necessary?
Structured	Open-ended	What is your occupation?
	Closed-ended	What is your age range? Under 18, 18–35, 36–50, more than 50

5.4.2   Questionnaires

Questionnaires provide another means of gathering user requirement information by asking participants to respond to questions or requests in a written format. One advantage of using questionnaires over interviews is the ability to collect information from more people at a time. In addition, questionnaires are superior to interviews when a researcher wants to avoid potential demand characteristics (see Chapter 2) during the interview. **Demand characteristics** are intentional or unintentional cues within the environment that suggest how a respondent should answer. For example, if the interviewer is nodding or smiling during some responses and not others, this might suggest to the respondent that he or she should answer a particular way. Using surveys removes the interaction between the interviewer and respondent, reducing the likelihood of demand characteristic effects.

Designing high-quality questionnaires that are statistically valid is a complex and iterative process. Care must be taken to ensure the items are interpreted the same by everyone, so the results are reliable and valid. (See Chapter 2 for a discussion on reliability and validity). In human factors, many questionnaires have been psychometrically validated and used in a variety of research domains. These include questionnaires to assess user satisfaction and perceived mental workload. Because questionnaires are used quite a bit in evaluation and research, we will review the process of questionnaire development.

To develop a survey, first one must determine what one wants to assess and then determine the types of items one wants to use. Just as with interviews, we can use open-ended and closed-ended questions or items. **Closed-ended questions** include forced choice (multiple-choice), two-alternative forced choice (e.g., Yes/No), Likert scales, and rankings. **Likert scales** range from a numerical low to numerical high value with diametric terms at each end of the scale (e.g., 1 = strongly agree, 5 = strongly disagree). See Box 5.1 for more information on Likert and Likert scales. **Open-ended questions** allow the user to write a free-form response (see Table 5.2). The open-ended item, "Explain how you felt during the driving

Table 5.2 Item types on questionnaires and surveys

Question types	Examples
Closed-Ended Questions	

Forced-choice	What is your preferred method of communication?
	    a. Phone
	    b. Text
	    c. In-person
	    d. Letter
	    e. Other

Two-alternative forced-choice	Are you 18 years of age or older? Yes or No Was the release valve on or off when you activated the heater? On or Off
Likert-type	Rate the extent to which you found the system easy to use: 1 = Extremely difficult to use to 5 = Extremely easy to use
Rankings	Rank the following processes in order of importance (1 = most important, 6 = least important) _____ Understanding the challenges of each team member's task. _____ Creating a backup plan in case of system failure. _____ Understanding how to perform at least two tasks other than your own. _____ Knowing the expertise of each team member. _____ Knowing the location of team members. _____ Developing a strategy for communicating in crisis situations.
Open-Ended Questions	Tell me about what you do when there is a system failure. Give me an example of when you need to submit a Change Request form.

simulation" allows the respondent to give many different responses even if the content is essentially the same. Because the analysis of open-ended items is more time intensive, this item could be written as a closed-ended item, "Rate how you felt during the driving simulation." The response options could be a Likert scale between 1 and 5, where 1 represents "horrible" and 5 represents "fantastic" and the other numbers represent feelings in between horrible and fantastic.

BOX 5.1:   EXAMPLE OF A LIKERT SCALE TO ASSESS A WEBSITE

Rensis Likert (pronounced lick-ert) was an organizational psychologist who developed his Likert scale to measure attitudes while completing his doctoral dissertation. The typical Likert scale has five benchmarks as shown in Figure 5.1. Instead of just giving individuals a scale from 1 to 5, we need to inform respondents what the anchors mean, and which are the *benchmarks*. Figure 5.1 shows a series of Likert scale statements assessing a website.

As the Likert scale is often modified to accommodate different situations using different anchors and benchmarks, these scales are referred to as **Likert scales**. Likert scale items are typically phrased as a statement to which the respondents' rate how they feel about something. The rating scale represents a continuum between the two extreme anchors, although individuals often must report or select discrete responses (e.g., 4, not 3.75). Likert-type scales also might differ in the number of anchors. The Likert scale is a 5-point scale and is labeled from 1 to 5. Another 5-point scale might start at 0 and end at 4. Likert-type scales might have as many as nine points and as few as three. It is recommended that Likert-type scales have five to nine anchors, as you do not want too many anchors whereby someone cannot make the distinction, for example, between a 26 and 27 on a 30-point scale. You also do not want too few anchors, as people might not feel a rating falls between 2 and 3 (e.g., "okay" and "excellent").

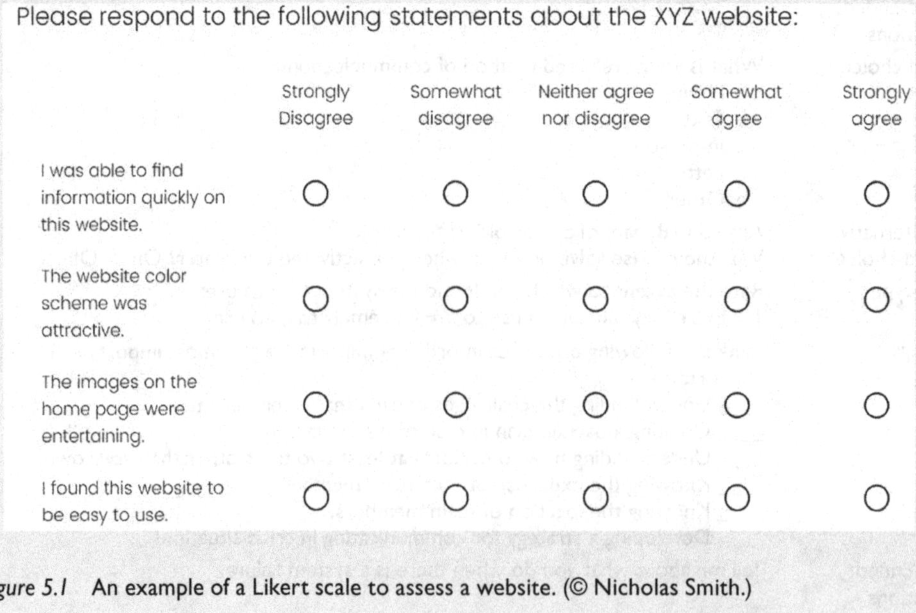

Please respond to the following statements about the XYZ website:

	Strongly Disagree	Somewhat disagree	Neither agree nor disagree	Somewhat agree	Strongly agree
I was able to find information quickly on this website.	O	O	O	O	O
The website color scheme was attractive.	O	O	O	O	O
The images on the home page were entertaining.	O	O	O	O	O
I found this website to be easy to use.	O	O	O	O	O

Figure 5.1   An example of a Likert scale to assess a website. (© Nicholas Smith.)

The benchmarks can also be changed to fit the statement as well as the number of response alternatives. Besides using "Strongly Agree" and "Strongly Disagree," one could use "Extremely Pleasant" and "Extremely Unpleasant" or "Completely Understandable" and "Completely Uninterpretable," depending on the wording of the statement being rated. One can also have either an odd number of response alternatives, which allows the respondent to indicate a neutral response, or an even number of response alternatives, which forces the respondent to respond for or against (i.e., agree or disagree) the statement. **Critical Thinking Questions:** A questionnaire was distributed among the faculty and staff users of the fitness center at a university. Individuals at the fitness center wanted to gain information to help make the facility better. One of the items was:

How often do you use the facility?

1	2	3	4	5	6	7	8	9

What problems can you identify with this survey item? How could you fix these problems?

Once the content is assessed and the types of questions or items have been determined, then the items are written. When writing the items, it is important to avoid loaded, leading, and double-barreled questions. The question "Do you think it is fair that those greedy owners of this company vote themselves pay raises?" is a loaded question. It is a **loaded question** because the question refers to owners as "greedy." A leading question might be, "You do not think the owners of this company are crooks, do you?" This is a **leading question** because there is the implied sense that you could not or should not believe the owners are crooks. Finally, the question, "Do you think high school students should be allowed to drive to school and have part-time jobs?" is considered a **double-barreled question** because it is asking two questions in one. These should be separated so you can get an answer to both questions.

For closed-ended items, it is also necessary to create response options and appropriate benchmarks for rating scales. If you are collecting demographic information about the students at Elvin's school, you might have items related to age or major. For example, how will age be listed? It is not reasonable to list every possible student age, so we might group the ages according to the likely demographics of the institution whereby we might have options such as 17, 18, 19, 20, 22, 23, 24–26, 27–30, 31–35, 35–40, 41–50, and over 50. Perhaps we should include a category for those younger than 18, as we sometimes have these younger students enrolled. How important age is for your evaluation might determine response options such as 18 or younger, 19–22, 23–30, and over 30. Of course, we could use an open-ended item and ask the respondents to write in their ages. Responding with actual age, though, might make a respondent feel more identifiable and less willing to respond, depending on the questions asked and if you use a small sample.

When using rating scales, benchmarks are needed. In the above example about one's experience with the driving simulator, a "1" represented "horrible" and a "5" represented "fantastic." These values and descriptors give the respondents benchmarks by which to judge where they will mark their responses. Problems can arise for respondents if the benchmarks are not clear or if there is no neutral point, as occurs with an even number of possible responses. In these cases, respondents might refuse to answer a particular item or to complete the survey, which we want to avoid. Yet, in some situations, we want to force individuals to choose one end of the scale, as is required when there is an even number of response items.

It is also important to consider whether the order of the questions matters. An item that targets specific information about the topic might influence the responses to a more general item or vice versa. Keep in mind, though, that if it is a paper-and-pencil survey or questionnaire, the respondents might go back and change a previous answer. If you need to be sure participants do not go back to earlier sections, you might want to give one part of the survey and then distribute the second part of the survey after the first part is submitted. The use of specific instructions on whether one should or should not go back to previous items can help, but it cannot prohibit someone from doing so. Given today's technology, it is easier to control the order of the items as well as whether respondents are allowed to go back to previous items when the survey is computerized.

A final issue to consider when using questionnaires includes sampling techniques. Recall that it is important to have a representative sample of the population (see Chapter 2). Ideally, a random sample should be considered. Sometimes, though, it is important to have a stratified sample. As discussed in Chapter 2, if your population includes 10% children and 40% elderly persons; 20% males and 80% females; 20% African American, 30% Hispanics or Latinos, and 50% Whites, then you often want to collect data in similar proportions. You would have a **stratified sample** when you randomly select from the population to ensure that you have close to 10% children, 50% young and middle-aged adults, and 40% elderly. You would also want to ensure that you had similar proportions of males and females as well as African American, Hispanic, and White respondents. Recall that if your random sample does not represent your population, it may be difficult to generalize the results to your population.

A couple of disadvantages to using a questionnaire include the inability to interact with the respondent during the completion of the survey and the lower response rates. When administering questionnaires, it is not possible to clarify respondents' questions about the survey and the researcher cannot ask follow-up questions if clarification is needed. This reiterates the importance of making sure your survey items are well-written. As many researchers will agree, writing quality questionnaires is not an easy task and can sometimes take several years to get it just right. Another disadvantage of the use of questionnaires is the tendency to get a poor response rate. Typically, less than 10% of the targeted population responds to a questionnaire. A response rate of 30% is considered to be quite good when administering organizational surveys. Yet, not all returned questionnaires are useful, as some might be incomplete or inaccurate. Incentives such as a gift card to those who complete the questionnaire or an opportunity to enter a drawing to win a prize are often used to encourage individuals to complete it.

5.4.3   Focus groups and contextual inquiry

Two other methods that human factors professionals employ to gather user requirements are focus groups and contextual inquiry. **Focus groups** target small sets of people (typically 6–8 people) who have something in common such as using the same machinery, process, or software. A facilitator discusses specific issues about the tools or processes with the participants to gain a better understanding of their feelings and attitudes toward that issue. The discussion topic or issue could be a current system or product, or a future product or conceptual idea.

While focus groups may appear to be casual conversations among participants, they are carefully planned and structured to elicit user perceptions and self-disclosure. Just as one prepares for interviews and the development of questionnaires, one must prepare the questions or discussion points for focus groups. The interviewer also must create an environment of trust by informing the participants about how the information will be used. In particular, the participants need to understand that their feedback is not identifiable, nor will it be used to evaluate them. Rather, the information will be used to enhance, develop, or change the tools or processes. By creating this trusting environment, focus groups are sometimes more

Figure 5.2 Focus groups are often used to gather user feedback and need requirements. (© IICD from The Hague, The Netherlands.)

effective than one-on-one interviews because of the small group environment. The participants can talk with one another about their feelings on a particular issue in a non-threatening and comfortable environment. Listening to what other people have to say often sparks more discussion than if each person were interviewed alone (see Figure 5.2).

Contextual inquiry is a method of gathering user requirements by observing the user in his or her work environment. Rather than using a structured interview, survey, or focus group, the researcher sits with the employee in the workplace and takes notes on the user's tasks, workflow, environment, and interactions with other individuals. Again, the researcher must create a trusting environment so the worker will perform the tasks as one normally would perform them. The researcher gains a much better understanding of the employee's tasks, interactions, and workflow by observing the actual work completed in the environment in which it is normally performed. The disadvantage of this method is that it is time-consuming to conduct the observations and analyze the data. As you can imagine, the observers in a contextual inquiry collect volumes of data that must be summarized.

5.4.4   Assimilating user-centered design information

As we discussed earlier, one might use an unstructured interview to gain more information to create a structured interview. We also can use interviews, questionnaires, focus groups, and contextual inquiry to inform other evaluation methods. For example, we might begin with a focus group to gather some understanding of the job to develop a better questionnaire. Perhaps we would need to observe or at least visit the work environment before the interviews or focus groups. As we gain a better understanding of the user requirements, we might use a previously used technique to hone in on some specific details. We also use this information to better understand the task. To fully understand the intricacies of a task, though, we should perform a task analysis.

5.5   TASK ANALYSIS

Besides understanding the user requirements, a thorough understanding of the tasks a user does is critical to the design process. The data collection techniques mentioned above, such as interviews, questionnaires, focus groups, and contextual inquiry, will provide the researcher with an understanding of the tasks that users typically perform with the system. Task analysis is a technique that allows the researcher to describe the tasks in further detail. Barry Kirwan and Les Ainsworth (1992) describe **task analysis** as the study of what an operator is required to do to achieve system goals. This includes all actions necessary for each step of the task as well as the cognitive processes necessary. In systems that involve technology, tasks are defined by what technology (e.g., controls, displays, mouse clicks, and switches) is used to complete each step.

A popular task analysis method called **hierarchical task analysis (HTA)** involves identifying the goals, operations, and plans of tasks. The *goal* of a task is simply what the operator wants to accomplish. For example, a goal in an email program may be to "Send an Email Message." This may then be broken down into sub-goals, which include "Open email software," "Compose Message," "Edit/Proof Message," and "Select Send button." *Operations* are the actions necessary to complete a goal or sub-goal. For a user to compose an email

#	TASK
1	Open browser
2	Enter URL address (www.hotmail.com)
3	Log in
4	Compose message
4.1	Determine Recipient
4.1.1	Find recipient address in address book
4.1.2	Enter recipient email address
4.2	Enter Subject
4.3	Type e-mail message
4.3.1	Format message (i.e., font, spacing, emoticons)
4.4	Review
5.0	Send email message

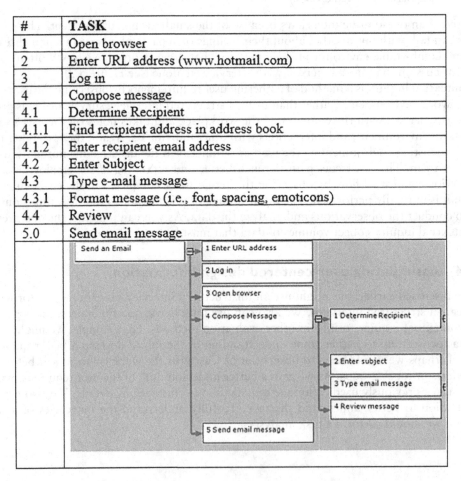

Figure 5.3   Hierarchical Task Analysis of the task to "Send an Email" in both tabular and graphical form. (© B. Chaparro.)

message, the user needs to click the mouse on a "Compose" button and then use the keyboard to type the message in the appropriate area on the screen. The top part of Figure 5.3 presents a sample task hierarchy for the task "Send an Email." In addition to the goals and sub-goals, *plans* are the method by which the goals are executed. The steps may follow a linear path, such as "Do Step 1, then Step 2, then Step 3" or they may be nonlinear or involve branching, or simultaneous steps. The output of a hierarchical task analysis is typically a tabular list or a diagram (see Figure 5.3).

The output of an HTA serves as the foundation of many additional methods designed to understand the user and the user's interactions with a system. Building from the information from an HTA, **cognitive task analysis (CTA)** is a technique that focuses on the cognitive and decision-making aspects involved in each task and subtask. In the task to "Send an Email," users might deliberate on finding the proper email address for the recipient or how best to format the message using the formatting tools available.

When observing users creating and sending an email, we also might observe users committing errors at various stages in the process. Users might have difficulty logging into the email program (i.e., they forgot their username or password, as Elvin did, or entered it incorrectly), or they might have difficulty finding a button to compose the email message once into the site. The HTA lays the framework for systematically identifying the potential for error. Using **human error identification** (HEI) techniques at the sub-task level of an HTA, experts can go through the HTA and identify all the sources of potential error at each task and subtask. Knowing where the problematic areas are allows designers to iterate on the design so that error potential decreases. In Elvin's case, if the designers knew that people were likely to have problems finding what their new ID was, some design changes could have reduced this problem.

5.6 USER PROFILE

Once the users' background information (i.e., user requirements) and job requirements (i.e., task analysis data) have been obtained, designers typically develop a user profile of the various target populations. The user profile describes in detail the critical characteristics of each user group, such as gender, age, domain experience, domain knowledge, technology experience, education level, physical characteristics, cognitive capabilities, and motivation level. Each of these characteristics is important to know when designing a new system or evaluating an existing system. Designers need to know what special considerations must be made in a design if the target population is in a particular age group (i.e., baby boomer, teenager, millennial, and Gen Z); a parent, or a person with visual, physical, or cognitive disability to determine what special considerations must be made in the design.

Less obvious features of the population, such as how motivated they are to use the system are also important to consider. If company employees are required to change to a new system to enter their personal contact information, they might be less motivated to learn how to use it than if it were an application that they chose to learn or purchased for their enjoyment. Table 5.3 shows a sample user profile questionnaire and possible data for a company website.

The user profile typically results in several different user groups based on all the user profile inputs. We often use the user profile to create **personas**, which bring life to the user profile data. For example, we might find that one segment of our target population for an e-commerce website is single females who have high levels of shopping experience both online (especially on their phones) and in physical brick-and-mortar stores. We could label this profile persona "Maria." Another persona might be based on an older adult population who might have low levels of technical experience and little online shopping experience. This persona might be represented by the name "Pat." By attaching names to the primary user groups, developers are better able to relate to their needs and capabilities and the kind of tasks they may want to do.

Table 5.3 Sample user profile for target users of a business website

Demographic	Range	Sample frequency
Gender	Male/Female	Male = 8 Female = 8
Age	**18–80**	18–23 = 4 24–35 = 4 36–50 = 4 51–65 = 3 66–80 = 1
Education Level	**High School—Professional Degree**	High School = 3 Associates Degree = 2 Bachelor's Degree = 7 Graduate Degree = 2 Doctoral Degree = 2
Occupation	**Variety of Professional Occupations and Student**	Consulting = 2 Education = 0 Engineering = 2 Finance = 2 IT = 0 Legal = 0 Management = 3 Manufacturing = 0 Media = 2 Public Relations = 0 Purchasing = 2 Sales & Marketing = 2 Student = 0 Other = 1
Own Computer	**Yes/No**	Yes = 16 No = 0
Frequency of Internet use	**Never to Daily**	Daily = 12 A few times per week = 3 A few times per month = 1 Less than once per month = 0 Never = 0
Reasons for using internet at home (check all that apply)	**Entertainment—Business**	Research = 10 News = 8 Business = 7 Buying = 3 Selling = 3 Investing = 3 Entertainment = 8 Communicating = 12 Other = 0
Reasons for using the internet at work (check all that apply)	Research/Communication	Research = 10 News = 10 Buying = 2 Selling = 2 Investing = 4 Communicating = 7 Other = 1

(Continued)

Table 5.3 (Continued) Sample user profile for target users of a business website

Demographic	Range	Sample frequency
Annual Income	$0–$200K+	$0–20,000 = 0
		$20,001–40, 000 = 4
		$40,001–60,000 = 3
		$60,001–80,000 = 4
		$80,001–100,000 = 2
		$100,001–150,000 = 1
		$150,001–200,000 = 1
		>$200,000 = 1

5.7 CARD SORTING

Given that evaluation methods often result in much data that is not in any form that is easy to analyze, we need to determine the meaning of these data, what data can be grouped, and the best method of data analysis.

A common method used to interpret the data and gain a better understanding of the conceptual model people use to organize the information of a system is the card sorting technique. During a **card sorting** task, participants sort and categorize related concepts into groups. To conduct a card sort, researchers present target users with a series of index cards, each containing the name of an object or piece of information (this activity also may be done electronically using virtual cards). Users are asked to group or categorize the cards in a way that makes the most sense to them. For example, the user shown in Figure 5.4 is sorting cards with the names of animals into piles based on how she categorizes them. Participants might be instructed to sort into a set number of groups, or the number of groups can be left undefined.

Once the cards are sorted, the participants are then asked to label the groups of cards. Although the researcher also could label the groups, it is good to get the participant's interpretation of the group name. For example, when developing the online enrollment system that Elvin encountered at the beginning of this chapter, users might be given a series of cards that reflect all aspects of the university, from coursework to campus life to university history, to upcoming events, news, and career opportunities on campus. Target users (which in this

Figure 5.4   A participant conducts a card sort activity. (© B. Chaparro.)

case include students, staff, faculty, and visitors) would group the cards in a manner that matches their conceptualization. Results from all participants are then analyzed statistically (using a technique called cluster analysis) to see where there is consensus as to which items are most related. This information is then used when designing the website architecture.

The clustered items from the card sort for web design could inform the designer as to which items should be displayed under the same menu category or grouped on the web page. Generally, different user groups will categorize the information in different ways. Students who are familiar with the campus and university life are likely to think of terms differently than a high school student who is just starting to consider attending college or a parent who is visiting the site to see if the school might be a good match for his or her child. Card sorting data allows designers to identify these discrepancies and what terminology might be interpreted differently across individuals and user groups.

5.8 DESIGN EVALUATION

While the methods described above will help designers create a product that meets the users' needs, it is still necessary to evaluate a design once it is created. It is recommended that designs be evaluated when they are in the initial prototyping stages. This allows developers to see where there may be design deficiencies early in the product life cycle before too much is invested in further design and development. There are many methods of evaluating whether a design is usable or not. We will cover two of the more common methods: heuristic evaluation and usability testing.

5.8.1 Heuristic evaluation

Jakob Nielsen (1993) proposed a set of ten usability heuristics that can be used to explain many problems in machine interfaces (see Table 5.4). To conduct a **heuristic evaluation,**

Table 5.4 Usability principles, or heuristics, proposed by Nielsen (1995)

Heuristic	Description
Simple and natural dialogue	The user interface should be as simple as possible and match the users' task. Only present the information the user needs to accomplish a task.
Speak the users' language	Use terminology (words, phrases, and concepts) that is familiar and that the user will understand.
Minimize the user's memory load	Allow the user to use recognition rather than recall in the interface (e.g., allow selection from a list rather than typing in key information).
Consistency	Allow the same commands, graphics, and terminology throughout the interface. Use similar commands, graphics, and terminology as other application interfaces that may be familiar to the user.
Feedback	Allow the interface to keep users informed about what is going on in the interface and how their input is interpreted.
Clearly marked exits	Allow users to easily exit an interface and not feel "trapped."
Shortcuts	Allow power users to use accelerators or shortcuts to perform frequent operations with the interface.
Good error messages	Provide intuitive and informative error messages that identify the problem and suggest a solution.
Prevent errors	Create an intuitive design that results in few errors.
Help and documentation	Provide help for users who need additional information beyond what can be gained from the interface itself.

several evaluators, typically usability or subject matter experts, individually evaluate the application and judge how well the interface adheres to a pre-defined set of usability principles or **heuristics**. Each evaluator generates a list of usability problems by principle. Lists from all evaluators are pooled to generate a summary list. It is recommended that more than one evaluator be used in a heuristic evaluation, as a single evaluator might only identify about 35% of the usability problems in an interface (Nielsen & Mack, 1994). Identification of usability problems increased to about 75% for five evaluators and 85% for 10 evaluators. The general recommendation is to use between three and five evaluators.

One way to determine how many evaluators to use is to weigh the costs versus the benefits of employing more evaluators in a **cost–benefit analysis**. Costs include the time to plan the evaluation, conduct the evaluation, and write up the results; fixed material costs; as well as variable costs such as the salary of the evaluator. The benefits constitute the usability problems that are discovered and the user time savings resulting from increased productivity should the usability problems be resolved. For applications that are used by many users, these savings add up very quickly (see Box 5.2 for an example of a cost–benefit analysis).

BOX 5.2: COST–BENEFIT ANALYSIS OF A USABILITY TEST EARLY IN DESIGN

To justify the use of user-centered design techniques to corporate executives, project managers, or program engineers, human factors specialists sometimes present a cost–benefit analysis. To do this, they add up all the costs associated with conducting the activity and estimate the savings that will be generated by fixing the issues found during the assessment. For example, in a financial trading company, a new application was developed for traders to use to log daily trades from all over the world. Early user testing was conducted on a prototype of the system. Approximately 50 problems were found with the design, some being as simple as reordering some of the fields on the screens so the trader could enter information into them more efficiently. It should be noted that the cost of making the corrections to the design was much cheaper for the developers to do early in the design process than later when the code was more rigid. The costs and benefits (estimated savings) are as follows:

1. Cost to conduct the user test (2 HF researchers each @ $1,500/day * 3 days) = $9000
2. Estimated savings from fixing the problems in terms of increased productivity, reduced errors, and reduced customer support for 20 users = $700,000
3. Overall Benefit from conducting test = $70,9000

As you can see, the benefits greatly outweigh the costs of the usability tests early in the design process. **Critical Thinking Questions**: Assume the usability test is not conducted early in the design process, but mid-way or near the end. What are the estimated costs for fixing the design problems? Would the benefits outweigh the costs (we would assume the estimated savings would be the same)?

The advantages of a heuristic evaluation are that it is an efficient, low-cost, and relatively quick technique for assessing an application's usability. The disadvantage is that it relies on expert opinion and not end-user experience (like usability testing which is discussed in 5.8.2). Heuristic evaluation also can be subjective and might not generate consistent results across evaluators. Some researchers have developed customized heuristics for domains with checklist items to provide more guidance to the reviewer of items to evaluate (see Box 5.3).

BOX 5.3: CUSTOMIZED HEURISTIC CHECKLIST FOR AUGMENTED AND MIXED REALITY

A heuristic evaluation checklist provides items under each main heuristic to provide guidance to the reviewer on specific items to evaluate. Researchers have created heuristic checklists for a variety of domains. One example is the Derby Dozen: A Toolkit for 12 Usability Heuristics for Augmented and Mixed Reality (Derby, 2023). This checklist was developed after an in-depth investigation into the literature and best practices for augmented and mixed reality applications and devices. This ranges from smartphones (i.e., Pokemon Go, retail try-in-your-space applications) to mixed-reality headsets like the HoloLens, Magic Leap, and the Meta Quest running applications for industry to augment the physical workplace. For example, an aircraft maintenance worker may use a HoloLens headset to display engine schematics or procedures virtually on the actual aircraft engine as they work. The 12 heuristics include Unboxing and Setting Up, Instructions, Organization & Simplification, Consistency & Flexibility, Integration of Physical & Virtual Worlds, User Interaction, Comfort, Feedback to the User, Intuitiveness of Virtual Elements, Collaboration, Privacy, and Device Maintainability. A total of 109 checklist items are included that allow the evaluator mark as "Yes," "No," "Somewhat," or "N/A." For example, one of the heuristics "Integration of Physical & Virtual Worlds" includes checklist items "Are physical (real-world) elements easily distinguishable from virtual elements?" and "Are virtual elements accurately placed on the real environment?" Evaluators walk through the application as a user might, conducting a series of tasks, and then use the checklist to assess the application usability. See Derby J Designing Tomorrow's Reality: The Development and Validation of an Augmented and Mixed reality Heuristic Checklist (PhD Dissertation Embry-Riddle Aeronautical University 2023).

5.8.2 Usability testing

Usability testing is an empirical method of measuring a product's ease of use (Rubin & Chisnell, 2011). Typically, people representing the target user population of the product are asked to complete a series of tasks with the product. Observational, user satisfaction, and performance data are collected, summarized, and used as the basis for review. Generally, one participant at a time works through the tasks with a facilitator. The facilitator provides instructions and tasks to the participant but is careful not to give any guidance as to how the tasks should be completed. Participants are instructed to "think aloud" as they step through the task. During the **think-aloud protocol**, participants say what they are thinking as they work through the task. The participant's verbal comments are recorded by the facilitator as well as by video cameras in the room (see Figure 5.5). In addition to the comments, other measures such as task success, time to complete each task, number of steps to complete each task, errors made on each task, user perceptions of ease of use and appeal, and overall user satisfaction with the product are collected. Many laboratories also use cameras and an observation area to allow other observers and/or product representatives to watch the exercise unobtrusively.

Figure 5.5 An example of usability testing configuration. A facilitator (a) observes a participant (b) as she completes tasks with a software application. Some labs use a one-way mirror (c) to separate an observation room from the test room or even the facilitator from the participant. (© B. Chaparro.)

One critical component of the usability test is the tasks, or scenarios, that are used. The data derived from a test is only as good as the tasks used to generate the data. Since human factors specialists are trained to work with the end users as well as the developers of a product, they are the ones who typically develop the task scenarios. The tasks should represent a set of common activities users would attempt to do with the product and may be a direct result of a task analysis exercise. For example, when testing the usability of a plug-and-play photo printer, users might first be asked to take the printer out of the box and connect it to a computer. Next, the user could be asked to install any relevant software and print photos of various configurations and sizes. During each of these activities, the researcher uses various data collection methods such as the think-aloud protocol and observation.

Another critical component of the usability test is the selection of users that serve as participants. The results of a usability test can be generalized to the target population only if the participants are *representative* of that population. This means that they have a similar user profile (e.g., background, capabilities, and goals) to the primary target user group.

Typically, usability studies involve between five and ten participants per user group. Suppose a product is targeted toward both a novice and a professional audience. In such a case, five to ten users from each user group would participate in the study, and their results would be analyzed separately to see if there are any differences in how they approach or complete the tasks.

Analysis of the data collected in a usability study involves the careful synthesis of both subjective and objective data that are evaluated with both inferential and descriptive statistical analyses where applicable (see Chapter 2). Task performance often is reflected by success, time on task, and errors committed. Although time on task and errors committed are **objective measures** because they can be measured without bias, success could be either objective (e.g., completed task? yes or no) or **subjective** (e.g., rating of performance by experts or rating of perceived performance by participants). The use of descriptive statistics could help us understand the percentage of participants who were successful, the distribution of time on task, or the mean number of errors committed. We also could assess whether the mean

VERY EASY	EASY	NEITHER EASY NOR DIFFICULT	DIFFICULT	VERY DIFFICULT
1	2	3	4	5

Figure 5.6 Participants are asked to rate the difficulty of tasks after they complete them. (© B. Chaparro.)

number of errors is significantly higher or lower than an expected number of errors when performing a competing design.

The research also could compare the number of steps for each task to the "optimal" number of steps that each task should have taken to give a measure of how "lost" the participant was. These data are summarized with the perceived difficulty (see Figure 5.6), perceived confidence that the task was completed correctly, satisfaction, and preference ratings to give an overall view of the product's usability. Without both objective and subjective data, the usability test is incomplete because the objective and subjective data do not always paint the same picture of usability. Participants might indicate that they "like" a product even though they perform poorly with it. The aesthetics of a product often plays a role in the subjective opinions of users. If a product looks attractive or "cool," users might say they like it, even if they cannot use it very efficiently.

BOX 5.4: SAMPLE USABILITY TEST PROTOCOL

Imagine that you wanted to conduct a usability study of the enrollment portion of the university website that Elvin experienced. Here is a sample protocol:

1. Recruit representative participants (undergraduate and graduate students with varied backgrounds in the use of the site).
2. Create tasks. These should be representative of the functions students most often do in enrollment. For example:
   a. Log into the site.
   b. Find a class to enroll in using the Search option.
   c. Add the class to your schedule.
   d. Find a class using course #.
   e. Add the class to your schedule.
   f. Change the section of the class to one on a different day/time.
   g. Delete/drop a class from your schedule.
3. Provide instructions to each user when they appear for the study. For example,

"Thank you for agreeing to participate in this study. Today, we are asking you to help us evaluate a website. To do this, we will ask you to complete several tasks with this site. Each task will be shown one at a time and we will walk through them. As you work through the tasks, we ask you to keep the following in mind:

Table 5.5 System Usability Scale (SUS).

System Usability Scale (All questions answered using the Likert scale: 1 = Strongly disagree, 5 = Strongly agree)
1. I think that I would like to use this system frequently
2. I found the system unnecessarily complex
3. I thought the system was easy to use
4. I think that I would need the support of a technical person to be able to use the system
5. I found the various functions in this system were well integrated
6. I thought there was too much inconsistency in this system
7. I would imagine that most people would learn to use this system very quickly
8. I found the system very cumbersome to use
9. I felt very confident using the system
10. I needed to learn a lot of things before I could get going with this system

From or adapted from Brooke (1986).

- *Please speak aloud your thoughts about the site, both positive and negative, so that we can gain a greater understanding of the strengths and weaknesses of the site.*
- *Because we are evaluating the site, I cannot help you with the tasks—I am primarily here as an observer.*
- *We are evaluating the website, not you—so if you cannot complete a task or find something, that is OK.*
- *After you complete a task, I will ask you to rate it on a scale as to how easy or difficult it was [show rating scale].*
- *We are not the designers of the site so feel free to speak candidly about your experience. You will not hurt our feelings!*
- *We expect this study to take about 30 minutes to complete.*

Do you have any questions?"
  Collect relevant data:

  - Task success, time to complete each task, and errors committed.
  - Perceived difficulty for each task.
  - Perceived usability of the site overall.
  - Subjective comments by users.

Critical Thinking Question: Think about how best to summarize the data gathered in this usability test. What visualizations might you use for each measure once it is analyzed?

5.8.2.1 Measures to assess usability

There is no single measure of product usability. Instead, researchers rely on data from many measures to collectively evaluate the usability of a product. Usability measures include time to complete a task, task success or failure, efficiency, perceived task difficulty, learnability, perceived mental workload, user satisfaction, and user comments on likes and dislikes. The goal of the evaluation and the goal of the product determine which measures may be more important overall. If the goal is for a product to be very easy to use the first time and require no training of the user, then efficiency, time on task, and first-time success would be important measures. On the other hand, if a product's goal is to be used efficiently over time (and requires some practice to master), then it would be important to measure learnability and user satisfaction.

**BOX 5.5:   SCORING AND APPLYING
THE SYSTEM USABILITY SCALE (SUS)**

The questions that make up the System Usability Scale (SUS) are in Table 5.5. To score the SUS, the responses of each item need to be converted and summed as follows (from Brooke, 1986):

1. Convert the rating of each item to be on a scale from 0 to 4. For items 1, 3, 5, 7, and 9, subtract 1 from the rating scale response. For example, if the rating for Item 1 was a 5 (Strongly Agree that they would use the system frequently), it will become a 4 $(5-1=4)$.
2. For items 2, 4, 6, 8, and 10, subtract the rating scale response from 5. For example, if the rating for Item 2 was a 1 (Strongly Disagree that the system was unnecessarily complex), it will become a 4 $(5-1=4)$.
3. Sum the converted scores.
4. Multiply the sum by 2.5. This will reveal the overall SUS score.

Critical Thinking Questions: Let's say you collected SUS data from university employees regarding the university website. The overall SUS score was 80 for faculty, 70 for staff, and 60 for students. How would you interpret these results? If this were Elvin's school, how would you apply these data? Administer the SUS to different groups in your class for any type of system (e.g., online registration and library book orders) and compare. What do you conclude?

To measure end users' perceptions of how usable a system or product is, specific measures have been developed to address the varying aspects which contribute to "usability." The System Usability Scale (SUS) is described as a "quick-and-dirty" assessment tool to evaluate users' perceptions of usability (Brooke, 1996; Brooke, 2013),=. Table 5.5 presents the SUS items. Scores on each item are summarized and reflected as a total score out of 100 points (see Box 5.5 for an explanation of how to score and apply the SUS).

Mental workload (MWL) is another important measure associated with product usability. **Mental workload** is the measure of how taxing product usage is (see Chapter 6 for a discussion on workload). If users feel they have to work hard mentally to use a product, this may impact their level of frustration and satisfaction. Time to complete a task, error rate, and efficiency tend to be positively correlated with perceived mental workload, but it is important to assess mental workload as a separate measure. Measures of mental workload are generally conducted with the use of physiological measures, such as heart rate, galvanic skin response, eye movements, or brain activity, or by subjective rating procedures.

The NASA Task Load Index (TLX) is one popular method for subjectively measuring MWL during or immediately after a user completes a task (Hart & Staveland, 1988; Stanton et al., 2005). Users rate the task on six dimensions: mental demand, physical demand, temporal demand, effort, performance, and frustration level using a low to high (1–20) scale. Next, the users are presented with all 15 pairwise comparisons of the six dimensions (e.g., mental demand versus physical demand) and asked to indicate which one contributed more to their overall workload. The NASA-TLX is used in many domains with a variety of products and complex systems. The multi-dimensional nature of the results is beneficial to researchers as they determine what works best or least within a product.

We also can learn a great deal about how an individual approaches, reviews, and scans a system with eye-tracking data. **Eye tracking** involves the electronic monitoring of eye gaze data while using a product. The use of eye tracking data provides an objective measure of where a user fixates on the product as they complete a task. For example, if we wanted to learn what portions of a website's home page were most salient or attention-grabbing in the first 10 seconds the page was viewed the first time, we could capture eye-tracking data to determine this (see Box 5.6).

BOX 5.6:   EYE TRACKING AND USABILITY TESTING

The use of eye-tracking data allows usability professionals to gather more information about what areas of a computer screen are: (1) most salient, (2) most informative, (3) most frequently ignored, and (4) most distracting to the observer. Eye tracking can be used to determine not only whether users fixate on a particular area of a screen, but also where and how they search the interface for a targeted object. This use of eye-tracking data as supplemental information to standard usability testing can be extremely informative to both user expectations and resulting design recommendations. For example, results from the use of eye tracking on web pages show that these data can be used to better understand how users initiate a search for a targeted link or web object. Frequency, duration, and order of visual attention to specific areas on a page are very helpful to designers in understanding user expectations and usage patterns. In Figure 5.7, we can see where people more often look or do not look. **Critical Thinking Questions**: How does the information from Figure 5.7 inform your design of a website? How else might you employ these data?

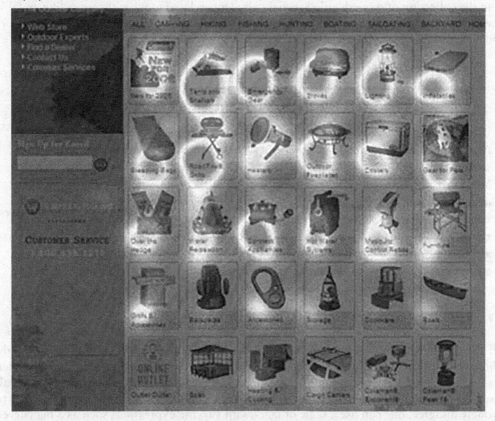

Figure 5.7 Heat map of eye gaze patterns of an e-commerce web page. The highest numbers of fixations are reflected by the brighter darker portions of the fixation circles. (© B. Chaparro.)

5.8.2.2 Variations to traditional usability testing

While traditional usability testing involves the observation of a single user with a product, human factors practitioners sometimes bring in two users at the same time to complete the tasks collaboratively. Like focus groups, this **co-participant usability testing** stimulates more interaction and discussion, which can help understand how users work with a product.

Another variation of the traditional usability testing that is primarily seen in the usability testing of computer software involves remote access to participants. In **remote usability testing**, researchers and participants may be in separate rooms, buildings, or even different parts of the world! Data are gathered via video conferencing software, phone, and/or web cameras.

These online tools allow researchers to gather data related to many users' online experiences at once. Data is gathered while the users interact with the software that is being evaluated. Users are given tasks to perform with the software; and task success, efficiency, and satisfaction data are collected as they complete the tasks. The advantage of this type of testing is that it allows researchers to collect data from a distributed population since they can be located anywhere in the world. This population also may be much more diverse when compared to what the researchers could access locally. One disadvantage is that it could limit the collection of more detailed verbal comments related to what the users like and dislike that typically come out of a face-to-face usability evaluation. See Box 5.7 to learn more about variations of usability testing.

BOX 5.7:   ASYNCHRONOUS VERSUS SYNCHRONOUS USABILITY TESTING

Usability testing can be conducted in different ways to allow researchers to collect extensive data. Asynchronous usability testing is done without a facilitator present during the test. Tasks are presented to the users, and they complete them on their own—sometimes the session is recorded. In addition, all mouse clicks and key presses also may be recorded. Synchronous usability testing is conducted with a facilitator present (either in person or remotely) as the user completes the tasks. This allows for follow-up to any key comments or observations. **Critical thinking questions**: If you wanted to evaluate the performance of different audiences with a new campus activity app, what type of data would you want to collect? How could you use both synchronous and asynchronous testing to help you collect the data? What methods of analysis would you use?

5.9   APPLICATION TO SPECIAL POPULATIONS

All the methods described in this chapter apply to any target population of users. Slight adjustments to the procedures may be made based on the population. For example, when evaluating a product with children, age-appropriate measurement materials are critical. Instead of numerical benchmarks on a Likert-type scale ranging from strongly agree to strongly disagree, one can use a scale of faces with varying expressions such as a big curved smile, then a slight smile followed by a straight line for the mouth to represent neutrality, and then a slight and big curved frown. When working with older adults, the text size of printed materials or the font size of online material might need to be enlarged. In addition, adaptations for product users with different physical and cognitive disabilities can be made. Box 5.8 describes an example of usability testing with a visually impaired user. It is important that proper task analysis and information gathering be conducted with each population to ensure their needs will be met.

BOX 5.8: TESTING SOFTWARE WITH A VISUALLY IMPAIRED USER

Evaluating products with the visually impaired presents some obvious challenges, especially when they are computer-related products. Having a visual impairment does not preclude a person from using computers. Many visually impaired individuals use products called screen readers to "view" what is on the computer screen. As they navigate through the software, the screens are read to them. Most users become very sophisticated with their computer keyboard and use different key sequences and shortcuts to navigate.

Evaluating software products with visually impaired users has revealed many issues that are not evident when testing sighted users. For websites, for example, the screen readers provide keyboard sequences that allow users to jump from link to link on a web page. This is one reason why programmers need to design link names to be as intuitive as possible. One common mistake seen on a webpage is a "Click here" link. The sighted user may be able to figure out what the "Click Here" link will do based on other content on the page. However, the visually impaired user who may only hear the links on a page will have difficulty figuring out what "Click Here" may mean.

Critical Thinking Questions: If a visually impaired individual wanted to purchase an airline ticket online, what potential problems like the "Click Here" example given above is the user likely to experience? How are ATMs designed to enhance the performance of the visually impaired?

5.10 SUMMARY

Human factors practitioners use a method called user-centered design to incorporate user needs into the design of products. This involves the gathering of key needs of the target end users of a product and a thorough understanding of the tasks users will do with it. This can be done through observation, interviews, and questionnaires. Even if this is done well, it is still necessary to evaluate a product's design with targeted users before the product is released. In addition, the design can be evaluated based on a series of key heuristics, or principles, known to influence end-user satisfaction and efficiency. Observing users working with a product also allows researchers to gain valuable insights.

Human factors practitioners often find themselves having to justify the early evaluation of product designs to the product engineers, marketing executives, and corporate executives. A simple cost–benefit analysis shows, however, that conducting user-centered design activities can pay off immensely. The time, effort, and cost associated with conducting a heuristic evaluation or a user test is typically much less than the time, effort, and cost of redesign after a product has been released.

LIST OF KEY TERMS

Card sorting
Closed-ended question
Cognitive task analysis
Content analysis
Contextual inquiry

Co-participant usability testing
Cost–benefit analysis
Demand characteristics
Double-barreled question
Explicit knowledge

Eye tracking

Focus group

Heuristic evaluation

Hierarchical task analysis

Human error identification

Interviews

Leading question

Likert scale

Likert-type scale

Loaded question

Mental workload (MWL)

Objective measures

Open-ended question

Participatory design

Persona

Probing question

Psychometrically

Questionnaire

Remote usability testing

Semi-structured interview

Stratified sample

Structured interview

Subjective measures

Tacit knowledge

Task analysis

Think-aloud protocol

Unstructured interview

Usability testing

User-centered design

User profile

SUGGESTED READINGS

Bias, R., & Mayhew, D. (2005). *Cost-Justifying Usability; An Update for the Internet Age.* San Francisco, CA: Elsevier, Inc.
   This book provides many examples demonstrating the cost-effectiveness of incorporating user-centered design into the development process.

Health & Human Services. (2006). *Research-Based Web Design & Usability Guidelines.* Washington, DC: U.S. General Services Administration
   This book provides research references that support web design and usability guidelines along with ratings of importance and research support.

Krueger, R., & Casey, M. (2000). *Focus Groups: A Practical Guide for Applied Research.* Thousand Oaks, CA: Sage Publications.
   This book is a nice practical guide for conducting focus groups.

Rubin, J. (1994). *Handbook of Usability Testing: How to Plan, Design, and Conduct Effective Tests.* Wiley Publishing.
   This book provides useful and practical guidelines for usability testing.

Ruel, E., Wagner III, W. E., & Gillespie, B. J. (2015). *The Practice of Survey Research: Theory and Applications.* SAGE Publications.
   This book is a practical reference for survey design.

Stanton, N., Salmon, P., Walker, G., Baber, C., & Jenkins, D. (2005). *Human Factors Methods: A Practical Guide for Engineering and Design.* Burlington, VT: Ashgate Publishing.
   This book is an excellent resource for a wide variety of human factors methods.

CHAPTER EXERCISES

1. What is user-centered design? Why is it used?
2. What is the difference between an interview and a focus group? What are the types of questions used in each? Give an example of each.
3. What is the difference between objective and subjective measures? Give an example of each.
4. You are exploring the use of cell phones among an older adult population. Design a 5-item questionnaire to survey this population on their cell phone usage. Include both closed-ended and open-ended questions.
5. You are exploring the use of a new app to capture people's daily exercise and eating habits and display them in a dashboard on their home screen. Assuming your

population has never worked with anything like this before, which method of requirements gathering would you recommend? Why?

6. You are interested in finding out more about how older adults use the social networking site Instagram. Create a user profile questionnaire for such users. How many target populations do you anticipate finding? Describe the characteristics of each.

7. What is card sorting? How is it used to understand users? Describe how you would use a card sort to design the architecture of a website.

8. Conduct a task analysis on making a bologna and cheese sandwich. List all tasks and subtasks necessary to complete this task.

9. List the tasks and subtasks necessary to extract $500 cash from your local ATM. Identify the potential errors that could occur at each subtask. What improvements, if any, would you recommend to the design of the ATM to reduce the potential for error?

10. Suppose you are interested in doing a usability study of your university's student club website. Define five tasks you would have users do to assess its usability.
    a. How many users would you recruit? What user characteristics would you include in your screening survey? Would you exclude any students from participating? If so, why?
    b. What measures would you log for each task and what does each tell you about the site's usability?

11. Conduct the study you defined in Question 10 with a minimum of five participants.
    a. Create a table of your results. How consistent is the feedback across participants?
    b. What recommendations do you have to make the site more usable and satisfying to your users? What changes would you make to your study to get more information?

12. Identify a system or product that might present a problem to someone (1) visually challenged, (2) hearing challenged, (3) mobility challenged, and/or (4) cognitively challenged. Describe the potential problems and present possible solutions.

REFERENCES

Albert, B., Tullis, T., & Tedesco, D. (2010). *Beyond the Usability Lab: Conducting Large-Scale Online User Experience Studies*. New York: Morgan Kaufmann.

Beyer, H., & Holzblatt, K. (1998). *Contextual Inquiry: Defining Customer-Centered Systems*. San Francisco, CA: Morgan Kaufmann Publishers Inc.

Bias, R., & Mayhew, D. (2005). *Cost-Justifying Usability; An Update for the Internet Age*. San Francisco, CA: Elsevier, Inc.

Brooke, J. (1996). SUS: A "quick and dirty" usability scale. In P. W. Jordan, B. Thomas, B. A. Weerdmeester, & A. L. McClelland (Eds.), *Usability Evaluation in Industry*. London: Taylor and Francis.

Brooke, J. (2013). SUS: a retrospective. *Journal of usability studies*, 8(2).

Derby, J. (2023). Designing Tomorrow's Reality: The Development and Validation of an Augmented and Mixed reality Heuristic Checklist (PhD Dissertation Embry-Riddle Aeronautical University).

Hart, S. G. (1988). Development of a multi-dimensional workload rating scale: Results of empirical and theoretical research. *Human Mental Workload*, 1, 39–183.Kirwan, B., & Ainsworth, L. K. (1992). *A Guide to Task Analysis*. Philadelphia, PA: Taylor & Francis.

Nielsen, J. (1993). *Usability Engineering*. San Diego, CA: Academic Press.

Nielsen, J. (1995). 10 usability heuristics for user interface design. https://www.nngroup.com/articles/ten-usability-heuristics/.

Nielsen, J., & Mack, R. (1994). *Usability Inspection Methods*. New York: John Wiley & Sons, Inc.

Nielsen, J. (2000). *Designing Web Usability*. Indianapolis, IN: New Riders Publishing.

Rubin, J., & Chisnell, D. (2011). *Handbook of Usability Testing: How to Plan, Design, and Conduct Effective Tests*. New York: John Wiley & Sons.

Stanton, N., Salmon, P., Walker, G., Baber, C., & Jenkins, D. (2005). *Human Factors Methods: A Practical Guide for Engineering and Design*. Burlington, VT: Ashgate Publishing.

Chapter 6

Attention, memory, and multitasking

Chapter Vignette

On a nice summer day, Roberta and Frank are strolling through the zoo engrossed in a conversation. During the walk, they navigated the grounds, and they avoided other zoo visitors, running into each other. Neither one seemed to have trouble doing these tasks: walking, navigating, and maintaining an appropriate distance while engaged in a conversation.

Jasmine works as an air traffic controller. As an air traffic controller, Jasmine must monitor multiple aircraft, be aware of other aircraft near the ones she is monitoring and listen for communication from the pilots and her co-workers. Because of these task demands, Jasmine sometimes has trouble responding to queries from co-workers, especially since she is new on the job.

Roberta does not seem to have trouble multi-tasking, but Jasmine does, at least on occasion. To understand these differences, let us consider for a moment the multiple task demands you experience when driving. As you drive, you must monitor your lane position, your speed, the position, and motion of vehicles in front, beside, and behind you, and the potential intent of pedestrians and other drivers. In addition, you must select a route to your destination and monitor your progress toward that goal. These behaviors are often performed while you are engaged in non-driving-related tasks, such as talking to a passenger, conversing on a cell phone, listening to music, or consulting a map. Despite our success at multi-tasking, it is apparent that we are more successful at juggling some tasks than others. Also, our ability to multi-task depends on whether we are just learning the task or if we are more experienced or expert at the task. For example, sometimes we drive home without really paying attention until getting close to our destination ("I don't remember driving from the corner of Hill St. to Catalina Boulevard!"). Other times, when traffic is greater or we are driving a friend's car with which we are not too familiar, we may pay closer attention to the task of driving. This chapter will explore the capacity of human operators to perform multiple tasks, which is affected by our information processing, attentional processes, and memory, as well as the task properties.

6.1 CHAPTER OBJECTIVES

After reading this chapter, you should be able to:

- Describe the major components of human information processing.
- Describe and explain the different components of how working memory.
- Identify the different types of attention.
- Identify different factors that influence our ability to multi-task.

DOI: 10.1201/9781003515463-6

6.2 WHAT ARE THE MAJOR STAGES OF HUMAN INFORMATION PROCESSING?

The ability to multi-task has received considerable attention recently due to the controversy surrounding the increased crash risk of distracted drivers using cellular phones. Research suggested that drivers using a cellular phone were four times more likely to be involved in a collision than when they drove with other typical distractions (i.e., talking to a passenger or listening to music; Redelmeier & Tibshirani, 1997). Laboratory studies using driving simulators have found that participants using a mobile phone are more likely to miss traffic signals and are slower to respond to the on-set brake lights (Strayer et al., 2003; Strayer & Johnson, 2001).

Concerns over distracted drivers have led to proposals for banning the use of cell phones while driving. Some 14 US states and the District of Columbia (DC) bar the use of hand-held cell phones while driving (Rudisill & Zhu, 2016), but allow the use of hands-free models (Strayer et al., 2003), despite evidence indicating no advantage of using hands-free cellular phones. Understanding the conditions and task properties that facilitate effective multi-tasking performance is an important goal of human factors. Before we can delve into these issues, we need to review the models of human information processing. Models of information processing emphasize the flow of information from sensory processes through hypothesized cognitive processes including attention, short-term memory, long-term memory, decision-making, and response-execution stages. These models conceive of human information processing in terms of a computer metaphor with distinct stages for input, processing, and output. That is, information processes are conceived of as a system as discussed in Chapter 1. The properties of each stage, including its capacity and the codes for representing information, are derived from many areas of research including experimental psychology, computer science, neuroscience, and physiology.

Figure 6.1 illustrates a model of human information processing. Diagrams like these are useful in illustrating the stages, internal flow of information, and relationships among hypothesized components of human information processing. Shown along the top of the diagram are the three general stages of human information flow: encoding, central processing, and responding. **Encoding** represents the registration of a stimulus on sensory receptors and its encoding supports detection and recognition of the input to the system, which is the human operator. **Central processing** is, of course, the processing performed by the operator. A **response** is the operator's output that can be in the form of a verbal or motor response. Below each of these stages, representative perceptual, cognitive, and motor processes are shown. The first stage of information processing that we need to consider is called the sensory register.

6.2.1 Sensory registers

As we navigate our environment, new sensory stimuli are continuously striking our senses. In the case of vision and audition, researchers have found evidence that an exact representation of the impinging information captured by the sensory system is available in a **sensory register**. That is, as you look at the visual scene in front of you, an impression is retained only briefly, so the sensory register avoids the equivalent of a double exposure by superimposing multiple impressions from successive stimuli. Although equivalent processes are presumed to exist for other sensory modalities, the visual (iconic) and auditory (echoic) sensory registers are the most studied and best understood. Below we review what is known about iconic and echoic sensory stores.

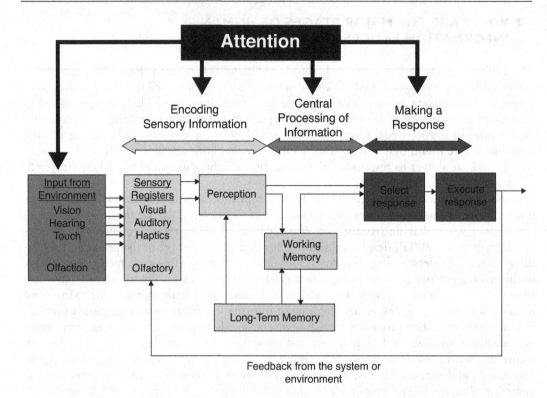

Figure 6.1 A schematic representation of human information processing illustrating the major stages (encoding, central processing, and responding), processes, information flow, influences of attention, and interrelations among processes. (Adapted from Wickens et al., 2004.)

6.2.1.1 Iconic register

Our understanding of the **iconic register** is in large part due to the classic experiments of George Sperling (1960), who measured the capacity and duration of information in the register. Sperling presented a 3×4 matrix of letters for a very brief period of time (i.e., 50 milliseconds or 0.05 seconds) and asked the participants to report as many of the letters as possible. Frequently, the participants could only report three to four letters from the matrix; however, they reported that many of the letters had been visible but had decayed before they could report them. In a subsequent experiment, he verified the participants' reports by having them only report a designated row of the matrix. Participants could reliably report any row of letters, thus confirming their phenomenological experience and demonstrating that the representation formed in the sensory register has a very short life span (approximately between 200 and 500 milliseconds or 1/5 to ½ second) before decaying completely.

These findings illustrate the temporary nature of the iconic register, and the difficulty observers may have maintaining complex visual images in mind when conditions allow only brief glances at a visual display. For example, many high-end automobiles have navigation aids including maps. To use the map, the driver looks at the map and then back at the road and traffic. Due to the rapid decay of the registered map image and overwriting of the image by subsequent visual impressions of the driving scene, the driver must repeatedly view the display. Thus, the driver must coordinate the driving tasks with monitoring other traffic and searching the navigation display.

6.2.1.2 Echoic register

There is an auditory analog of the iconic register wherein auditory information persists for a brief period following an auditory stimulus. Like the iconic register, the **echoic register** preserves auditory information allowing the listener to defer attending to the auditory stimulus temporarily (Crowder, 1982; Crowder & Morton, 1969). The persistence of echoic sensory store is longer than the iconic store, lasting between 250 milliseconds (1/4 of a second) and 4 seconds. This is why you are often able to answer someone's question even though you asked, "What?" immediately after hearing the question.

Presenting information aurally offers certain advantages over visual presentation. For instance, auditory presentation allows the user to move about to perform other tasks; the listener is not constrained by having to view a display to obtain information. Because of its transitory nature, though, auditory information demands the user's immediate attention (unless the information can be repeated); otherwise, the information is lost.

Information in the iconic and echoic registers is susceptible to decay or erasure by subsequent stimuli if it is not transferred to a more durable form of memory. To retain the information, it needs to be transferred to another more durable stage of processing, such as short-term memory.

6.2.2 Short-term memory

To get information into short-term memory, we usually must attend to it and rehearse the information otherwise; it will be lost usually in less than 20 seconds. The representation of information in **short-term memory** (STM) tends to be short lived, and the capacity of STM is severely limited. Most of us have experienced being lost and asking for directions to our destination. We can maintain the directions indefinitely in STM—long enough to arrive at our destination—if we rehearse them. Our experience also reveals that information may be lost if we are momentarily distracted, if the list of items is long, or if there is interference from similar information.

The ability to retain information in STM is greatly reduced if a participant engages in another task, interfering with the rehearsal of the information. In two studies, participants were instructed to maintain a trigram (a set of three consonants such as the letters T R M) in STM while counting backwards by threes, a task chosen to prevent rehearsal of the trigram. Participants' ability to recall a trigram declined as the duration of the backward counting task increased (Brown, 1958; Peterson & Peterson, 1959). Furthermore, recall accuracy was reduced to 7% after just 18 seconds of counting backwards! The contents of STM may decay even with rehearsal—especially if the list is long or if the items take too long to generate. Lists of items with more syllables take longer to rehearse, reducing the number of items that can be recalled later. Likewise, the driver soliciting directions would have greater success in recalling a list of short street names rather than a list of longer street names.

Suppose we had to remember a list of grocery items or as in the example above, a set of street names, how many items can we maintain? More specifically, is there a limit to the capacity of STM? In a classic study, George A. Miller (1956) concluded that STM has a **capacity** of seven plus or minus two items (7±2). As noted above, this capacity limit depends on the complexity of the individual items (i.e., words having fewer syllables), how similar sounding the items are, and the coding strategies we employ. The memory span for similar sounding letters like DVPBGCET is smaller than if more dissimilar sounding letters like DFKMOIZQ are used (Conrad & Hull, 1964). We are also better able to remember a list of letters if we "chunk" the letters into meaningful groups. Imagine trying to recall the following list of letters: BIKURDCFO, your recall would improve if you organized the letters

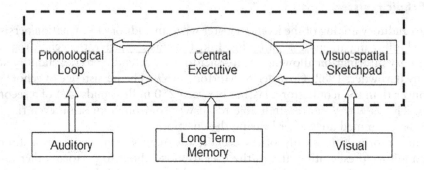

Figure 6.2 The schematic of Alan Baddeley's model of working memory. (© Alex Chaparro.)

into the words BUICK and FORD. So, rather than having to remember nine individual letters you have to recall just two words! The same strategy can be used to chunk numbers into meaningful sets, thereby effectively increasing the capacity of working memory. This is something you may have done with phone numbers. That is, for people who live in your area, the area code becomes one chunk, and the prefix becomes another chunk. So, you only need to actively encode the last four digits. Interestingly, other findings suggest that participants in experimental studies use strategies like chunking to maximize their performance in tests of STM capacity and that the true capacity of STM may be as small as 3±1 items (Cowan, 2001).

To this point, STM has been described as a single process whose function is the temporary maintenance of information. But what role does STM play in supporting complex behaviors like flying, driving, and decision-making? In 1974, Allan Baddeley and Graham Hitch proposed a reconceptualization of STM as part of a larger dynamic, multi-component process that they called **working memory (WM)**. This is illustrated in Figure 6.2.

6.2.3 Baddeley's model of working memory

Baddeley and Hitch (1974) proposed that working memory consists of three subsystems: a central executive, a visuo-spatial sketch pad, and a phonological loop. This tripartite model incorporates the capacity limitations and rehearsal functions described above as part of two systems—the **phonological loop** and **visuo-spatial sketchpad**—that interface with a supervisory system, called the **central executive**. The phonological loop consists of two parts: a **phonological store** that contains unrehearsed verbal or acoustic information that decays with time, which is akin to echoic memory that we discussed above, and an **articulatory control process** that generates the **subvocal** rehearsal process used to maintain a grocery list, a telephone number, or directions. The visuo-spatial sketchpad is used to maintain visual or spatial information, such as when you attempt to remember the positions of pieces in a game of chess (Baddeley, 1992) or imagine the appearance of the letter "B" after it has been rotated counterclockwise 90 degrees.

Of the three components in this model, the central executive is the least studied and understood. In Baddeley's model, the central executive serves as an interface between long-term memory and the phonological loop and visuo-spatial sketchpad. In this capacity, the central executive is hypothesized to function as the controller, coordinating the functions of the phonological loop and visuo-spatial sketchpad. The central executive is also hypothesized to fulfill multiple roles including planning behavior and selecting operational strategies as well as maintaining and switching of attention as task demands vary. We will come back to this, as it is important in current models of multi-task behavior.

6.2.4 Long-term memory

The transfer of items into long-term memory (LTM) ensures the integrity of the information. Information is transferred from WM to LTM through repetition, elaboration, or by relating it to information already stored in LTM. Unlike STM, **LTM** is believed to have unlimited capacity with evidence of little or no decay of memory over time. Failure to retrieve information from LTM may represent a failure of recovery processes rather than the permanent loss of information.

In the case of driving, information about the meaning of traffic signs, how to shift gears, and the purpose of your current trip are believed to be retained in long-term memory. These different types of information may be stored by distinct memory systems that organize and store information in different ways (Tulving, 1985, 1989). A driver's memory of the meaning of street signs, what to do at a blinking red light, and the location of the nearest gas station rely on semantic memory. **Semantic memory** represents the storehouse of general knowledge, rules, and facts about the world that are not unique to an individual. Your ability to name the capitals of different countries is one example of a semantic memory. This form of memory is different from either episodic memory or procedural memory. **Episodic memories** consist of individual experiences where the context and time of when the information was learned are important. Your ability to recall what you ate for lunch yesterday, the last three cities that you have visited, or what material was covered in the last class lecture are examples of episodic memories. Later, an additional form of memory called **procedural memory** was proposed (Tulving, 1985), which explains the acquisition of new skills like riding a bicycle, parallel park (or not), or how to coordinate the use of the clutch with the stick shift in a car with a manual transmission. Procedural memories are also classified as a type of implicit memory because the information is learned and used unconsciously unlike semantic memories (e.g., learning the symbols for the chemical elements) that are encoded intentionally and effortfully.

Recalling what you ate for dinner a week ago today is much more difficult than attempting to recall what you ate yesterday or the names of the three ships used by Christopher Columbus when he discovered the new world. These cases illustrate several important qualities of episodic and semantic memories. Unlike semantic memories, episodic memories are often learned after a single exposure with no special effort on part of the individual, but they are more susceptible to loss or interference from new information. In contrast, semantic memories, such as knowing the names of the planets, are established via conscious effort at memorization and are more stable across time, although they may be activated infrequently. Like semantic memories, procedural memories are acquired through extensive practice; but, unlike either semantic or episodic memory, procedural memories may not be verbalizable. Can you articulate how to ride a bike, tie your shoes, or walk? Despite these differences, procedural memories are quite robust and show little decay across time. This property is illustrated by the familiar idiom "It's like riding a bicycle"—which acknowledges that the ability to ride will come back to you once you get back on the bicycle.

Semantic, episodic, and procedural memories focus on our memory for past events, yet memory for future events, which has received less attention in the published literature, is just as important (Herrman et al., 1999). **Prospective memory** is a type of episodic memory that involves remembering to perform something in the future. Researchers distinguish between two types of prospective memory: **event-based** (e.g., remembering to tell your friend the latest gossip when you see them) and **time-based** (e.g., remembering to take the cookies out of the oven in 20 minutes). Examples of prospective memory include remembering to send an email, remembering to go by the grocery on the way home from work, or remembering to complete a task that was interrupted.

Frequently, failures of prospective memory only cause embarrassment, but in some instances, they represent a significant potential for harm as in the case of aviation where prospective memory failures have contributed to a number of aircraft mishaps (Dismukes & Nowinski, 2007). Failures of intentions are more likely under conditions of stress or high workload where other task demands contend for our limited attention. To avoid these failures, people employ different strategies including using alarms, to-do lists, posting notes on calendars, as well as strategically positioning reminders such as putting books by the door to remember to return them to the library. In aviation, pilots follow detailed checklists to ensure the aircraft is configured appropriately for takeoff and landing. For a strategy to be effective, the cues must be associated with a particular intention and have a high probability of being spotted. The pilot checklist is very specific. On the other hand, tying a string around one's finger is sometimes ineffective; while it is likely to be noticed later, it is a poor cue of what exactly you were to remember.

The performance of everyday tasks relies on prospective memory and activation of other types of memory (see Squire & Zola, 1996; Tulving, 1985). The simple task of walking across campus to the library requires that you access a variety of procedural, semantic, and episodic memories that allow you to recall motor acts (walking and opening doors), interpret the meaning of signs (handicap, exit and entrance signs), and recall where you previously found the psychology-related journals and books. Even though these activities involve several types of memory, all the information is brought into WM to deal with a particular situation. Because information in WM can be easily replaced with new information, you need to attend to the information to accomplish a specific task.

6.3 ATTENTION

Although we use the term "attention" in everyday conversation and assume that others understand our intended meaning it turns out that defining "attention" is somewhat of a challenge. This difficulty derives from a lack of specificity in the use of the word "attention" as reflected in the often-contradictory processes it is used to refer to. We use **Attention** to refer to processes underlying our ability to focus on the performance of a single task (**selective attention**), the performance of multiple concurrent tasks (**divided attention**), or maintaining attention for an extended period (**sustained attention**). Likewise, the target of attention may be a particular feature of an object, an entire object, a grouping of objects, or a region of space where a target is expected to appear. For the purpose of discussion, we will define attention as a selective process by which attended information is processed more efficiently than non-attended information (Fernandez-Duque & Johnson, 2002, p. 154). Attention improves our ability to detect, identify, and distinguish between objects. Attending to information also increases the likelihood that we will remember it, as unattended information is unlikely to make a lasting impression on memory that can be recovered later. Finally, when we attend to a task or stimulus, we tend to respond faster than when our attention is diverted.

So, how do task-related demands, our experiences, and situational factors affect our ability to attend to task relevant information? As a driver gains experience, they learn to deploy their attention to search strategically to identify the most likely potential hazards (Summaia et al., 1996). For instance, the driver learns to scan intersections and crosswalks, as well as their blind spot. When individuals deply attention based on prior experience (e.g., knowing where to look for potential hazards), this is known as **top-down** control. Episodic memories are a type of top-down knowledge. Episodic memory of the location of a pedestrian, the parked police car you just passed, or the location of a pothole that damaged your tire may

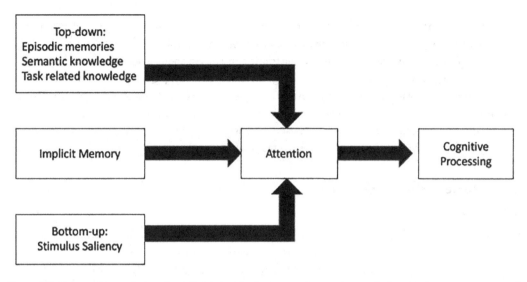

Figure 6.3 Types of knowledge that affect the deployment of attention including implicit memories, and top-down and bottom-up influences.

influence attentional deployment. Implicit memories in the form of unconscious learned associations can also influence attention. One of the authors once found a hundred-dollar bill near a curb at a shopping mall. They found themselves looking in the same area the next time they parked in the same location.

In many situations, though, we must be ready to attend to unexpected, surprising events (**bottom-up or data driven**) as in the case of a child running from behind a parked vehicle or a driver opening a car door in the right of way. Driving safety and efficiency would be compromised if other salient but irrelevant stimuli like swaying of tree branches, flashing roadside business signs, or traffic flow in a parking lot were to capture our attention. Figure 6.3 summarizes the different types of information known to influence the deployment of attention.

BOX 6.1   PLAYING VIDEO GAMES CAN IMPROVE VISUAL ATTENTION SKILLS!

For parents, seeing their children play computer games raises some concern as they fear that this may be a wasteful indulgence for a child or at worst may stunt the development of their social skills. There is some good news for parents and video game aficionados—playing video games may improve not only motor skills but also visual selective attention! Shawn Green and Daphne Bavelier (2003) of the University of Rochester found that video game players possessed improved visual attention skills including enhanced selective attention and the ability to allocate attention across the visual field. They were less affected by increases in the difficulty of visual attention tasks. However, not all computer games may have these effects. Non-video gamers asked to play Medal of Honor—an action game requiring distributed attention and the switching of attention around the visual field—for 1 hour a day for 10 days improved their performance on several tests of visual attention. In contrast, training on Tetris, a game that demands focused attention on one object at a time, did not. The finding that skills refined by playing video games translated to other tasks is interesting as it may represent a means for cost-effective training for complex tasks such as driving, piloting, or air traffic control. An added advantage is that trainees may be willing to volunteer for more training! **Critical Thinking Questions**: Do you think that playing video games may offer a cost-effective way to train personnel for some complex real-world tasks such as driving, piloting, or air traffic control? Can you identify characteristics (e.g., the game requires the player to develop strategies and divide their attention) of specific computer games that might translate best to tasks like driving?

6.3.1 Varieties of attention

6.3.1.1 Selective attention

In contrast to drivers who must be alert for various events within the environment, sometimes we focus on just one activity. Note that as you read this text, you are sitting comfortably somewhere, and your attention is devoted to the text (hopefully!). Until it is called to your attention, you probably have been unaware of the sensations produced by the shirt you are wearing, the shoes on your feet, or the watch on your wrist. Without **selective attention** to focus on task relevant information, these other sensations would easily exceed our cognitive processing capacity. That is, selective attention is a critical property of behavior given the mismatch between the limited capacity of our cognitive processes and the amount of potential information available to consciousness through the different sensory systems.

Several factors including expectancy, value, salience, and effort (Wickens et al., 2003) influence what we select to attend to. The role of expectancies in directing attention was made obvious to one of the authors recently when he could not locate the plastic cover of a tub of margarine. After searching the counter surface where he usually placed the cover, he unexpectedly found it lying on top of a can of coffee grounds located in front of him. His attempts to find the cover were frustrated by searching in the wrong locations even though the cover was distinctive in color and size from the cover of the coffee grounds container. Through experience, we learn about the probabilities of events and where information is likely to be found. This knowledge in turn can be used to guide our attention when we are engaged in a task. Performance can be degraded when these expectancies are violated, as illustrated by a study showing that drivers are less likely to detect a stop sign even though it is very visually salient when it was positioned in the middle of a city block rather than at an intersection (Shinoda et al., 2001).

The relative value or cost of information can influence the likelihood of an operator monitoring information available in the channel even if the probability of the event is low. For instance, there is a significant potential cost of failing to monitor your rear-view mirror or blind spot when driving. Value and expectancy constitute two types of top-down influences on the allocation of attention.

Similarly, salience represents the influence of bottom-up or stimulus-based factors on selective attention. Auditory, visual, and tactile stimuli including loud sounds, abrupt visual on-sets (flashing light) or unique colors, and movement are particularly effective in drawing observers' attention (Yantis, 1998), as are tactile-vibratory cues like those produced by cellular phones. Figure 6.4 shows examples of visual cues that have been shown to produce **attentional capture**. These salient cues are used intentionally to draw attention in the case of alarms or advertisements. The drawing of attention away from a primary goal or task by an irrelevant (e.g., advertisement) or unexpected event (e.g., fire alarm) is called attentional

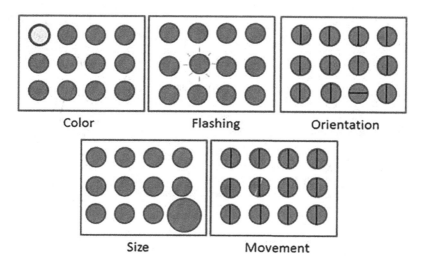

Figure 6.4 Illustration of the visual stimuli shown to produce attentional capture. (Adapted from Wolfe and Horowitz, 2004.)

capture (Most & Simons, 2001). While stimulus properties that produce attentional capture can enhance the effectiveness of alerts, they can also be distracting once the operator is apprised of the situation. In some cases, operators have been known to intentionally turn off alarms to minimize distractions while doing other tasks.

Finally, **effort** can also influence selective attention by biasing the likelihood of engaging in certain visual behaviors, such as checking the rear-view mirror or making larger physical movements, which would be necessary to check the blind spot on the driver's side.

The effectiveness of selective attentional processes appears to rely on two independent mechanisms: (1) a perceptual process that filters irrelevant information under conditions of high perceptual demands and (2) cognitive top-down processes that filter irrelevant stimuli guided by expectancies and knowledge of target properties (Downing, 2000; Lavie et al., 2004). The first mechanism works at the early stages of processing and is illustrated by research (Lavie & Cox, 1997) using the **flanker compatibility task** illustrated in Figure 6.5. This task measures the effects of a to-be-ignored distracter on a target task. Participants are instructed to report whether a target letter (X or V) is present among a group of letters forming a circle while at the same time ignoring a distracter letter positioned outside the ring of letters. The difficulty of the letter search task was varied by using either homogeneous (all O's) or heterogeneous non-targets (i.e., MKZHW) to form a circle of letters. A distracter letter was positioned outside the circle, and this letter could be identical to the target letter (compatible conditions c and f), an alternate target letter (incompatible conditions a and d), or a neutral letter that did not belong to either the target set or the letters forming the circle (conditions b and e). Keep in mind that the participant does not have to attend to the distracter letter at all to perform the task.

The results of the study showed that the effect of the distracter was dependent on the processing demands of the target task. That is, the distracter (whether compatible or incompatible) had negligible effect on the time to find the target letter when the search was made more difficult by using heterogeneous distracters. In contrast, the incompatible distracter (e.g., V distracter and X target) significantly slowed search when using homogeneous distracters (i.e., condition a). It is hypothesized that in the case of the latter "easy" condition, spare attentional capacity "spills over" to process the distracter, thereby interfering with

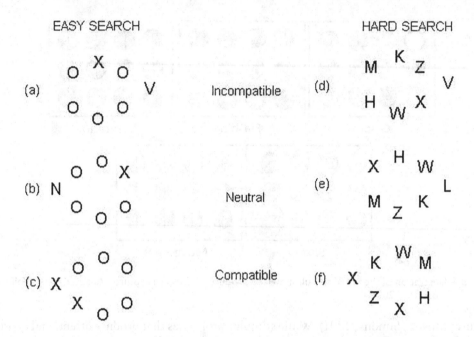

EASY SEARCH   HARD SEARCH

(a)   Incompatible   (d)

(b)   Neutral   (e)

(c)   Compatible   (f)

Figure 6.5 Examples of the easy (a–c) and hard (d–f) search tasks employed by Lavie and Cox (1997).

and slowing the execution of the search task. Processing of the distracter is prevented only in cases where processing the target task is challenging enough that all available resources are committed to the target. Thus, when our attention is engaged in another demanding task, salient stimuli may be less effective in capturing our attention. This is consistent with findings showing that drivers are more likely to fail detecting important driving-related information (also see Attentional Blindness Box 6.3) when talking on a cellular phone (Recarte & Nunes, 2003; Strayer et al., 2003).

Higher-level cognitive processes also help regulate the selectivity of attention. Allan Baddeley proposed that the central executive is critical to our capability to focus attention and to switch attention from one object or task to another. Other findings (de Fockert et al., 2001) support the view that directing attention to task-related information requires that the observer maintains information relevant to the task at hand. Like the traveler at an airport who is attempting to find their rental vehicle in a parking lot or searching for a friend in a crowd, we must maintain information about the target of our search to avoid being distracted by non-target stimuli.

This hypothesis was put to the test by having participants classify the written names of famous pop stars or politicians while ignoring images of faces that served as distracters (de Fockert et al., 2001). The load on verbal working memory was varied by having the participants remember the same series of digits (low load condition) or a new set (high load condition) prior to each trial. It was anticipated that rehearsal of new digit strings under the high load condition would interfere with the ability to maintain information specifying properties of objects that the observers are searching for. It was predicted that participants would be slower in categorizing the names as belonging to pop stars or politicians in trials under the high load condition. The data supported their hypothesis, as the response times were significantly longer in the high load condition. More importantly, the distracter images had a greater effect under the high load condition, demonstrating that effective selective attention depends on WM to maintain stimulus priorities for attention. Similarly, the car renters' search for his or her vehicle may be impeded by attempts to maintain a conversation or answer questions that would interfere with the maintenance (by the phonological loop) of the target vehicle's make and color.

Just as someone searching for a car, drivers also need selective attention. Selective attention has been found to be one of the best predictors of crash rate among young drivers between 22 and 32 years of age (Kahneman et al., 1973). In this study, drivers were divided into three groups based on their accident history: accident-free, accident-prone, and an intermediate group. The auditory selective attention test required participants to attend to one of two (different) messages. A different message was presented to each ear and a tone indicated to which ear the driver was to attend. Participants were instructed to report the message presented in the indicated ear as soon as it was heard and to disregard the message in the other ear. The number of errors made in this task was found to be predictive of the participants' crash rate (classified as accident-free, accident-prone, and intermediate). Drivers with poor selective attention had more accidents. See Box 6.2 for recent work linking visual attention to crash risk.

BOX 6.2   DRIVING AND VISUAL ATTENTION SKILLS

Driving is a visual task, and one might expect that standard tests of visual ability including visual acuity, color vision, and stereoscopic vision might be predictive of crash risk. However, research over the last several decades has shown that measures of visual function (i.e., acuity, color vision, and stereovision) do not predict crash risk. Why might this be the case? One reason is that tests like the test of visual acuity performed by an eye care specialist assess your ability to resolve minute details using small, high contrast letters (black letters on a white background). Whereas visual acuity might explain a drivers' inability to read a street sign or the license plate of another vehicle, it likely plays a minor role in daytime accidents involving large and often clearly visible vehicles. It is more likely that the failure to attend to another vehicle, rather than an inability to see the other vehicle, has been a factor in many such accidents. Drivers often report that the other vehicle "came out of no-where," suggesting that they failed to attend to it (Rumar, 1990). This is corroborated by an analysis of traffic accidents that sought to identify the different failures or errors that played a part in 72 accidents (Malaterre, 1990). The most frequently cited error was the driver not noticing other road users, signs, or signals, or becoming aware of them too late to make an appropriate response (33%). The important role of visual attention in driving is also evidenced by the findings of Karlene Ball and Cynthia Owsley (Ball et al. 1993; Owsley et al., 1998). They developed a test called the Useful Field of View (UFOV), which measures visual processing speed as well as divided and selective attention. The UFOV test measures the extent of the visual area from which a driver can extract information from a briefly presented display while dividing attention between two tasks: identifying a central figure and detecting a target among distracters (selective attention). They reported that the UFOV was a significantly better predictor of accident risk among older drivers than were measures of visual function like visual acuity. Drivers who showed a 40% or greater reduction in the extent of the UFOV were twice as likely to be involved in a car crash. **Critical Thinking Question**: Why are some measures of performance like visual acuity, color vision, and reaction poor predictors of a driver's crash risk or crash history?

6.3.1.2 Divided attention

Like selective attention, the ability to divide attention is critical to the performance of many tasks including driving a car, dribbling a basketball, or maintaining a conversation while walking without stumbling or running into the other conversant. **Divided attention** is the ability to distribute attention among two or more tasks. Multi-tasking performance depends not only on mastery of the individual component skills that must be performed concurrently, but also on higher level cognitive skills associated with coordinating multiple tasks, selecting the relative emphasis to place on different task components, and choosing when to shift attention between tasks (Kramer et al., 1995). These are functions believed to be mediated by the hypothesized central executive of Baddeley's model of working memory.

Harold Pashler (1984) has provided convincing evidence that the decision processes (labeled Central Processing in Figure 6.1) represent an obstacle or "bottleneck" to effective multi-tasking. His studies of participants performing multiple tasks have demonstrated that two tasks cannot be performed in parallel, but rather they must be performed sequentially when both tasks require that a choice to be made. Imagine having to decide which way to turn at an unfamiliar intersection while also trying to reply to a passenger's query. According to Pashler, the response to the passenger cannot be initiated until after selecting which way to turn at the intersection, or vice versa. Both decisions cannot be performed concurrently because cognitive processes involved in the decision stage appear capable of supporting only one decision at a time.

These findings might first appear to contradict our phenomenological experience and observations of other people doing multiple things at one time, such as walking and talking. That is, people are quite capable of executing a response to one stimulus while selecting a response to another, such as turning the steering wheel while selecting a reply to the passenger. Our effortless performance of multiple tasks reflects the human ability to execute multiple sets of well-practiced behaviors like talking and walking so long as the critical cognitive operations do not overlap (Fagot & Pashler, 1992). Similarly, in instances where the mapping between a stimulus and response is very natural, as in the case of repeating words as you hear them, tasks can be executed in parallel (McLeod & Posner, 1984).

In many circumstances, we can eliminate the need to make multiple decisions at the same time by delaying the response, for example, by asking our passenger to wait a moment while we decide which direction to turn. In the case of driving, passengers may also regulate their conversation in anticipation of the cognitive demands placed on the driver by changing driving circumstance they are also viewing, which is something the other conversant on a cellular phone cannot do. Not only may passengers vary their conversation in anticipation of the cognitive demands on the driver, but they may also alert the driver to hazards or missed signals.

6.3.2 Controlled versus automatic processes

One reason there are performance decrements when individuals are multi-tasking is because they are dividing their attention between two or more tasks that require a certain amount of cognitive effort. In a driving situation, the performance decrement tends to be greater for inexperienced drivers than experienced drivers because the task of driving requires less cognitive effort for the experienced driver. That is, the efficient performance of complex tasks requires that individual task components be mastered before the behaviors can be integrated into coordinated wholes. Consider the case of riding a bicycle. The bicyclist must learn the individual tasks of balancing, turning, braking, and pedaling while also monitoring their path. The ability to coordinate riding a bicycle while maintaining a lookout for obstacles is initially severely limited because the component skills have not been mastered, requiring considerable cognitive effort. In fact, it is not uncommon to observe young children so focused on the task of riding a bicycle that they ride straight into an obstacle.

It is possible to reduce the necessary cognitive effort with extensive practice, as these component skills become automatic and are integrated into more complex behavioral patterns that can be performed with little effort or oversight. Thus, experience allows the individual to perform multiple actions with minimal interference. These processes are referred to as **automatic processes**. In contrast, **controlled processes** are not practiced enough to become automatic (i.e., parallel parking, backing up) or they lack consistent mapping between the stimulus and the response (see below; Schneider & Shiffrin, 1977; Shiffrin & Schneider, 1977). Processes executed under conscious control require more effort, are slower to execute, and interfere with other tasks to a greater extent than do automatic processes.

An important prerequisite for automaticity is consistent mapping between a stimulus and a response. To study the development of automaticity, researchers employed a search paradigm wherein participants are asked to search a set of test stimuli (probe items) and report whether any of the probe items matches a target set. The participant first memorizes four letters of a **target set**. During the test trial, **probe items** (single letters) are presented individually, and the participant is asked to report whether the probe item (e.g., L, G, M, and X) matches any of the items in the target set (e.g., N or K), and the response time is recorded. Probe items that do not match the target set are distracters. Experimenters manipulate the number of distracter items and whether a target item on one trial may become a distracter on a subsequent trial. **Consistent mapping** (CM) occurs when letters forming the target set are never used as distracters in later trials; in **variable mapping** (VM), letters forming the target set are sometimes used as distracters in subsequent trials.

Research has shown that consistency in mapping affects the degree to which the tasks become highly automatic, requiring little or no attention (Schneider & Fisk, 1982). Furthermore, training that employs consistent mapping (CM) tends to produce shorter reaction times, searches that require little effort, and is relatively unaffected by the number of distractors. CM supports better multi-tasking, and consistently mapped tasks are less susceptible to the effects of stress than are variably mapped tasks. VM has been found to produce slower response times, require more effort, and is more affected by the number of distracters.

6.3.2.1 Interference

The data discussed above demonstrates that the ability to effectively divide attention between tasks depends on familiarity with the component tasks and the consistency in the mapping between the stimulus and response. Repeated practice and consistent mapping significantly reduce the need for constant monitoring of step-by-step execution, freeing attention for other tasks. For young drivers, operating a vehicle is an effortful, controlled process that is susceptible to interference from other behaviors like talking on a cell phone. Tasks may interfere with one another when they compete for a limited resource like attention, or when they require the use of a cognitive process like response choice that cannot be subdivided.

The youngest drivers (Redelmeier & Tibshirani, 1997) may have the highest risk of being in an accident while using a cell phone. This would not be surprising given their limited driving experience and the attention-demanding nature of cell phone conversations. It is important to note that even experienced drivers, for whom driving is more of an automatic process, also experienced an increased crash risk when using a cellular phone. Finally, a related finding reported by Redelmeier, Tibshirani, and others is that users of hands-free cellular phones have similar accident rates to those who use hand-held models. This suggests that the risk posed by cellular phones is not due to handling the phone, but rather to its interference effects on attentional processes.

The ability to multi-task is also expected to be dependent on the degree to which the two tasks place demands on identical cognitive processes and response methods

(Wickens & Hollands, 2000). As introduced above, one might expect greater interference between tasks such as recalling a set of directions while dialing a telephone number, which place loads on the same WM structure (i.e., phonological loop) than tasks that place loads on distinct processes, such as recalling directions while monitoring a traffic signal (one task is phonological, the other visual). Similarly, two tasks that compete for access to the same sensory system (visually monitoring the driving environment while trying to view a GPS moving map display) or response mode (e.g., steering or shifting gears while manipulating keys on a cell phone) are expected to interfere more than those that place demands on separate systems (vision and audition, verbal response and motor response).

BOX 6.3   FAILURES OF ATTENTION: INATTENTIONAL BLINDNESS AND CHANGE BLINDNESS

If asked, most observers would express confidence in their ability to detect stimuli within their field of view or in their ability to detect changes in stimuli they recently had scrutinized. Nevertheless, researchers (Mack & Rock, 1998; Rensink et al., 1997; Simons & Chabris, 1999) have provided ample evidence that our attentional abilities are severely limited, restricting our awareness of the visual world. These limitations are beautifully illustrated by two phenomena called Inattentional blindness and change blindness.

Inattentional blindness is the failure to detect an otherwise visible object when attention is directed or engaged somewhere else. This was powerfully demonstrated by Daniel Simons and Christopher Chabris (1999) who showed subjects a 75-second video wherein six students formed a circle and three of the students wore white t-shirts and three wore black t-shirts. The students with the different colored t-shirts were intermingled, and each group had a basketball that they passed among other similarly attired team members. In one condition, participants were instructed to count the number of passes between members of the white team or black team. Unbeknownst to the subjects, after 45 seconds, a gorilla or a woman with an umbrella would pass through the middle of the circle of players. Surprisingly, approximately half of the subjects failed to notice the gorilla or the women with the umbrella even though the event was highly salient and directly in their field of view. This and related findings illustrate how observers can show dramatic failures in detecting unattended features of the visual environment. However, even attending to an object does not necessarily ensure we will detect changes in features of the object.

Studies of **change blindness** show that observers are not necessarily any better at detecting changes in scenes or objects that they attended to just previously. Change blindness involves the failure to detect visually salient changes in a visual scene, between eye movements or changes in viewpoint like what one experiences watching a TV show or movie. Figure 6.6 shows two pictures that look identical, but which differ in a number of ways. See if you can identify the differences. These failures in attention are not limited to laboratory demonstrations. Similar effects have been reported for army (Durlach & Chen, 2003) and naval operators (DiVita et al., 2004) performing monitoring and control functions using tactical displays illustrating friendly and unfriendly troop positions, as well as the location and movements of aircraft or other assets. These findings illustrate the importance of considering not only the cognitive demands of tasks but also how work interruptions may increase the operator susceptibility to failures of attention. **Critical Thinking Questions**: Do you think that inattentional blindness and change blindness may explain instances where drivers failed to detect hazards like other cars of pedestrians? Can you identify some situational factors that would exacerbate the effects of inattentional blindness and change blindness?

Figure 6.6 Example of spot the difference game. This game commonly found in children's magazines illustrates the change blindness phenomena. Observers find it difficult to identify the differences between the two pictures despite their spatial proximity. (From https://en.wikipedia.org/wiki/Spot_the_difference.)

6.3.3 Sustained attention (vigilance)

Besides selective and divided attention, there is also another form of attention called sustained attention, which is also known as vigilance. **Vigilance** is the ability to sustain attention across an extended period. Vigilance is a characteristic of many tasks, including identifying defective parts on a conveyor belt, scrutinizing X-rays of passenger carry-on baggage for prohibited items, maintaining a lookout for hazards while driving on a highway, and monitoring swimmers in a pool to identify individuals in distress.

Initial interest in vigilance performance among psychologists was driven by a real-world problem. During WWII, Norman H. Mackworth was approached by the Royal Air Force to identify the optimal watch duration for radar operators. The question was motivated by evidence that radar operators on aircraft performing anti-submarine patrols had failed to detect potential Nazi U-boat contacts (see Davies & Parasuraman, 1982).

BOX 6.4   VIGILANCE AND HUMAN INSPECTION PERFORMANCE

Monitoring and inspection tasks such as reviewing X-rays, scrutinizing images of luggage, or the surface of structures to identify damage or corrosion that may compromise structural integrity or ubiquitous. Human monitoring or inspection often represents the last line of defense against defects in manufactured components or the deterioration of existing structures that can lead to catastrophic failure.

Although inspectors are called on to perform a variety of inspection tasks, field evidence and experimental findings demonstrate that they may not perform them particularly well. Craig (1984) summarized the reported miss rates for a wide variety of real monitoring or inspection tasks including X-rays, inspection of glassware and metal fasteners. In some instances, the miss rates were low (i.e., 1%–9%), but higher miss rates of ~20% to 40% were typical of most of the reported studies. Reports cited in the popular press (USAToday, Oct 18, 2007) of tests on TSA

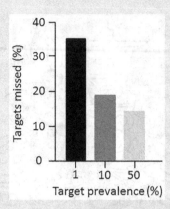

Figure 6.7 Graph of data from the Wolfe et al. (2005) showing how target prevalence affects the likelihood of missing a target when there were 18 objects in a display. The black, gray, and light gray bars represent data for different experimental conditions where target prevalence (i.e., the likelihood that a target is present on any given trial) was very low (black bar, 1% prevalence), slightly higher (gray bar, target present on 10% of the trials), and common (light gray bar, targets were present on 50% of the trails). (Adapted from Wolfe et al., 2005.)

screeners at major airports cite miss rates of 60%–75%. Even the comparatively low miss rates (1%–9%) cited above are unacceptable for socially important tasks, where "misses" have significant negative consequences (e.g., baggage screening, medical care, and aircraft maintenance).

Earl Wiener (1984) noted that in many cases that the prevalence of targets for many real-world tasks (enemy attack, aircraft engine failures, bombs in luggage, etc.) is vanishingly small. Recent laboratory studies have shown that very low target prevalence (i.e., the number of defects per set of inspected items) has a pronounced effect on detection performance (Wolfe et al., 2005, 2007). Jeremy Wolfe and colleagues created an artificial baggage screening task and had participants search for "tools" positioned among other objects. Figure 6.7 shows how target prevalence affects the likelihood of failing to detect the target when there were as many as 18 objects in the display. The blue, yellow, and red bars represent data from different experimental conditions where target prevalence (i.e., the likelihood that a target is present on any given trial) was very low (blue bar, 1% prevalence), slightly higher (yellow bars, target present on 10% of the trials), and common (red bars, targets present on 50% of the trails). An example of one of the displays is also shown. The data reveal that the number of misses or failures to detect a target increased significantly from 15% to 37% for target prevalence conditions corresponding to 50% and 1%, respectively. Thus, as the likelihood of a target decreased, participants were significantly less likely to detect the target when it was present. This is true even when the target is clearly visible, and the operators' attention is drawn to them. Prevalence may also play a role in another phenomena called Satisfaction of search (SOS). Imagine a physician reviewing an X-ray to identify all the broken bones in an image. Research shows that when there are multiple targets (i.e., multiple broken bones), finding one target can reduce the likelihood of detecting another target (i.e., another broken bone) in the image. Evidence suggests that SOS errors occur in a range of radiological exams including abdominal radiography and skeletal radiography (Berbaum et al., 1994, 1996). **Critical Thinking Questions**: Besides prevalence, what other factors may explain why human operators have such difficulty with inspection tasks? Are there ways in which the inspection task could be redesigned to improve performance such as including frequent breaks and alternating inspection with other tasks for instance?

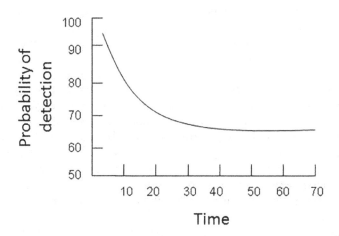

Figure 6.8 Illustration of the vigilance decrement. (© Alex Chaparro.)

To study vigilance performance, Norman Mackworth developed the classic "Clock Test," which consists of a rotating pointer that moves in discrete single steps once a second. The participants' task was to monitor the pointer and to report when the pointer made a larger double jump. Double jumps of the pointer occurred 12 times per 30-minute interval. Using this and other tasks, Mackworth found evidence of deterioration in detection performance after less than 30 minutes on the task! (Figure 6.8) This effect is not solely a visual phenomenon, as a vigilance decrement is also observed using auditory vigilance tasks.

Over the years, several hypotheses have been offered to explain the vigilance decrement including diminished sensory sensitivity, decreased arousal, and individual differences in motivation. The **sensitivity hypothesis** attributes the vigilance decrement to a decrease in the observer's ability to discriminate between signals and non-signals, whereas the **arousal hypothesis** attributes the vigilance decrement to a reduced ability to maintain a high level of alertness due to perceptual habituation to the repetitive stimulation. Finally, the **motivational hypothesis** proposes that the decrement reflects the averaging or pooling of data across at least two types of participants: those who are highly motivated and do not exhibit a vigilance decrement, and those who are lesser motivated participants who respond more sporadically (Davies & Parasuraman, 1982). Based on detailed analysis of some 138 vigilance studies, researchers (See et al., 1995) have concluded that the experimental evidence supports the notion that vigilance tasks produce a robust and reliable decrement in sensitivity.

The vigilance decrement is a significant concern given the many every-day tasks that require vigilance including lifeguarding, air traffic control, and inspections that require an observer to sustain attention for extended periods of time. What can be done to minimize the effects of the vigilance decrement? The most frequently recommended intervention is to minimize the duration of the task and/or to allow the observer frequent rest breaks. Training on the task may also offer some additional benefits. Training programs that maximize the development of automatic processing rather than controlled processing of the vigilance task might eliminate or reduce the magnitude of the vigilance decrement (Fisk & Schneider, 1981). Fisk and Schneider (1981) recommend where possible that training provide extensive practice and exposure to targets and present consistent mapping (CM) between critical signals and targets to facilitate the development of the automatic detection of targets.

6.4 APPLICATIONS TO SPECIAL POPULATIONS

The findings of Green and Bavelier (2003) hold considerable promise for a segment of older drivers who experience diminished attentional skills that may compromise safe driving. If the attentional skills of young adults can be improved by playing games, then it is possible that the attentional skills of older drivers may also be improved. Computer games may serve as a training tool to help older drivers maintain their skills at a high level of proficiency as well as a means of developing new skills that could aid them in compensating for other age-related changes. It should be stressed that many older drivers are safe drivers but that a small segment of this population may be more crash prone. An important futures step in this research is the demonstration that improvements in these attentional skills transfer to real-world driving. In other words, it is possible to demonstrate that game-related changes in performance assessed using laboratory tests of attentional skill translate to improved on-the-road driving performance.

Attentional training programs have broad application including young and older drivers as well as in rehabilitation for individuals who have brain injuries who want to regain some measure of independence. The independence afforded by driving is an important quality of life issue for many older adults. Loss of driving privileges is associated with increased isolation and symptoms of depression.

6.5 SUMMARY

Accident databases offer too many examples of mishaps that result in part from tools or environments whose design failed to consider the capabilities and limitations of the human user. In this chapter, we have reviewed the capacity of human operators to attend, process, and remember important task-related information, as well as the theories that attempt to explain the basis of these abilities. Psychological theories like those reviewed above are powerful tools that the human factors specialist can employ when evaluating new designs to identify the potential impact of design features, thereby maximizing the operators' performance and minimizing the likelihood of errors.

LIST OF KEY TERMS

Articulatory control process
Attention
   Divided attention
   Selective attention
   Sustained attention
   Change blindness
   Inattentional blindness
Automaticity
Automatic processes
Central executive
Capacity (working memory)
Change blindness
Chunking
Clock test
Consistent mapping (CM)
Controlled processes
Decision-making

Distracters
Flanker compatibility effect
Memory
   Echoic memory
   Episodic memory
   Iconic memory
   Long-term memory (LTM)
   Procedural memory
   Prospective memory
   Semantic memory
   Short-term memory (STM)
   Working memory (WM)
Multi-tasking
Phonological loop
Phonological store
Probe
Rehearsal (working memory)

Response execution
Sensory register
Variable mapping (VM)
Vigilance
  Vigilance decrement

Arousal hypothesis
Sensitivity hypothesis
Motivation hypothesis
Visuo-spatial sketchpad
Visual search

CHAPTER EXERCISES

1. Compare and contrast the properties of the echoic and iconic stores.
2. Describe the different types of long-term memories and give examples of each type.
3. Describe the different strategies you have used to remind yourself to do something in the future (i.e., prospective memory). What kind of cues did you use to remind yourself? Were they time- or event-based prospective memories? From your experience, were time- or event-based prospective memories more prone to failure?
4. Describe Baddeley's model of working memory. What are the functions of the different components of Baddeley's model?
5. Explain the difference between bottom-up and top-down influences on the deployment of visual attention.
6. Describe the differences between inattentional blindness and change blindness. Which of these forms of attentional failure may be most closely related to the attentional failure's drivers' experience? Explain.
7. Experimental evidence suggests that young drivers are more at risk of accidents when talking on a cellular phone. Explain how the theory of automatic versus controlled processing might account for the challenges young drivers face.

SUGGESTED READINGS

Green, S. G., & Bavelier, D. (2003). Action video games modifies visual selective attention. *Nature, 423,* 534–537.
  This paper was one of the first published studies documenting the effects of playing video games on different visual attention skills.
Lavie, N., Hirst, A., de Fockert, J. W., & Viding, E. (2004). Load theory of selective attention and cognitive control. *Journal of Experimental Psychology: General, 133*(3), 339–354.
  This paper provides a new account of the relationship cognitive load and the selectivity of attention.

REFERENCES

Baddeley, A. D. (1992). Is working memory working? The Fifteenth Bartlett Lecture. *Quarterly Journal of Experimental Psychology, 44A,* 1–31.
Baddeley, A. D., & Hitch, G. J. (1974). Working memory. In G. A. Bower (Ed.), *The Psychology of Learning and Motivation,* pp. 47–89. New York: Academic Press.
Ball, K., Owsley, C., Sloane, M. E., Roenker, D. L., & Bruni, J. R. (1993). Visual attention problems as a predictor of vehicle crashes in older adults. *Investigative Ophthalmology & Visual Science, 34*(11), 3110–3123.
Berbaum, K. S., El-Khoury, G. Y., Franken, E. A., Kuehn, D. M., Meis, D. M., Dorfman, D. D., Warnock, N. G., Thompson, B. H., Kao, S. C., & Kathol, M. H. (1994). Missed fractures resulting from satisfaction of search effect. *Emergency Radiology, 1*(5), 242–249.
Berbaum, K. S., Franken, E. A., Jr., Dorfman, D. D., Miller, E. M., Krupinski, E. A., Kreinbring, K., Caldwell, R. T., & Lu, C. H. (1996). The cause of satisfaction of search effects in contrast studies of the abdomen. *Academic Radiology, 3,* 815–826.

Briggs, G. E., & Johnsen, A. M. (1973). On the nature of central processes in choice reactions. *Memory & Cognition, 15*, 91–100.

Brown, J. (1958). Some tests of the decay theory of immediate memory. *Quarterly Journal of Experimental Psychology, 10*, 12–21.

Conrad, R., & Hull, A. J. (1964). Information, acoustic confusion and memory span. *British Journal of Psychology, 55*(4), 429–432.

Cowan, N. (2001). The magical number 4 in short-term memory. A reconsideration of mental storage capacity. *Behavioral and Brain Sciences, 24*, 87–185.

Craig, A. (1984). Human engineering: The control of vigilance. In J. S. Warm (Ed.), *Sustained attention in human performance*, pp. 247–291. New York, NY: Wiley.

Crowder, R. G. (1982). Decay of auditory memory in vowel discrimination. *Journal of Experimental Psychology: Learning, Memory, and Cognition, 8*, 153–162.

Crowder, R. G., & Morton, J. (1969). Precategorical acoustic storage (PAS). *Perception & Psychophysics, 5*, 365–373.

Davies, D. R., & Parasuraman, R. (1982). *The Psychology of Vigilance*. New York: Academic Press.

de Fockert, J. W., Rees, G., Frith, C. D., & Lavie, N. (2001). The role of working memory in visual selective attention. *Science, 291*(5509), 1803–1806.

Dismukes, K., & Nowinski, J. (2007). Prospective memory, concurrent task management, and pilot error. In A. F. Kramer, D. Wiegmann & A. Kirlik (Eds.), *Attention: From Theory to Practice*, pp. 225–236. Oxford: Oxford University Press.

DiVita, J., Obermeyer, R., Nygren, T. E., & Linville, J. M. (2004). Verification of the change blindness phenomenon while managing critical events on a combat information display. *Human Factors, 46*, 205–218.

Downing, P. E. (2000). Interactions between visual working memory and selective attention. *Psychological Science, 11*(6), 467–473.

Durlach, P. J., & Chen, J. Y. C. (2003). Visual change detection in digital military displays. Paper presented at the *Proceedings of the Interservice/Industry Training, Simulation, and Education Conference*, December 2003. Orlando, FL.

Fagot, C., & Pashler, H. (1992). Making two responses to a single object: exploring the central bottleneck. *Journal of Experimental Psychology: Human Perception and Performance, 18*, 1058–1079.

Fernandez-Duque, D., & Johnson, M. L. (2002). Cause and effect theories of attention: the role of conceptual metaphors. *Review of General Psychology, 6*(2), 153–165.

Fisk, A. D., & Schneider, W. (1981). Control and automatic processing during tasks requiring sustained attention: a new approach to vigilance. *Human Factors, 23*(6), 737–750.

Green, C. S., & Bavelier, D. (2003). Action video games modifies visual selective attention. *Nature, 423*(6939), 534–537.

Herrman, D., Brubaker, B., Yoder, C., Sheets, V., & Tio, A. (1999). Devices that remind. In F. T. Durso, R. S. Nickerson, R. W. Schvaneveldt, S. T. Dumais, D. S. Lindsay & M. T. H. Chi (Eds.), *Handbook of Applied Cognition*, pp. 377–407, New York: John Wiley & Sons, Ltd.

Kahneman, D., Ben-Ishai, R., & Lotan, M. (1973). Relation of a test of attention to road accidents. *Journal of Applied Psychology, 58*(1), 113–115.

Kramer, A. F., Larish, J. F., & Strayer, D. L. (1995). Training for attentional control in dual task settings: a comparison of young and old adults. *Journal of Experimental Psychology: Applied, 1*(1), 50–76.

Lavie, N., & Cox, S. (1997). On the efficiency of visual selective attention: efficient visual search leads to inefficient distractor rejection. *Psychological Science, 8*(5), 395–398.

Lavie, N., Hirst, A., de Fockert, J. W., & Viding, E. (2004). Load theory of selective attention and cognitive control. *Journal of experimental Psychology: General, 133*(3), 339–354.

Mack, A., & Rock, I. (1998). *Inattentional Blindness*. Cambridge, MA: MIT Press.

Malaterre, G. (1990). Error analysis and in-depth accident studies. *Ergonomics, 33*, 1403–1421.

McLeod, P., & Posner, M. I. (1984). Privileged loops from percept to act. In H. Bouma & D. G. Bouwhuis (Eds.), pp. 55–66. *Attention and Performance X*. London: Lawrence Erlbaum Associates.

Miller, G. A. (1956). The magical number seven, plus or minus two: some limits on our capacity for processing information. *Psychological Review, 63*, 81–97.

Most, S. B., & Simons, D. J. (2001). Attention capture, orienting, and awareness. In C. Folk & B. Gibson (Eds.), *Attraction, Distraction, and Action: Multiple Perspectives on Attentional Capture*, pp. 151–173. Amsterdam: Elsevier.

Owsley, C., Ball, K., McGwin, G., Sloane, M. E., Roenker, D. L., White, M. F., & Overley, E. T. (1998). Visual processing impairment and risk of motor vehicle crash among older adults. *Journal of the American Medical Association*, *279*, 1083–1088.

Pashler, H. (1984). Processing stages in overlapping tasks: evidence for a central bottleneck. *Journal of Experimental Psychology: Human Perception and Performance*, *10*, 358–377.

Peterson, L. R., & Peterson, M. J. (1959). Short-term retention of individual verbal items. *Journal of Experimental Psychology: Applied*, *58*, 193–198.

Recarte, M. A., & Nunes, L. M. (2003). Mental workload while driving: effects on visual search, discrimination, and decision making. *Journal of Experimental Psychology: Applied*, *9*(2), 119–137.

Redelmeier, D. A., & Tibshirani, R. J. (1997). Association between cellular-telephone calls and motor vehicle collisions. *The New England Journal of Medicine*, *336*(7), 453–458.

Rensink, R. A., O'Regan, J. K., & Clark, J. J. (1997). To see or not to see: the need for attention to perceive changes in scenes. *Psychological Science*, *8*, 368–373.

Rudisill, T. M., & Zhu, M. (2016). Who actually receives cell phone use while driving citations and how much are these laws enforced among states? A descriptive, cross-sectional study. *BMJ Open*, *6(6)*, e011381C.

Rumar, K. (1990). The basic driver error: late detection. *Ergonomics*, *33*(10/11), 1281–1290.

Schneider, W., & Fisk, A. D. (1982). Degree of consistent training: improvements in search performance and automatic process development. *Perspectives in Psychophysics*, *31*(2), 160–168.

Schneider, W., & Shiffrin, R. W. (1977). Controlled and automatic human information processing, Vol. I. Detection, search, and attention. *Psychological Review*, *84*, 1–66.

See, J. E., Howe, S. R., Warm, J. S., & Dember, W. N. (1995). Meta-analysis of the sensitivity decrement in vigilance. *Psychological Bulletin*, *117*(2), 230–249.

Shiffrin, R. M., & Schneider, W. (1977). Controlled and automatic human information processing: perceptual learning, automatic attending, and a general theory. *Psychological Review*, *84*, 127–190.

Shinoda, H., Hayhoe, M. M., & Shrivastava, A. (2001). What controls attention in natural environments. *Vision Research*, *41*, 3535–3545.

Simons, D. J., & Chabris, C. F. (1999). Gorillas in our midst: sustained inattentional blindness for dynamic events. *Perception*, *28*(9), 1059–1074.

Sperling, G. (1960). The information available in brief visual presentations. *Psychological Monographs*, *74*(11), 1–29.

Squire, L. R., & Zola, S. (1996). Structure and functions of declarative and nondeclarative memory systems. *Proceedings of the National Academy of Sciences*, *93*, 13515–13522.

Strayer, D. L., Drews, F. A., & Johnston, W. A. (2003). Cell phone-induced failures of visual attention during simulated driving. *Journal of Experimental Psychology: Applied*, *9*(1), 23–32.

Strayer, D. L., & Johnson, W. A. (2001). Driven to distraction: dual-task studies of simulated driving and conversing on a cellular phone. *Psychological Science*, *12*, 462–466.

Summala, H., Pasanen, E., Raesaenen, M., & Sievaenen, J. (1996). Bicycle accidents and drivers' visual search at left and right turns. *Accident Analysis & Prevention*, *28*, 147–153.

Tulving, E. (1985). How many memory systems are there? *American Psychologist*, *40*(4), 385–398.

Tulving, E. (1989). Remembering and knowing the past. *American Scientist*, *77*, 361–367.

Wickens, C. D., & Hollands, J. G. (2000). *Engineering Psychology and Human Performance* (3rd ed.). Upper Saddle River, NJ: Prentice-Hall Inc.

Wickens, C. D., Lee, J. D., Liu, Y., & Gordon-Becker, S. (2003). *Introduction to Human Factors Engineering* (2nd ed.). Englewood Cliffs, NJ: Prentice Hall.

Wiener, E. L. (1984). Vigilance and inspection. In J. S. Warm (Ed.), *Sustained Attention in Human Performance*, pp. 207–246. Chichester, UK: Wiley.

Wolfe, J. M., & Horowitz, T. S. (2004). What attributes guide the deployment of visual attention and how do they do it? *Nature Reviews Neuroscience*, *5*, 1–7.

Wolfe, J. M., Horowitz, T. S., & Kenner, N. (2005). Rare items often missed in visual searches. *Nature*, *435*, 439–440.

Wolfe, J. M., Horowitz, T. S., Van Wert, M. J., Kenner, N. M., Place, S. S., & Kibbi, N. (2007). Low target prevalence is a stubborn source of errors in visual search tasks. *Journal of Experimental Psychology: General*, *136*(4), 623–638.

Yantis, S. (1998). Control of visual attention. In H. Pashler (Ed.), *Attention*. East Sussex, UK: Psychology Press.

Chapter 7

Decision-making

Chapter Vignette

John is the owner of a small company who used his personal airplane to fly several company executives to a meeting in another state. The meeting is running late, and he and the other executives are anxious to return so that they do not have to spend the night away from home. The weather briefing at the airport indicates a potential for increasing clouds and thunderstorms along the route home. This is of concern because John is only licensed to fly under visual flight rules. That is, he can only fly when visibility meets certain minimum requirements (e.g., 3 miles when taking off from an airport). Thus, he cannot fly at night and should not fly into clouds. He has begun training to obtain a rating that would allow him to fly in poorer weather using only instruments, but he is not yet licensed to do so.

John reviews his options again and realizes that he must quickly decide now: go or don't go. If they do not leave soon, they will not arrive home before sundown and the weather may catch them before they land. The other passengers, his executives, though, are eager to go and remind John of the added expense of staying another night. After evaluating the alternatives, John makes the decision to go.

Did John make the correct decision? When making decisions we are often faced with "facts" (such as the danger of the potential storm) but also other information (e.g., the executives' desire to get home and the trip cost) that can influence our decision-making process. Studies of decision-making reveal that decision-makers employ rules of thumb and systematic tendencies called heuristics and biases, respectively, that reduce the cognitive demands of the decision-making process. It is important to understand how heuristics and biases influence the outcomes of decision process, as well as the impact of task and situational demands, on decision outcomes. We also need to consider how the decision process changes as individuals gain experience as evidence suggests that the decision-making processes of experts differ considerably from those of novices. This chapter reviews the role of biases, stress, expertise, and other factors on decision-making and explores the use of design of aids to support decision-making.

7.1 CHAPTER OBJECTIVES

After reading this chapter, you should be able to:

- Describe the two major approaches to decision-making.
- Describe the different types of heuristics and decision biases that are characteristic of human decision-makers.

DOI: 10.1201/9781003515463-7

- Describe some of the attributes of experts and non-expert decision-makers.
- Explain what decision aids are and how they can improve decision-making.

7.2 WHAT IS DECISION-MAKING AND WHY DO WE STUDY IT?

Decision-making is generally defined as a task requiring an individual to choose among several alternatives with some uncertainty regarding which choice is the best one. Typically, this process is extended requiring a choice to be made on the order of minutes or longer, using whatever information may be available. By understanding how individuals make decisions, what information they use, and what decision-related tasks they find difficult or perform poorly, the human factors expert can determine how to change the environment, so individuals make better decisions. If you recall from Chapters 3 and 4, there are optimal ways to present information to individuals using displays. Research findings can inform decisions regarding what information should be displayed to the decision-maker and its best format. The research also provides useful information regarding which tasks to allocate to automated computer systems, which to assign to the human operator, and the design of decision aids to support better decision-making.

Consider the situation John faced in our chapter-opening scenario. What decision should John make? With the potential for a plane crash and the loss of lives, perhaps it appears as though John would obviously opt to stay the night and to fly home the next day. Unfortunately, situations like John's occur all too frequently, and the decision-makers often make decisions that lead to catastrophic consequences. Although only about 2% of general aviation (GA) accidents are weather-related, between 1990 and 1997, they accounted for 11% of GA fatalities (Weigmann & Goh, 2000). One of the most common factors in these accidents is a pilot rated for visual flight rules (VFR) who flies into poor or deteriorating weather conditions. These decision errors are a major causal factor in 30% of GA accidents (Weigmann & Shappel, 1997).

Decision-making is also a significant concern in air ambulance operations used to pick up accident victims or to transport patients from rural to urban hospitals that are better equipped to address some emergency medical conditions. A review of accident statistics indicates that 55 accidents involving Emergency Medical Services (EMS) aircraft occurred between January 2002 and January 2005 (NTSB, 2006). The hazards associated with these flights vary considerably due to the time pressure, the wide range of environmental conditions in which pilots may be asked to fly, and the lack of familiarity that pilots may have with the destination.

It is hard to imagine any other conscious behavior that absorbs as much of our day-to-day existence and is so critical to our welfare as decision-making. Despite the importance of decision-making, we usually receive little formal training on how to approach important life- or work-related decisions. How did you decide which college to attend? Which factors did you consider and weigh most heavily in the final decision? Consider the case of a patient with a life-threatening condition who must decide among multiple treatment options with different side effects or the case of a fire chief who must select a firefighting strategy during rapidly changing and stressful circumstances. What strategies do patients or firefighters adopt to avoid being overwhelmed by the decision process? How effective are these strategies? What strategies do experts employ? Can these expert strategies improve the decision-making of novices? These questions are addressed below.

7.3 APPROACHES TO DECISION-MAKING

There are several different approaches to the study of decision-making, but they can be generally categorized as belonging to either the normative, descriptive, or naturalistic approaches.

A normative approach or model of decision-making specifies what an individual *should* do given the facts of the situation. In contrast, the descriptive approach to decision-making attempts to describe how individuals *actually* make decisions.

7.3.1 Normative approach to decision-making

The **normative approach** is synonymous with a rationale or deliberate process of selecting among alternatives. That is, individuals using the normative approach rationally evaluate and weigh the available information to determine the optimal decision. In addition, it would be assumed that there are no time or memory constraints on this decision-making process. This is the method presumed to characterize the decision-making styles of scientists and physicians, and we are often encouraged to emulate it. By contrast, guessing or relying on one's gut is often characterized as evidence of an undisciplined mind and a decision-making style that should be avoided. The popular character Sherlock Holmes, created by Sir Author Conan Doyle, states "I never guess. It is a shocking habit—destructive to the logical faculty" (Doyle, 1987, p. 93).

The normative decision-making process is described as consisting of four components: (1) identification of choice alternatives (e.g., go or not go on a flight); (2) events that might or might not transpire (e.g., thunderstorms might appear, thunderstorms might not appear); (3) potential outcomes (e.g., accident, diversion to another airport, stay in local hotel, and arrive home); and (4) an assessment of the probability of each potential outcome. These considerations are illustrated in Table 7.1.

In our flight example (see Table 7.1), we have two alternative decisions: go or don't go. However, many decisions require that we consider several different alternatives and the probabilities that certain outcomes will occur because of those alternatives. The decision we make depends on the possible outcomes of those alternatives, the probability of those outcomes, and how much we value each outcome. That is, if we decide to fly home now, we consider the probability that there will be a thunderstorm and we will have an accident. Then, we essentially weigh the probabilities and the values to make a decision. If an event we do not want is likely to occur (e.g., an accident), we are less likely to select that alternative, but if the alternative leads to a high probability of an event we want (e.g., go and get home today), we are more likely to select that alternative.

Although problems often exist in determining the probabilities and values people associate with various outcomes, an important aspect of normative theories is the desirability or undesirability that is associated with the potential outcomes. The utility may be an objective or physical gain (e.g., money) or a gain of something more subjective (e.g., getting home today). The normative model of decision-making that combines probabilities and values to identify the best decision is called **expected value theory**.

Table 7.1 Four components of normative decision-making

Alternatives	Possible events	Possible outcomes	Probability of possible outcomes
Go	Thunderstorms	Accident, rough flight	%?
	No Thunderstorms	Nothing, get home today	%?
Don't go	Thunderstorms	Nothing, get home day later and have to stay in a hotel	%?
	No Thunderstorms	Nothing, get home day later and have to stay in a hotel	%?

7.3.1.1 *Expected value theory*

Recall that normative models describe how to optimally combine information (probabilities and values) to arrive at the best decision, and they assume we are not constrained by time or memory capacity. Because the utility of the outcomes in our flight example is more subjective, let's first look at an example of gambling in which the utility is more objective.

In gambling, the utility of an outcome corresponds to the monetary value of potential gains. Although the subjective utility of an outcome for a gambling addict may well exceed its monetary values, we will just consider the monetary gains. When deciding whether to gamble or not, we may consider both the probability of an outcome (winning or losing) and the utility (e.g., monetary value) of winning and losing. Probability and utility values can be used to calculate the expected value of an outcome.

The **expected value** is calculated by multiplying the probability of each potential outcome by its value and summing the products. In the gambling example, the expected value is represented by the average amount of money gamblers may expect to win or lose when they bet. Let's consider another example of expected value. Imagine you go to the carnival and there is a game akin to a Wheel of Fortune, where six different sectors of the wheel are randomly labeled 1 through 6. If you spin the wheel and the pointer stops on 6, you win $10; however, there is a catch. You must pay a $2 entry fee to play the game. Would you play the game? A rational decision-maker would use the equation below to calculate how much money on average one could expect to win. To calculate the expected value (EV), you need to enter quantities for the monetary value of winning, V(W), the probability of wining, P(W), the monetary value of losing, V(L), and the probability of losing, P(L).

$$EV = V(W) \times P(W) - V(L) \times P(L)$$

$$= \$8 \times \frac{1}{6} - \$2 \times \frac{5}{6}$$

$$= 1.33 - 1.66$$

$$= -0.33 \text{ (or a loss of 33 cents)}$$

In this example, the probability of a win is 1/6 (only one of the six sectors represents the $10 win) and the potential winning is $8 ($10 minus the $2 entry fee). By contrast, the probability of losing is 5/6, and the value of the loss is the $2 dollar entry fee. Given the calculation shown above, the expected value of the game is a loss of 33 cents or, in other words, on average, players will lose 33 cents when they play the game.

Now that you know that the expected value of this carnival game is negative, are you likely to play? The expected value of other gambling games is also negative. Will people stop gambling? Not likely. Gambling tasks or scenarios are frequently used to study decision-making because the probabilities of different alternatives can be readily calculated, and the monetary winnings or losing's represent values. These models can serve as a useful benchmark or standard against which to compare the real behavior of decision-makers.

How well does expected utility theory predict decision-making behavior? If the participants in the experiment were making decisions using the outcomes of an expected value model, then they should choose not to gamble. Considering the popularity of gambling casinos and carnival games, this suggests that expected value theory represents a poor description of human decision-making. Given that the probabilities of different outcomes are easily calculated, it follows that the monetary values associated with the different options must not really capture their utility to the decision-maker.

Returning to our flight example, it is much more difficult to determine an individual's value or utility rating of an accident, returning home today or returning home in the morning. Although we could assume that the utility of an accident is normally lower than the utility of returning home today or tomorrow, we do not have exact values. Nevertheless, these subjective values tend to influence our decisions. You do not have to be a pilot considering bad weather to experience the influence of subjective values. For example, your instructor has announced that surprise quizzes are scheduled throughout the semester. Your choice to study is likely to be influenced by several factors including: (1) your estimates of the probability of having a quiz and (2) the consequences of studying or not studying. Studying for the quizzes may improve an already strong grade or it may improve a marginal one. You could use the time to study for another class where you are not performing to expectation or, alternatively, you could use the time to catch up with friends.

7.3.1.2 Subjective expected utility (SEU)

In response to the discrepancy between predictions based on expected value theory and actual human behavior, theorists proposed the notion of **subjective expected utility** (Savage, 1954). This approach acknowledges that the values individuals assign to potential outcomes are not strictly dependent on monetary gains and losses but are influenced by the goals of an individual. Calculations of subjective utility are like expected value, but estimates of subjective utility determined for each individual are multiplied by the probability that the outcome obtains.

7.3.2 Descriptive models of decision-making

One attractive aspect of normative decision models described above is their systematic approach to deciding; however, one might question how often we engage in such deliberate processes. Even if one were tempted to use a normative strategy, the limited capacity of working memory severely restricts such processes (see Chapter 6). Situational factors and our limited capacity to process information may not allow the use of such slow, cognitively demanding decision processes required by the normative model of decision-making.

One of the strengths of human decision-makers is the ability to adapt to new circumstances and demands. Rather than employing only one strategy or heuristic, human decision-makers appear to be quite flexible in selecting an appropriate decision strategy from a repertoire of methods.

Descriptive models of decision-making seek to describe how individuals arrive at decisions and offer explanations for the discrepancy between the predictions of normative models of decision-making and actual human behavior. Researchers (Simon, 1957; Tversky & Kahneman, 1974) have proposed that experienced decision-makers employ a variety of heuristics, shortcuts, or rules of thumb that can be extremely effective in some situations but result in biases that lead to poor decisions in other situations.

Individuals use different decision strategies and heuristics to reduce cognitive demands and to speed up the decision-making process. **Heuristics** are mental shortcuts or rules of thumb that allow an individual to make a quick decision or choice requiring little cognitive effort. The selection of a decision-making method for a given situation appears to be based on an evaluation of the tradeoff between accuracy and effort. John Payne and colleagues

(Payne, 1976; Payne et al., 1988) conducted a series of experiments demonstrating factors that influence strategy selection. Payne, for instance, found that decision-makers are more apt to select a normative decision process that evaluates all the pertinent information and considers potential tradeoffs between one or more valued attributes when the decision problem involves two or three alternatives. Tradeoffs among attributes represent a form of **compensatory decision-making** where positive attributes can compensate for negative ones. By contrast, Payne (Payne, 1976; Payne & Braunstein, 1978) found that when faced with problems that involved six to twelve alternatives, the decision-makers were more likely to adopt non-compensatory strategies like elimination by aspects (Tversky, 1972) or satisficing (Simon, 1957), which are discussed next.

7.3.2.1 Non-compensatory strategies

One non-compensatory strategy decision-maker's use is the **elimination by aspects** heuristic. In 1972, Amos Tversky proposed that we make decisions by eliminating those alternatives that are least attractive. This approach assumes that individuals evaluate the attributes (i.e., aspects) of different alternatives and eliminate from further consideration of those alternatives that are not judged to be attractive. For instance, imagine you want to purchase a computer and you have $2,000 to spend. You might begin by eliminating computers that cost more than this amount. You are also interested in having powerful graphics card, a large hard drive, and a high-resolution monitor. As you continue to identify desired attributes and reject those computers that do not meet your minimum requirement, you will eventually eliminate most alternatives. According to Tversky's (1972) model, the likelihood of selecting an attribute for consideration depends on its importance. If you are independently wealthy, the price of the computer might be a minor consideration; consequently, it is evaluated later in the sequence. This is important because the final choice depends critically on the order in which the different attributes are evaluated.

Instead of eliminating those options that do not meet our minimum requirements, we might settle for a "good enough" rather than the perfect choice. Time constraints or a limited capacity to identify and evaluate alternatives can restrict our ability to identify the best choice. In such instances, we may be more likely to select an "acceptable" alternative. This strategy is called **satisficing** which is an amalgam of "satisfying and "sufficing" (Anderson et al., 2000). For example, a highly selective university is unlikely to conduct a detailed review of each applicant's file, rather they might set cutoffs pertaining to GPA and SAT scores and only review those that meet the cutoffs.

7.3.2.2 Heuristics

Satisfying and elimination by aspects are not the only tricks that decision-makers have at their disposal to handle the demands of decision-making. Decision-makers employ a diverse set of heuristics or rules of thumb that support rapid decision-making. It should be stressed that, in many cases, the heuristics described below represent shortcuts that serve the decision-maker well but may fail them in new or unfamiliar circumstances.

Unlike in the case of satisficing or elimination by aspects discussed earlier, the decision-maker does not always eliminate the worst alternatives or select an option that is good enough. With the **availability heuristic**, the decision-maker is influenced by information that is recalled or available at the time the decision is made. For example, doctors anticipating potential complications from a medical procedure or pilots choosing to fly to a distant destination are influenced by their judgment of the relative likelihood of certain outcomes. They may select an action or make contingencies for those outcomes that they believe to be most likely; however, these estimates of the probabilities or frequency of different outcomes are influenced by the memory retrieval process and the relative ease with which certain examples come to mind. That is, our pilot might recall one or two exceptional examples of pilots flying into bad weather and successfully reaching their destinations, which influences our pilots' decision to fly into bad weather. Unfortunately, our pilot might not recall other examples in which pilots had flown into bad weather and had encountered severe consequences, which might have led to a different decision.

What happens in the case of the availability heuristic is that we tend to estimate the likelihood of events that readily come to mind to be higher than those that do not. Some decisions we might make can include choosing to drive rather than fly, deciding not to enter the water because of a fear of sharks, or standing outside to watch a lightning storm. In each case, it is likely that we have not made a good decision. It is well known that flying is considerably safer than driving and that you are more likely to die from a lightning strike than being bitten by a shark ("Death Odds," September 24, 1990). Our decisions, though, were likely influenced by our recollections of what was safest. Perhaps recent news reports impacted this recollection, which influenced our decisions.

In his book on judgment and decision-making, Scott Plous (1993, p. 3) cited the following examples to illustrate the effects of the availability heuristic:

"Which is a more likely cause of death in the United States: (1) being killed by falling airplane parts or (2) by a shark?

Which of the following causes of death claim more lives: (1) diabetes or homicides or (2) stomach cancer or car accidents?"

Accident statistics indicate that you have a 30-fold greater chance of dying due to falling aircraft parts than dying from a shark attack ("Death Odds," September 24, 1990). Similarly, deaths due to diabetes and stomach cancer account for roughly twice as many deaths as homicides and car accidents do. In each of these choices, our decisions may be influenced by the availability heuristic.

People may fear flying in an airplane or shark attacks because instances of these events easily come to mind due to the widespread coverage they receive in on-line news sources and the evening news. Keeping in mind that events that easily come to mind are more frequent and that events that are difficult to conjure up are rare is often a good rule of thumb. This strategy probably served well our ancestors who had to rely on their previous experience to identify where to find game. However, in a modern society, information about rare medical conditions found in one part of the world can make our evening news and bias our estimates of their likelihood of occurring, and this, in turn, can lead us to change our behavior. The pilot who reads about an aircraft accident caused by a pilot striking a migrating bird might take precautions to avoid migratory flight paths but might not take precautions for more common and likely accident causes. Likewise, the mechanic attempting to troubleshoot a problem in a car might waste precious time looking for a rare but memorable cause rather than a more likely but less memorable problem.

7.3.3 Naturalistic decision-making

Klein and colleagues (Klein, 1998; Zsambok & Klein, 1997) have argued quite persuasively that traditional models of decision-making bear little resemblance to how experienced decision-makers behave under real-world conditions, where the need to respond rapidly is great, and alternatives may be uncertain, events are dynamic and evolving, and competing goals may exist and change as a function of time. In these types of situations, there may not be a right decision, and the decision one chooses can change by the minute. For instance, a fire chief must make a series of decisions that are influenced by a variety of factors, including the rate at which a fire is spreading, the size of the fire, and the risk it poses to other structures or people (Klein et al., 1993). There may be little time to enumerate and consider all options, their probabilities, and their utility and then sum their products. Naturalistic Decision-Making (NDM) studies individuals in real-world settings and seeks to describe how they make decisions rather than defining how they should make decisions.

Klein and his colleagues (Klein et al., 1993, 1996) have proposed the *Recognition Primed* decision model (RPM) of NDM to describe the decision-making behavior of experts working in time-sensitive environments like aviation, the military, or firing fighting. The name of the theory stems from the fact that the model presupposes that the individual making the decision is an experienced professional who, when confronted with a situation, can readily generate one or two alternative solutions based on his or her experience with similar situations and rules stored in memory. By recognizing similarities between the current situation and past encounters, the professional can identify an appropriate response. In more complex circumstances, they may rely on mental simulation. **Mental simulation** allows the decision-maker to make inferences about the positive and negative outcomes associated with different decisions by using his or her knowledge of a system, its components and their interrelationships to predict its response. In this way, much like a chess player, a fire chief may evaluate alternative moves by considering his opponents counter moves (i.e., a prediction of the fire's likely behavior) and the resulting strategic advantages and disadvantages to identify a workable solution rather than an optimal one. Note that this approach is consistent with Simon's concept of satisficing described earlier.

The process of comparing and matching the current situation to past encounters is hypothesized to occur relatively automatically with little conscious awareness on the part of the decision-maker and without the evaluation and generation of alternatives. For example, a fire captain confronting a new fire recognizes the situation as representative of a type of fire (i.e., electrical, wood, and oil/gas) and chooses a promising course of action derived from her experience without considering multiple options. This process of assessing the current circumstances and identifying a course of action occurs very quickly while also limiting the cognitive demands of the task. The approaches of NDM and PRM have been used to study professionals in several domains including firefighting, aviation, and the navy.

BOX 7.1   DECISION-MAKING, EXPERTISE AND THE EFFECTS OF STRESS

The mishap involving U.S. Airways Flight 1549 (see Figure 7.1) that crash landed in the Hudson River is an excellent example of operators performing admirably under acute stress. The aircraft struck geese at low altitude after taking off from LaGuardia airport in New York City and lost power in both engines. Emergencies like this pose a significant challenge to decision-makers because they involve sudden events, requiring near or immediate responses whose outcomes pose a serious threat to property, life, or limb. Acute stressors, defined as "sudden, novel, intense, and of relatively short duration" (Driskell & Salas, 1991), can provoke a complex set of physiological, cognitive, and emotional responses in operators. The physiological responses can include increased heart rate, blood pressure, as well as changes in respiration, while emotional reactions may include tension, anxiety, or fear. Stressors can have both negative and positive effects on decision-making processes. Research shows that decision-makers consider fewer

Figure 7.1 Photo of Flight 1549 that landed in the Hudson River after striking birds on takeoff. (Photo from https://en.wikipedia.org/wiki/US_Airways_Flight_1549#/media/File:Plane_crash_into_Hudson_River_(crop).jpg.)

alternative solutions, evaluate a smaller set of cues, and are less systematic in evaluating alternatives when faced by time pressure (Janis, 1982). In addition, decision-makers may become less flexible, adopting a problem-solving strategy and persisting with that strategy even when it may not lead to a solution (Janis & Mann, 1977).

It also important to consider the flip side of this "coin" and acknowledge that stress can have positive effects. Stress can serve as a motivator to act, increase attentional focus, and reduce distraction from irrelevant stimuli. Gary Klein (1996), one of the founders the field of NDM, has argued that in the case of experienced decision-makers that many of the documented reactions to stress are adaptive rather than dysfunctional. In response to time constraints or a threat, decision-makers adaptively respond by narrowing their field of attention, selecting a smaller set of information sources to monitor, and employing simpler, less cognitively demanding strategies. The use of simpler strategies is adaptive because compensatory decision strategies are simply not viable when faced with goals that shift as an event unfolds, as well as when there is insufficient information and what information is available may be ambiguous.

Importantly, Klein argues that these adaptive responses make experienced decision-makers less vulnerable to the negative effects of stress because they can quickly assess a situation, identify similarities to cases they have seen in the past, and pick a course of action. These recognition processes are very rapid and require little cognitive effort. When more complex or novel situations arise, the individual can use mental simulation to evaluate alternative courses of action which could make them more susceptible to the effects of stress.

In the case of the crew of flight 1549, the captain and first officer's responses were honed by years of experience and training that exposed them to a broad range of emergency situations allowing them to practice identifying different types of emergencies, executing emergency procedures, and coordinating their responses. They had a detailed emergency checklist that spelled out each step to perform thus reducing demands on memory retrieval processes at this critical juncture.

Unlike experts, novice operators lack the range of experiences that experts can bring to bear on a problem. **Critical Thinking Question**: How do you think the effects of stress are qualitatively different for novices versus experienced operators—consider how stress might affect cognitive process including memory retrieval, working memory, etc.?

7.4 PROBLEMS IN DECISION-MAKING

As is apparent from the previous discussion, individuals employ a variety of heuristics to manage the demands of decision-making. While effective under some circumstance, these tricks may also result in systematic errors or biases.

7.4.1 Decision biases

Experimental psychologists have identified several common systematic tendencies or what are often called **biases** that derive from the use of heuristics. These biases, including confirmation bias, hindsight bias, anchoring and adjustment, and overconfidence, are discussed in the following section.

7.4.1.1 Confirmation bias

One of the strengths of human operators is their ability to identify solutions for novel situations or to troubleshoot systems when they fail. This is quite unlike automated computer systems, which cannot fix themselves and that cannot handle novel circumstances for which they were not programmed. The problem that arises when individuals attempt to solve novel problems is a bias in hypothesis generation like that which might occur when you try to identify why your computer is not booting up or your car is not starting. The bias is called the **confirmation bias** and refers to the finding that human decision-makers are more likely to entertain hypotheses that confirm a belief rather than ones that would disprove it.

Consider the following example from a classic experiment performed by Peter Wason in (1960). In the experiment, the participants were shown three numbers (e.g., 2, 4, and 6) and were asked to discover the rule that the experimenter used to generate the numbers (Wason, 1960). Can you guess the rule? Paraphrased in Table 7.2 is an example of the typical questions made by participants before arriving at a rule and the experimenter's response.

The rule governing the set of numbers 2–4–6 is "any set of ascending numbers." Wason (1960) observed that participants adopted the strategy of trying to confirm the rule (e.g., responses numbered 1, 2, and 3 in Table 7.2) rather than attempting to disconfirm the

Table 7.2 Typical responses made by participants in Watson's (1960) study

Examples of common participant responses	Experimenter's responses
1. The numbers 8, 10, 12	Yes, the numbers conform to the rule
2. The numbers 14, 16, 18	Yes, the numbers conform to the rule
3. The numbers 20, 22, 24	Yes, the numbers conform to the rule
4. The numbers 1, 3, 5	Yes, the numbers conform to the rule
5. Participant states the rule as "numbers increasing by 2"	No, that is not the answer. Please continue

rule (e.g., selecting the numbers in the order 10, 8, 6). This inclination is referred to as a confirmation bias. Participants find this task to be surprisingly difficult and typically fewer than 20% of participants identify the correct rule. The confirmation bias is not a case of an individual trying to find evidence to support a strongly held view, but rather an example of a heuristic that often works but that fails us when confronting certain kinds of problems (Klayman & Ha, 1987). The confirmation bias may slow system troubleshooting in some emergency situations where a disconfirming rule might be more effective. In such cases, educating technicians about different strategies for generating hypotheses may improve trouble-shooting performance.

7.4.1.2 Hindsight bias

If our decision-making is sometimes compromised by biases and other weaknesses, can we learn to avoid these mistakes by reviewing the errors of other individuals? For instance, it is the practice of some aviation publications to publish a summary of recent general aviation accidents and identify their probable cause with the goal that pilots might learn from them. An example of an accident summary is shown below:

Aircraft: RV-6
Location: Cloverdale, Calif.
Injuries: 1 Fatal
Aircraft damage: Destroyed

What reportedly happened: The pilot, who had approximately 1,800 hours, was attempting to land on runway 14 although the winds favored runway 32. Witnesses said the wind was blowing from the northwest at 15 knots with gusts to 25 knots, so the aircraft had significant tail-wind. The plane came down hard 10 feet short of the runway threshold and burst into flames.

Probable cause: The pilots' inadequate in-flight planning/decision and selection of the wrong runway and inadequate compensation for the gusting tail wind conditions, which resulted in a stall/mush and subsequent crash (GA News, June 23, 2006).

Upon reading the summaries, it is often difficult not to conclude that the pilot was "an accident waiting to happen" or that the mishap was inevitable. The question is whether this is a case of smugness or of hindsight being 20/20. How does knowing the outcome affect our judgment of the case?

The research literature suggests that knowing the outcome does cause people to believe that they could have predicted it, thus, presumably that they would not have suffered the same consequence. This is known as the **hindsight bias**. The following example illustrates how this type of judgment heuristic is studied (Pohl, 2004). In the experiment, participants were first asked several questions, one of which was to estimate the number of books written by Agatha Christie. The average estimate was "51." Later, participants were told the correct number, which was "67," and were then asked to recall their original estimate. The mean of the recalled estimates was "63," which is considerably closer to the correct answer than their original estimates. In these circumstances, participants are apt to convey with increased confidence that they knew the answer all along (Fischhoff, 1975).

Hindsight bias highlights the importance of seeking independent feedback when doing things such as asking for a doctor's second opinion or having your house or car appraised. Consider the case of a doctor evaluating the symptoms presented by a patient to decide whether the condition requires surgery or not. To confirm his judgment, the doctor asks a colleague to have a look. Should the doctor wait to tell the colleague his conclusion until after the colleague makes his judgment? The data suggest that it would be best to withhold the information until the colleague has a chance to make an independent report; otherwise, the doctor's information or conclusion might bias the answer. This is of particular concern if the first evaluation is dubious or just wrong.

Hindsight bias is also relevant for those investigating accidents because knowledge of the circumstances and the related outcome may make the accident appear more predictable or more avoidable and the actors more negligent. It is important to note that the relationship between the cues available at the time of the accident and the eventual outcome may only be obvious after the fact, and it is not clear that, given the same circumstances, anyone would have acted differently.

Does informing people of hindsight bias and urging them to avoid it reduce its effects? Baruch Fischhoff (1977) suggests that simply knowing about hindsight bias is not enough. Rather, she found that only when participants reflected on how the outcomes might have turned out differently was hindsight bias reduced. Through this process, participants can develop more reasonable judgments regarding the likelihood of an event taking place and the probability of a similar outcome occurring in the future.

7.4.2 Anchoring and adjustment

There are other ways in which knowing or seeing a value, even an irrelevant one, can influence one's judgment. Consider the posturing that often precedes negotiations between unions and company representatives or lawyers representing the two parties in criminal cases. The employees' union may outline new salary and benefits demands that are considerably higher than the existing benefits packages. Likewise, the prosecutor may state that she will seek a maximum 40-year sentence for the crime committed by a defendant. The intent of using such extreme values is to influence the company's and judge's or jury's decisions. The question is whether awareness of these values affects the outcome of the sentencing or the final benefits package.

Studies investigating this phenomenon called **anchoring** demonstrate that the effects are robust and that it influences many types of judgments (Mussweiler et al., 2004). The classic demonstration of anchoring was performed by Amos Tversky and Daniel Kahneman (1974) wherein participants were first asked to spin a wheel of fortune. The wheel of fortune was rigged so that the needle stopped on either the number 10 or 65. The participants were randomly assigned either to the condition where the needle stopped on 10 or to the condition where the needle stopped on 65. After spinning the wheel of fortune, the participants were then asked two questions: First, is the percentage of African countries in the United Nations greater or less than the number indicated on the wheel of fortune? Second, what is the exact percentage of African countries in the United Nations? The results of the experiment revealed that those participants who were presented the higher anchor (i.e., 65%) estimated the number of African countries in the United Nations to be roughly 45%, whereas those who received the lower anchor (i.e., 10%) estimated the number to be roughly 25%. This is true although the number (e.g., 10% and 65%) was randomly chosen. Thus, given a particular anchor, people tend to adjust their decisions, so they are closer to the given anchor. (Note: There are 195 countries in the world and 192 are represented in the UN. Africa accounts for 45 of the countries in the UN or 24%.)

The effects of anchors are observed in novices as well as experts in task relevant domains, even when participants' motivation is manipulated by offering cash rewards for accuracy. Surprisingly, anchoring behavior is observed even when participants are given an absurdly high or low anchor, such as being told that Mahatma Gandhi was 9 years old or 140 years old when he died and then later being asked to estimate his age at the time of his death (Strack & Mussweiler, 1997).

Whether information influences decisions like an anchor depends on several factors. For instance, when integrating multiple cues or pieces of evidence, decision-makers are more strongly influenced by information that is presented first than information presented later. Thus, a doctor's diagnosis may be unduly influenced by the first lab reports when attempting to generate hypotheses regarding the cause of a patient's symptoms. In addition to this **primacy effect**, decision-makers are also influenced by **cue saliency** wherein information or cues that stand out by virtue of a feature that makes them more salient may be given greater weight in decisions. These effects demonstrate that *how* information is presented can be as important as *what* information is presented in influencing the decision outcome and represents an important design consideration.

7.4.3 Overconfidence

Even though the data indicate that individuals make errors in decision-making and judgments, we are often quite sure of our decisions and judgments. **Overconfidence** represents a discrepancy between one's perceived and actual ability and skill. Let's consider the question, "How would you rate your driving skills: average or above average?" Often, people respond that they are an above-average driver. Obviously, it is not possible for most people to be "above average?" Although most of us have confidence in our abilities, our estimate of our abilities or our confidence in our judgments is often wrong. The consequences of overconfidence are well documented by historians and by recent events including the Challenger and the Columbia space shuttle disasters. Prior to the Challenger space shuttle catastrophe, NASA estimated the risk of a catastrophic accident to be 1 in 100,000 launches (Feynman, 1988, February). In retrospect, it seems extraordinary that anyone took such estimates seriously in light of the fact that this would be equivalent to launching a Shuttle a day and experiencing one accident in 300 years (Plous, 1993).

NASA later revised calculations and estimated the risk to be 1 in 100 (CNN, 2/07/2006), a 10,000-fold increase in crash risk. Studies of decision-makers indicate that NASA's overconfidence is typical of judgments made by novices and experts, although there is some evidence that experts make more conservative evaluations of the accuracy of their decisions or predictions (Levenberg, 1975).

Because overconfidence can possibly lead to poor decisions, it is important to determine when an individual's tendency for overconfidence can be tempered. Fortunately, overconfidence can be reduced through training and providing feedback about the accuracy of decisions. There is experimental evidence (Camerer & Johnson, 1991) that professional bridge players and weather forecasters—who receive prompt feedback regarding the quality of their judgments—show little or no overconfidence. In contrast, doctor evaluations of the accuracy of their diagnoses of pneumonia and skull fractures can be off the mark (Camerer & Johnson, 1991).

7.5 DECISION QUALITY AND STRATEGY

The earlier discussion of research findings indicating that decision-makers rely on various shortcuts and that judgments show evidence of bias has possibly left you with a sobering impression of human decision-making. After all, the use of less systematic decision strategies like heuristics by individuals might be expected to result in less satisfactory decisions. This, however, is not always the case. Payne and his colleagues (1988) used a computer to simulate both normative and non-normative decision processes under different time constraints. If, at the end of the time period, the computer had not converged on a satisfactory choice, the program was designed to choose randomly among the remaining options.

Comparisons of the decision choices of Payne et al.'s computer program indicated that, under conditions of time pressure, non-normative strategies (e.g., satisficing, elimination by aspects) produce decisions that are comparable or better than those produced by normative decision processes. One explanation for this finding is that heuristics allow decision-makers to winnow poorer choices relatively quickly. In contrast, normative processes are slower. Consequently, the simulation ended up picking randomly from a larger set of alternatives, including some non-optimal choices. Payne's findings also demonstrated that the most efficient decision heuristic varied depending on time constraints and differences in the probabilities of different outcomes. Recent evidence also indicates that, in some circumstances, deliberate consideration of options reduces the quality of some decisions (See Box 7.2).

BOX 7.2   TOO MUCH THINKING CAN RESULT IN BAD DECISIONS

A group of Dutch researchers found that too much thinking about which car to buy can result in a poor choice, especially if the choice is complicated. Dijksterhuis, Bos, Nordgren, and van Baaren (2006) had 80 participants select one of four cars. Participants were provided either a short (4) or long (12) list of attributes describing each car (leg room, age, gas mileage, transmission type, etc.) and asked to select the best of the four cars. The attributes describing each of the four cars were carefully chosen such that, for one car, most of the attributes (75%) were positive (the optimal choice); for two cars, 50% were positive attributes; and, for one car, 25% were positive. After reading the attributes, the participants were instructed that they would be asked to select a car in 4 minutes. For the next 4 minutes, half of the participants were instructed to sit and think about their choice, whereas the other half was distracted by having to solve anagrams. When the number of attributes was small (i.e., 4), participants were more likely to select the better vehicle after conscious deliberation. In contrast, when there were more attributes to consider, participants were more likely to select the better vehicle if they had been distracted for 4 minutes. It seems commonsensical that evaluating more attributes should take longer but why should this process result in a poorer choice rather than a better one?

The authors proposed two reasons why thinking about alternatives can result in poor choices. The first is that conscious processes have low capacity due to limitations associated with cognitive processes like working memory (see Chapter 6) that allow the decision-maker to evaluate only a subset of the relevant information. The second is that conscious processes can result in the inappropriate weight assignment of some cues and neglect of other important cues. These results suggest that, when confronted with difficult decisions, it may be a good idea to first evaluate one's choices and then sleep on it before deciding. **Critical Thinking Question**: Usually, we think of distraction as negative and detrimental to performance. How might distraction be strategically employed to enhance performance?

7.6 EXPERT VERSUS NOVICE DECISION-MAKERS

Throughout this discussion on decision-making, differences between expert and novice decision-makers have often been noted. Because there are some striking differences between experts and novices, we will consider these differences in a bit more detail here.

We often solicit the opinions of "experts" when confronted with difficult problems or when seeking confirmation of decisions. **Experts** are believed to possess skills that allow them to identify correct, more efficient, or alternative solutions quickly and reliably. Studies of experts in other domains, including chess and medicine, have identified several skills that distinguish experts from novices. For instance, expert chess players can consider more options and their potential outcomes (Charness, 1981) than novices, and they can remember the positions of chess pieces even after a brief view of the board (Chase & Ericsson, 1982; de Groot, 1978). Randomly positioning the pieces, though, nullifies the experts' advantage, indicating that experts were recognizing patterns in the positions of the chess pieces that facilitated later recall. That is, rather than memorize the position of each chess piece, experts can relate the pattern of chess pieces to familiar moves or strategies they can easily reproduce.

Experts' memories are related to their more efficient encoding of information in long-term memory (LTM), which circumvents capacity constraints at earlier stages (i.e., short-term memory) of information processing (Chase & Ericsson, 1982), as discussed in Chapter 6. Avoiding short-term memory (STM) capacity limitations enables experts to resume a task after interruptions from unrelated activity, which would otherwise interfere with the maintenance of information in STM (Charness, 1991). This may be the basis of expert pilots' ability to handle multiple task demands effectively.

K. Anders Ericsson and Neil Charness (1994) cite evidence that experts store and index information in LTM in a qualitatively different manner than novices, allowing experts to readily recognize patterns by comparing the present circumstances to exemplars in memory. This skill could be vital when time-consuming deliberation of options is impossible and makes performance less susceptible to disruption by stress or other tasks. Like expert chess players, experienced pilots exhibit similar memory skills that allow them to store flight-related information without decay (Endsley, 1995). Interestingly, the improved memory skills of experts appear to be domain-specific, and they do not generalize to other tasks (Ericsson & Charness, 1994). That is, an expert chess player may be mediocre at recalling information from other board games, which reduces their ability to make good quick judgments in these situations.

Experts' ability to select only task-relevant information facilitates their management of multiple demands because they do not burden memory processes with less important information. For example, medical experts outperform medical students in identifying and recalling important pieces of presented information (Boshuizen & Schmidt, 1982; Chase & Ericsson, 1982), whereas medical students could recall more information in general. Studies of pilots corroborate these findings. Other researchers (Beck as cited in Wiggins & O'Hare, 2003; Rockwell & McCoy, 1988) have found that experienced pilots were more efficient in acquiring and evaluating details of weather-related information. These and other studies suggest that novices often lack an understanding of the utility of different cues (Schvaneveldt et al., 2000; Wiggins & O'Hare, 2003).

Finally, experts employ different metacognitive skills in monitoring their own thinking processes. **Metacognition** refers to individuals' monitoring, controlling, and organizing of their own cognitive processes. Experts are more apt to check their progress toward a goal, evaluate their accuracy, make decisions about the most effective use of their time and mental effort, and search prior experience to find instances of situations like the current

BOX 7.3   UNSKILLED AND UNAWARE OF IT

The title of this box is taken from a similarly titled article (Kruger & Dunning, 1999) reporting the result of research on people's perception of their competence and how this influenced their ability to reach good decisions and evaluate their performance. They had participants estimate the percentile ranking of their logical reasoning ability compared to other students in their psychology class and then had the same students complete a logical reasoning test. After the test, the students were also asked to estimate how many questions they believed they had gotten right. Kruger and Dunning found that students that performed in the lower 12th percentile (i.e., 88% of the students performed better than they) believed their reasoning ability was above average and estimated it to be in the 68th percentile. They also believed they had answered more problems correctly than they really had. Interestingly, students performing in the top quarter tended to underestimate their ability and actual performance by a smaller margin. These findings are very important because they suggest that inexperienced individuals are not only less competent and thus perform poorly, but they also lack the ability to recognize how bad they are! **Critical Thinking Questions**: Can you identify other ways in which technology and feedback might be used to improve novices' understanding of their abilities thereby improving their decisions? Consider the implications of these finding for the flying scenario described at the start of this chapter.

one (Cohen et al., 2000; Halpern, 1998). They can refine their understanding and knowledge through active learning strategies. Individuals become better thinkers and learners by developing the habit of monitoring their understanding and judging the quality of their learning.

The findings described in Box 7.3 demonstrate that experience benefits the individual in several ways that can improve their decision-making skills. As a result of experience, decision-makers not only derive greater knowledge of the types of problems they might confront and strategies for handling them, but they are also better able to recognize situations, match them to prior experiences, and evaluate their own performance. What specific kinds of experience do you think might benefit the novice pilot the most? Why?

7.6.1 Limitations of experts

It is important to acknowledge that there are limitations to expertise. Experts are not infallible and often fall short of optimal performance (Shanteau, 1992). Comparisons of the performance of novices and experts, where an expert was defined as "a person who is experienced in making predictions in a domain and has some professional or social credentials" (Camerer & Johnson, 1991, p. 196), show experts were better at generating hypotheses and developing sophisticated decision rules and guided by their superior knowledge, and were more effective in searching for information. However, the benefits of these superior skills were typically modest. That is, experts usually outperformed novices (but not always), and experts were often outperformed by simple statistical models. One explanation for the modest advantage of experts was that they were inconsistent in their weighting of different cues and made errors in adding them together (Camerer & Johnson, 1991).

BOX 7.4   IF A LITTLE INFORMATION IS GOOD, IS MORE INFORMATION BETTER?

Modern information technologies such as the web provide us with access to unprecedented amounts of information. Does the ready access to ever greater amounts of information improve the quality of our decision-making? Consider the case of a potential investor looking to invest in a company. The web offers timely access to a plethora of information, including regulatory information filed by publicly held companies, information on corporate websites, company press statements, annual reports, news stories, and analyst research reports, just to name a few. But is the quality of an investors' stock selection improved by reviewing more information? In other words, if a little information is good, is more information better? The answer to this question is—probably not.

Camerer and Johnson (1991), in their review of decision-making by experts, discuss several issues that bear out this point. They cite research findings showing that, although experts readily identify task relevant cues, they often employ simplistic models of the underlying system or processes. Namely, experts may not differentially weigh the cues and might treat the cues as if they were equally reliable, even though some cues are more predictive than others and some cues are reliable predictors only in certain outcomes. Also, when experts assign cues of differential importance (i.e., weights), they do so inconsistently and are prone to errors when combining them. Consulting additional sources for cues could have the negative consequence of increasing the complexity of integrating the information. In fact, experts typically rely on less—not more—information, although they still have difficulty with the problems cited above. **Critical Thinking Questions**: Do you think that more information would alter the decision of the pilot described in the opening scenario of this chapter? What situational factors do you think conspire against a more conservative choice in this case?

James Shanteau (1992) also analyzed the performance of experts and identified some of the characteristics of the tasks that were related to reliable and unreliable expert performance. Shanteau noted that tasks that require decisions about things—where stimuli and cues are static, and decision-makers receive feedback regarding the accuracy of their judgments—tend to yield better performance. Some of the types of experts who performed reliably better were weather forecasters, accountants, and chess masters. Table 7.3 lists examples of task conditions supporting better performance and fields where experts demonstrate good and poor performance.

7.7 IMPROVING DECISION-MAKING

7.7.1 Training

Although studies have demonstrated that novices can acquire some of the sophisticated cognitive abilities and skills of experts, to date, no one has developed training strategies that support the rapid development of these abilities. Typically, extended practice is necessary to approach performance levels comparable to that of experts in restricted domains. Achieving the level of competency of an expert may take years of experience.

Table 7.3 Task properties with good (left side) and poor (right side) expert performance

Task properties associated with good or poor performance	
Good expert performance	Poor expert performance
Static stimuli	Dynamic (changeable) stimuli
Decisions about things	Decisions regarding behavior
Experts agree on stimuli	Experts disagree on stimuli
Predictable problems	Unpredictable problems
Tasks tend to be repetitive	Tasks are unique
Feedback available	Feedback unavailable
Objective analysis available	Subjective analysis only
Fields with good or poor performance	
Good expert performance	Poor expert performance
Weather forecasters	Clinical psychologists
Accountants	Polygraph administrators
Astronomers	Personnel selectors
Test pilots	Student admissions personnel
Physicists	Court judges
Chess masters	Behavioral researchers

Source: Adapted from Shanteau (1992).

7.7.2 Decision aids

One way to improve human decision-making is to provide support in the form of decision aids such as simple protocols or computer-generated lists of options. **Decision aids** are tools that help the human decision-maker deliberate upon two or more options (Bekker et al., 2003). Protocols may consist of an enumerated list of options to consider when purchasing a car (e.g., price, gas mileage, and financing) or a decision tree, like that shown in Figure 7.2, which provides a guide to selecting an appropriate statistical test for an experiment. Alternatively, computer-based decision aids may implement sophisticated algorithms to generate a list of options based on inputs that are automatically sampled by the computer or based on user responses to computer queries. Decision aids are widely used in the medical profession. One database (http://decisionaid.ohri.ca) alone lists over 400 medical decision aids, covering everything from aids for selecting among treatments for breast cancer, choosing whether to have cataract surgery, and deciding whether to have hair loss treated medically.

Medical professionals use decision aids in a variety of roles including the monitoring of patient information and the critiquing of medical decisions. In the monitoring role, a medical decision aid may monitor test results and alert the doctor of abnormal lab results. **Critiquing systems** have been employed to confirm a doctor's diagnosis and treatment plan by reviewing the pertinent information including symptoms and test results stored in a computer database. The decision aid alerts the doctor of drug–drug interactions, drug allergies, or alternative treatments or diagnoses. One advantage of decision aids is that they do not suffer from the biases (e.g., confirmation and availability) and limitations of human cognitive processing that affect which hypotheses we entertain, how many options we consider, and what information we monitor.

7.7.2.1 Bias in use of decision aids

Even though decision aids help eliminate or reduce some biases, they can give rise to others. In aviation, automated decision aids have been developed that recommend flight plans that minimize fuel consumption and flight time (Lauber & Foushee, 1981) and that generate

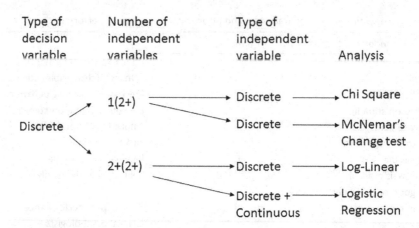

Figure 7.2 An example of a decision tree used to select an appropriate statistical test to analyze data. (Adapted from Tabachnick & Tidell, 2001.)

flight trajectories to landing sites during in-flight emergencies (Chen & Pritchett, 2001). The appropriate use of these aids tends to improve performance and results in fewer errors than when human operators work without the aid. Yet, research has demonstrated that operators can develop an **automation bias**, or tendency to disregard or fail to search for contradictory information when a computer aid generates options assumed to be correct (Parasuraman & Riley, 1997). Thus, the operator can become complacent, over-reliant on the aids, and more likely to defer to the choices or options provided by the aids even when they are incorrect.

An example of an incident where automation bias played a role was the mistaken downing of an American F/A-18 and a British Tornado aircraft by an American Patriot missile battery during operation Iraqi Freedom, which killed three crew members (Defense Science Board Task Force, 2005). In this instance, the operators of the missile battery were given 10 seconds to veto the identification of an aircraft by the missile system computers as a friend or foe.

Operators were trained to trust the system even though the missile system software, which identified an aircraft as a friend or foe, was known to perform very poorly. Confusing displays and insufficient training in the complex system further compromised operator performance. This example highlights one of the challenges of training users to use sophisticated decision aids, namely, how to instill the users with a sense of caution and awareness that these systems sometimes are in error even though they may be correct most of the time.

7.8 APPLICATIONS TO SPECIAL POPULATIONS

Some elderly face significant challenges making decisions in everyday circumstances like deciding among alternative medical treats for medical conditions that are more common among older adults. Even the most capable among us can be overwhelmed by an avalanche of facts, statistics, potential complications, testimonials, and medical jargon. Nevertheless, as this book makes clear an understanding of decision-making, information presentation, and cognitive abilities and limitations (i.e., wording, visual design) can be used to guide the design of information design and delivery (paper, computer-based decision aid) in ways that are more meaningful and readily interpretable. Improving decision-making through better information presentation benefits all users, not just older adults.

Medication adherence is a significant issue with other adults who are often prescribed multiple types of medication with different dosing schedules. Research suggests that medication adherence (i.e., remembering to taking medication at the time) may increase when older adults are provided explicit, well-organized instructions and external aids including organizational charts detailing which medications to take and when, or a plastic medical organizer with individuals bins containing the medications to take in the morning, noon, evening, or night of each day (Park et al., 1991, 1992, 1994). Importantly, the design of information displays, memory aids, and organizational charts that support decision-making and compensate for some of the changes in working memory processes and resources more common among older users stand to benefit all users regardless of age.

7.9 SUMMARY

In this chapter, we focused on models of decision-making and how decision-making under real-world conditions often differs from the models. The human decision-maker is flexible using different strategies depending on task and situational demands. For instance, decisions-makers employ a variety of heuristics that reduce the demands of choosing among alternatives, but which can result in poor choices when applied in some situations.

We also discussed the decision-making processes of experts, how they differ from novices, and how decision aids can improve the quality of human decisions. These considerations are central to human factors evaluations and recommendations for the design of crew stations. Our understanding of decision-making helps us determine what and how much information to display, the best format of the display to facilitate understanding, and which design of aids to use to improve the quality of individuals' final decisions.

LIST OF KEY TERMS

Adjustment

Anchoring

Availability heuristic

Automation bias

Bias

Compensatory decision-making

Confirmation bias

Critiquing

Cues

Cue primacy

Cue saliency

Decision aids

Decision-making styles

Descriptive decision models

Elimination by aspects

Expected value theory

Expert

Hindsight bias

Heuristics

Mental simulation

Metacognitive skills

Monitoring

Naturalistic decision-making

Normative decision models

Non-compensatory decision-making

Overconfidence

Recognition primed model

Satisficing

Subjective expected utility (SEU)

Utility

SUGGESTED READINGS

Kruger, J., & Dunning, D. (1999). Unskilled and unaware of it: how difficulties in recognizing one's own incompetence lead to inflated self-assessments. *Journal of Personality and Social Psychology*, 77, 1121–1134.

The paper offers a very readable scientific investigation of the differences between novices and experts and the discrepancy between their assessment of their abilities and their performance.

Ericsson, K. A., & Charness, N. (1994). Expert performance: its structure and acquisition. *American Psychologist*, 49(8), 725–747.

The paper offers a detailed overview of how experts differ from novices and the behaviors and experiences that allowed them to become experts.

Shanteau, J. (1992). The psychology of experts: an alternative view. In G. Wright & F. Bolger (Eds.), *Expertise and Decision Support*. New York: Plenum Press.

This is a seminal article in the area of expertise and provides a description of the task conditions supporting better performance and fields where experts demonstrate good and poor performance.

CHAPTER EXERCISES

1. Describe the difference between normative and non-normative decision-making.
2. Given the method shown for calculating expected value shown on page 7, calculate what the monetary value of winning would need to be such that expected winnings and losses would balance out (i.e., equal 0).
3. Explain the difference between compensatory and non-compensatory methods of evaluating the attributes of decision alternatives.
4. Describe how expert decision-makers differ from novices. How do experts reduce the effects of workload on decision-making?
5. What task characteristics are related to reliable differences between expert and non-experts in decision-making?
6. What are decision aids and how can they improve human decision-making? Give an example.
7. What is automation bias?

REFERENCES

Anderson, J. R., Reder, L. M., & Simon, H. A. (2000). Applications and misapplications of cognitive psychology to mathematics education. *Texas Educational Review*.

Bekker, H. L., Hewison, J., & Thornton, J. G. (2003). Understanding why decision aids work: linking process with outcome. *Patient Education and Counseling*, 50, 323–329.

Boshuizen, H. P. A., & Schmidt, H. G. (1982). On the role of biomedical knowledge in clinical reasoning by experts, intermediates, and novices. *Cognitive Psychology*, 16, 153–184.

Camerer, C. F., & Johnson, E. J. (1991). The process-performance paradox in expert judgment: how can experts know so much and predict so badly? In K. A. Ericsson & J. Smith (Eds.), *Towards a General Theory of Expertise: Prospects and Limits*, pp. 195–217. Cambridge, UK: Cambridge University Press.

Charness, N. (1981). Search in chess: age and skill differences. *Journal of Experimental Psychology: Human Perception and Performance*, 7, 467–476.

Charness, N. (1991). Expertise in chess: the balance between knowledge and search. In K. A. Ericsson & J. Smith (Eds.), *Toward a General Theory of Expertise: Prospects and Limits*, pp. 39–63. Cambridge, UK: Cambridge University Press.

Chase, W. G., & Ericsson, K. A. (1982). Skill and working memory. In G. H. Bower (Ed.), *The Psychology of Learning and Motivation*, Vol. 16, pp. 1–58. New York: Academic Press.

Chen, T. L., & Pritchett, A. R. (2001). Development and evaluation of a cockpit decision-aid for emergency trajectory generation. *Journal of Aircraft, 38*, 935–943.

Cohen, M. S., Thompson, B. B., Adelman, L., Bresnick, T. A., & Shastri, L. (2000). *Training critical thinking for the battlefield. Volume II: Training system and evaluation* (No. 00–2). Ft. Leavenworth, KS: U.S. Army Research Institute.

de Groot, A. (1978). Thought and Choice and Chess. The Hague, The Netherlands: Mouton.

Death Odds. (September 24, 1990). *Newsweek, 10.*

Defense Science Board Task Force (2005). *Patriot System Performance: Report Summary.* Washington, DC: Office of the Under Secretary of Defense for Acquisition, Technology, and Logistics.

Dijksterhuis, A., Bos, M. W., Nordgren, L. F., & van Baaren, R. B. (2006). On making the right choice: the deliberation-without attention effect. *Science, 311*(5763), 1005–1007.

Doyle, A. C. (Ed.). (1987). *The Sign of the Four.* Cutchogue, NY: Buccaneer Books.

Endsley, M. R. (1995). Measurement of situation awareness in dynamic systems. *Human Factors, 38*(1), 65–84.

Ericsson, K. A., & Charness, N. (1994). Expert performance: its structure and acquisition. *American Psychologist, 49*(8), 725–747.

Feynman, R. P. (1988, February). An outsider's inside view of the challenger inquiry. Physics Today, 26–37.

Fischhoff, B. (1975). Hindsight foresight: the effect of outcome knowledge on judgement under uncertainty. *Journal of Experimental Psychology: Human Perception & Performance, 1*, 288–299.

Fischhoff, B. (1977). Perceived informativeness of facts. *Journal of Experimental Psychology: Human Perception & Performance, 3*, 349–358.

G. A. News, (June 23, 2006). Choice of wrong runway fatal for California pilot, p. 41.

Halpern, D. F. (1998). Teaching critical thinking for transfer across domains. *American Psychologist, 53*(4), 449–455.

Klayman, J., & Ha, Y. (1987). Confirmation, disconfirmation, and information hypothesis testing. *Psychological Review, 94*, 211–228.

Klein, G. (1998). *Sources of Power: How People Make Decision.* Cambridge, MA: Massachusetts Institute of Technology.

Klein, G., Calderwood, R., & Clinton-Cirocco, A. (1996). Rapid decision making on the fire ground. *Proceedings of the Human Factors Society-30th Annual Meeting, 30*(6), 576–580.

Klein, G., Orasanu, J., Calderwood, R., & Zsambok, C. E. (Eds.). (1993). *Decision Making in Action: Models and Methods.* Norwood, NJ: Ablex.

Kruger, J., & Dunning, D. (1999). Unskilled and unaware of it: how difficulties in recognizing one's own incompetence lead to inflated self-assessments. *Journal of Personality and Social Psychology, 77*, 1121–1134.

Janis, I. L. (1982). Decision making under stress. In L. Goldberg, & S. Breznits, (Eds.), *Handbook of Stress: Theoretical and Clinical Aspects*, pp. 69–87. New York: Free Press.

Janis, I. L., & Mann, L. (1977). *Decision Making: A Psychological Analysis of Conflict, Choice and Commitment.* New York: Free Press.

Lauber, J. K., & Foushee, H. C. (1981). Guidelines for line-oriented flight training Volume 1. *Paper presented at the NASA/Industry Workshop*, Moffet Field, CA.

Levenberg, S. B. (1975). Professional training, psychodiagnostic skill, and kinetic family drawings. *Journal of Personality Assessment, 39*, 389–393.

Mussweiler, T., Englich, B., & Strack, F. (2004). Anchoring effect. In R. F. Pohl (Ed.), *Cognitive Illusions: A Handbook of Fallacies and Biases in Thinking, Judgment and Memory.* New York: Psychology Press.

National Transportation Safety Board. (2006). *Special Investigation Report on Emergency Medical Services Operations.* Special Investigation Report NTSB/SIR-06/01. Washington, DC.

Parasuraman, R., & Riley, V. (1997). Use, misuse, disuse, abuse. *Human Factors, 39*, 230–253.

Park, D. C., Morrell, R. W., Frieske, D., Blackburn, A. B., & Birchmore, D. (1991). Cognitive factors and the use of over-the-counter medication organizers by arthritis patients. *Human Factors, 33*(1), 57–67.

Park, D. C., Morrell, R. W., Frieske, D., & Kincaid, D. (1992). Medication adherence behaviors in older adults: effects of external cognitive supports. *Psychology and Aging, 7*(2), 252–256.

Park, D. C., Willis, S. L., Morrow, D., Diehl, M., & Gaines, C. L. (1994). Cognitive function and medication usage in older adults. *Journal of Applied Gerontology, 13*(1), 39–57.

Payne, J. W. (1976). Task complexity and contingent processing in decision making. An information search and protocol analysis. *Organization Behavior and Human Performance, 16*, 366–387.

Payne, J. W., Bettman, J. R., & Johnson, E. J. (1988). Adaptive strategy selection in decision making. *Journal of Experimental Psychology: Learning, Memory and Cognition, 14*(3), 534–552.

Payne, J. W., & Braunstein, M. L. (1978). Risky choice: an examination of information acquisition behavior. *Memory and Cognition, 6*, 554–561.

Plous, S. (1993). *The Psychology of Judgment and Decision Making*. Philadelphia, PA, McGraw-Hill.

Pohl, R. F. (2004). Hindsight bias. In R. F. Pohl (Ed.), *Cognitive Illusions: A Handbook on Fallacies and Biases in Thinking, Judgment and Memory*. pp. 293–308. New York: Psychology Press.

Rockwell, T. H., & McCoy, C. E. (1988). *General Aviation Pilot Error: A Study of Pilot Strategies in Computer Simulated Adverse Weather Scenarios*. Cambridge, MA: United States Department of Transportation.

Savage, L. T. (1954). *The Foundations of Statistics*. New York: Wiley

Schvaneveldt, R., Beringer, D. B., & Lamonica, J. (2000). Priority and organization of information accessed by pilots in various phases of flight. *The International Journal of Aviation Psychology, 11*(3), 253–280.

Shanteau, J. (1992). The psychology of experts: an alternative view. In G. Wright & F. Bolger (Eds.), *Expertise and Decision Support*. pp. 11–23. New York: Plenum Press.

"Shuttle Crew faced 1-in-100 chance of dying." Retrieved June 27, 2006, from www.cnn.com/2006/TECH/space/06/27/shuttle.risk.ap/index.html.

Strack, F., & Mussweiler, T. (1997). Explaining the enigmatic anchoring effect: mechanisms of selective accessibility. *Journal of Personality and Social Psychology, 73*, 437–446.

Tabachnick, B. G., & Fidell, L., S. (2001). *Computer Assisted Research and Design and Analysis*. Needham Heights, MA: Allyn & Bacon.

Tversky, A. (1972). Elimination by aspects: a theory of choice. *Psychological Review, 79*, 281–299.

Tversky, A., & Kahneman, D. (1974). Judgment under uncertainty: heuristics and biases. *Science, 185*, 1124–1131.

Wason, P. C. (1960). On the failure to eliminate hypotheses in a conceptual task. *Quarterly Journal of Experimental Psychology, 12*, 129–140.

Weigmann, D. A., & Goh, J. (2000). *Visual Flight Rules (VFR) Flight into Adverse Weather: An Empirical Investigation of Factors Affecting Pilot Decision Making*. Washington, DC: Federal Aviation Administration

Weigmann, D. A., & Shappell, S. A. (1997). Human factors analysis of post accident data. *International Journal of Aviation Psychology, 7*, 67–82.

Wiggins, M. W., & O'Hare, D. (2003). Expert and novice pilot perceptions of static in-flight images of weather. *International Journal of Aviation Psychology, 13*(2), 173–187.

Zsambok, C. E., & Klein, G. (Eds.). (1997). *Naturalistic Decision Making*. Hillsdale, NJ: Lawrence Erlbaum Associates.

Chapter 8

Motor skills and control

Chapter Vignette

Antonia is playing her favorite mobile video game. In this game, colorful balloons fall from the top of the screen at random intervals and locations, and Antonia has to "pop" them by tapping them with her fingertip. The smaller and faster balloons are worth more points than the larger or slower ones. If she lets too many balloons fall to the bottom of the screen, she will lose the game—yet as the game progresses, the balloons get smaller and move faster. How far can she go before missing too many and losing? Today, she is going for a new high score and needs one last balloon, a very small blue one that is racing to the bottom of the screen. Can she pop it in time?

Antonia's ability to pop the balloon in time depends on several factors. In part, it depends on her ability to see the balloon clearly, despite her finger hovering over the touchscreen of her tablet. But it mainly depends on her ability to quickly and accurately locate the balloon on screen. This is the human factor: people have a limited capacity to move quickly—and sometimes when we do move quickly, we can be prone to error. Antonia is faced with a situation where both speed and accuracy of response are of utmost importance. In other tasks, sometimes only speed is important, in others, perhaps only accuracy is needed. In this chapter, we will see how understanding human limits on speed and accuracy can be used in the design of games, other forms of computer software, and the physical controls used in industry, military, and consumer technologies. Even in the more mundane tasks that Antonia performs on her tablet, such as tapping on the icon to open her game or swiping upward to view more content in her social media feed, we can examine human factors. For example, how does she know that in order to view social media content that is displayed at the bottom of the screen, she should swipe upward to bring it better into view? Is this mapping of the movement of the control (her finger) to the movement of the text and images on screen the best mapping? Ultimately, the design goal is to maximize performance and minimize error. While scrolling through social media may not seem like a performance-critical situation, a confused user is surely an annoyed user who may abandon a technology. But there are many other time-critical and accuracy-critical tasks that crucially depend on designing with human movement skills in mind.

8.1 CHAPTER OBJECTIVES

After reading this chapter, you should be able to:

- Describe aspects of control design and its effects on performance.

DOI: 10.1201/9781003515463-8

- Describe aspects of control layout and its effects on performance.
- Describe the factors that influence the speed of a motor response.
- Describe the factors that influence the accuracy of a motor response.
- Outline the theoretical perspectives on motor functioning.
- Outline the tracking loop and the role of control order in tracking.
- Understand the role of development and aging in motor functioning.

8.2 WHAT ARE MOTOR SKILLS AND CONTROL?

Human factors specialists must consider many aspects of human behavior in the design and analysis of a person–technology interface. The list can be long and involve the visual and attentional capabilities of the person, as well as their decision-making tendencies. Yet, there is one aspect of human behavior so ubiquitous that it is easy to forget but must be considered in nearly every application of person-centered design. This aspect is movement.

We have to move in order to operate controls. We need to be able to select the right control—sometimes there are many controls at our disposal, such as when driving a car. There are numerous stories in the news about individuals who pressed the gas pedal instead of the brake and rear-ended someone or drove into a storefront! This demonstrates that there are numerous limitations in our ability to move quickly and accurately in certain circumstances. Before we can address issues related to the manual operation of controls, we must first understand basic principles of human motor control and motor skill. These principles will involve constraints on the human ability to move quickly and accurately, as well as our theoretical understanding of how skill and coordinated movement are achieved.

8.3 RAPIDITY OF CONTROL

Recall the situation confronted by Antonia, who needs to tap on small moving targets in her video game. The question is whether she will be able to do this quickly enough to avoid losing. Several aspects of the movement determine the rapidity of her response. The first is **reaction time**, which corresponds to the time between when a stimulus is first noticed and when a response to that stimulus is initiated. For Antonia, this is the time between when she first notices a target appear at the top of the screen, and the time that she initiates her hand movement toward it. The second is the speed of the movement itself, or movement time, which is the time it takes for her finger to travel from its current location to the target. As we will see, there are a number of variables that can influence both reaction and movement time.

8.3.1 Reaction time

More than 100 years ago, Franciscus Donders (1869/1969) began the first studies into the measurement of the speed of human reactions. He used a clever combination of experimental conditions to tease apart the durations of multiple phases of the response to a stimulus. Consider the complexity of this process. First, we notice the presence of a stimulus using one of our sensory systems (in the case of Antonia, the relevant sense is vision). Second, we have to decide whether a response is required, and what that response should be (again, in Antonia's case, the response would be to tap the target). Finally, we have to initiate the response by sending a signal from the brain to the appropriate muscles.

In Donders' first experiment, he presented observers with a stimulus and asked them to press a response key as quickly as they could once they noticed the stimulus. This task was a highly simplified version of most response tasks—the observers did not have to decide whether a response is required, or choose which response to make, they merely had to notice the stimulus and respond as quickly as they could. The latency, or delay, in making the response after the presentation of the stimulus was thus labeled **simple reaction time.** Several studies have determined that human simple reaction time is on the order of 200 ms, or 0.200 seconds (Keele, 1986).

Next, Donders added a level of complexity to the experiment by presenting one of two possible stimuli. Observers were instructed to respond with a key press only if one of the stimuli appeared and to make no response to the other. This task required them to make a decision on whether to respond, and the subsequent response latency was longer than the previously measured simple reaction time. By subtracting the simple reaction time from this new response time, Donders claimed to have measured the duration of the decision-making process, or **decision time.**

The final step was to add a second possible response. In this version of the task, one of two possible stimuli could appear on any given trial, as in the decision time experiment, but now observers were instructed to vary their response according to which stimulus was presented. This task was slightly more complex than the decision task because it required a response selection in addition to a decision. After measuring the resultant response latency and subtracting both decision time and simple reaction time from it, Donders claimed to have measured the duration of the **response selection** process.

Returning to Antonia, if we assume there is only one type, say blue balloon, in the game, then her reaction need not entail a response selection phase. However, what if the game designers added in special balloons that have negative effects—perhaps green balloons that, if you pop them, make all the blue balloons start dropping faster, or maybe a red balloon that if popped causes her to instantly lose. Antonia's ability to pop just the blue balloons will now be slower, because her response must include the extra decision time, to tap or not to tap. What if the game designers also added in an alien ship that randomly appears and shoots at the blue balloons, which also makes them fall faster—but there is also a response to this event; if Antonia performs a swiping motion on the alien ship, she can swipe if off the screen. At this point, her reaction time to blue balloons is even longer, because in addition to the decision time, she must also select between tapping and swiping motions, and her reaction time will include this extra response selection phase.

Because the primary factor that determines simple reaction time is neural transmission speed, then one might expect it to be fixed; however, it turns out that a variety of factors can affect it. Researchers have found that simple reaction time depends on the intensity of the stimulus, with stimuli near threshold (i.e., faint stimuli that are difficult to discern) exhibiting longer reaction times than stimuli above threshold. Once the stimulus is sufficiently visible though, little is gained by increasing the intensity. For example, if the brightness settings on Antonia's tablet display are low, or if there is glare from an overhead lamp, she may have difficulty seeing the targets.

8.3.1.1 Hick–Hyman law

Recall that at various points in the game, Antonia may have multiple balloons on screen, maybe a single blue balloon target and multiple red and green non-targets. An interesting finding regarding the number of possible stimuli is that total reaction time increases with the logarithm of the number of choices, known as the **Hick–Hyman law** (Hick, 1952; Hyman, 1953).

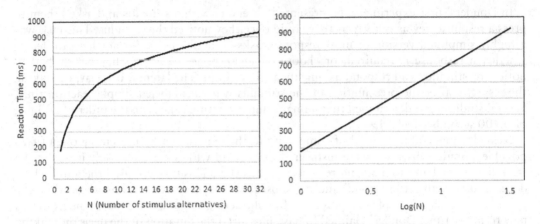

Figure 8.1 The Hick–Hyman law. Starting with an average simple reaction time of 180 ms when only one target is present, each doubling of the number of targets results in a 150 ms increase in reaction time, resulting in a logarithmic curve as shown on the left. Taking the log of N results in a linear function, as shown on the right. (© Daniel S. McConnell.)

Starting with a simple reaction time of about 180 ms, Hick and Hyman independently found that with every doubling in the number of stimuli, total reaction time increased by a fixed amount (about 150 ms; see also Keele, 1986). The logarithmic relation between the number of stimuli and reaction time is illustrated in Figure 8.1.

8.3.2 Movement time

After Antonia has noticed a blue balloon and initiated a tapping action, we next need to consider what happens during the movement of her hand to the target. How rapidly can she propel her fingertip toward the balloon? The speed of her movement, in conjunction with the distance she has to travel, determines her **movement time**.

8.3.2.1 Speed/accuracy tradeoff

The primary factor that constrains the speed of reaching and pointing movements such as those required by Antonia is the accuracy required of the movement. We need to consider how much room there is for error in hitting the target, as we generally become less accurate the faster we move. This is entirely the challenge in Antonia's game, as it progresses to higher levels, the targets get smaller, but they also drop faster so she has to make faster movements which are also less accurate. While the designers of Antonia's balloon game have installed this challenge on purpose, in many other contexts, the goal may be to minimize difficulty and increase performance. For example, in an industrial setting where a critical incident has occurred and a worker needs to hit an emergency kill switch button to shut down a dangerous, over-heating machine, there is nothing to be gained by making this button very small or nesting it within an array of multiple other buttons. Likewise, on mobile devices, icons on the touchscreens vary in size, and often, the smaller ones are more difficult to press accurately. While selecting these icons may not always be time-critical, users may still find it annoying when they repeatedly miss when trying to perform a task on their device. Clearly, small targets increase task difficulty, and difficulty can either increase errors or decrease speed, either of which may be counter-productive to successful task performance.

The **speed/accuracy tradeoff** is just as its name suggests: moving quickly and moving accurately tend to be inversely related. Therefore, as you increase speed, you normally decrease accuracy. Similarly, when you decrease speed, accuracy tends to increase. This principle applies to many aspects of human behavior, including pressing buttons, using a computer mouse to point to on-screen icons, hammering a nail, and even walking through a doorway or parking a car. In fact, this phenomenon has been studied so much that the mathematical expression of the speed/accuracy tradeoff has become one of the most established principles in the study of human movement: Fitts' Law.

8.3.2.2 Fitts' law

The speed/accuracy tradeoff suggests that increases in movement speed come at a cost in movement accuracy. If we need to be accurate, we must slow down. The obvious implication is that if a task requires rapid actions, as in the case of emergency shutoffs, the difficulty constraints of the task need to be low. In 1954, the psychologist Paul Fitts discovered that two factors, target width and movement amplitude, or distance, determine the difficulty of such tasks. He designed a task that required participants to point back and forth between two rectangular targets (See Figure 8.2) as quickly as possible while still being accurate.

In his experiment, Fitts manipulated these two factors and discovered that movement time (MT), defined as the time taken to move from one target to the other, was longest for the smallest targets and the longest distances. Because distance (D) and target width (W) were inversely related to each other, they could be expressed as a ratio, D/W, which represented the difficulty of the task. One could further assume that participants ideally aimed at the center of the target, so that the actual room for error at the end of the movement was really one-half the target width (i.e., the distance from the target center to the edge), which is $W/2$. Thus, the ratio can be refined to $D/W/2$, which could be expressed more simply as $2D/W$. Fitts called this ratio the **index of difficulty** (ID).

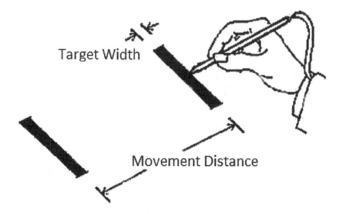

Figure 8.2 Fitts' (1954) classic experiment designed to test the relationship between target width, movement distance, and movement time. Participants held a stylus connected to an electrical circuit and were instructed to tap back and forth between the two targets (the dark rectangular bands) as rapidly as possible without missing. The targets were strips of copper plating connected to the same circuit as the stylus; thus, each time the stylus touched down, it would close the circuit and a timer would record this event. From the timer data, Fitts could obtain the durations of the movements. Sliding occluders allowed him to manipulate both the width of the targets and the distance between them. (Adapted from Fitts (1954).)

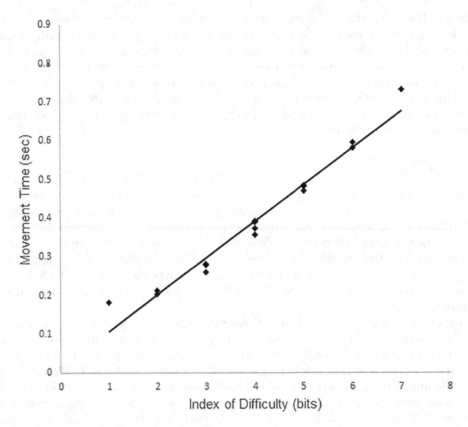

Figure 8.3 The movement time data from the 1 oz. stylus condition (Table I, Fitts, 1954). The index of difficulty ($\log_2$ 2D/W) strongly correlated to the duration of the pointing movement. The best fitting
line accounted for over 96% of the variance in movement time.

When Fitts plotted the *MT* data as a function of $\log_2$ of *ID*, as shown in Figure 8.3, he
found that there was a strong linear relation between the two, described by the linear equation: $MT = a + b$ (ID). This relationship is known as **Fitts' Law**, and using it, researchers can
predict *MT* by knowing the distance of the required movement and the width of the target.
Fitts' Law has been verified with tasks involving different body parts and several types of
tasks such as underwater movements, movements viewed through a microscope, pointing
and clicking on a computer screen, or target selection in virtual reality (VR).

Fortunately, the designers of controls are aware of Fitts' Law, and we can conclude that
emergency kill switches in industrial settings are generally large, located close to where
workers are likely to be, and easily discriminated from other controls. Less fortunate for
Antonia, the game designers pitted Fitts' law against her as they increased the difficulty of
her game.

It is also sometimes possible that controls will be too easy to activate, leading to errors.
Accidental activation of a kill switch in a factory might lead to costly downtime and wasted
resources. Other controls might have more disastrous consequences if activated at the wrong
time (think of a missile launch button on a military aircraft or a button that retracts the
landing gear on an aircraft). Thus, a balance must be found between the need to operate certain controls quickly when needed but also prevent their accidental activation. One
example is the use of button covers and guards. These are typically clear plastic hinged covers that require the user to lift them before the control can be activated. These can be useful

in preventing accidental control activation, but they carry the risk of slowing access to the control in time-critical situations.

8.4 ACCURACY OF MOVEMENT

Having considered the speed of pointing movements, we next address their accuracy. We have already seen that the most accurate movements are executed relatively slowly. In addition, we make fewer errors when we can see the hand and target. Thus, visual control and monitoring of the movement is important. There are numerous theories designed to explain the visual control of reaching, pointing, and other movements. These models can be divided into two groups: those that emphasize closed-loop feedback processing and those that focus on open-loop control.

8.4.1 Closed-loop control

As mentioned in Chapter 1, **closed-loop** control means that movements are guided by a feedback loop that continuously monitors the state of the movement and generates modifications to the trajectory as errors are detected. The information evaluated during the feedback loop is the position of the hand relative to the desired target. In a closed-loop system, this feedback continuously informs the motor commands issued to the limb. If the effects of these commands are perceived to be accurately propelling the limb to the appropriate place, no new commands are issued. But if an error is detected, new motor commands are subsequently re-issued to change direction. This looping of motor output and sensory input continues until the movement is successfully accomplished. It is important to note that processing and evaluating sensory feedback takes time, which causes some delay between when an error in the trajectory is detected and when a correction to the movement is initiated.

According to closed-loop theories, feedback information is most important during the later stages of the movement, when the hand is nearing the target. For this reason, and because of the delays involved in the feedback loop, movements that approach the target at high speed do not leave sufficient time at the end of the movement to process the feedback and make any necessary corrections. In other words, there is a "point of no return" beyond which feedback-based corrections cannot be executed. At this point, any remaining errors in the trajectory will remain uncorrected and the movement will be inaccurate. In slower approaches, the hand can get relatively closer to the target and there would still be time left to process feedback, and such movements are correspondingly more accurate than fast movements. Hence, the speed/accuracy tradeoff can be understood as a function of how close the hand can get to the target and still have time left to process feedback and correct for error.

This theoretical perspective places an obvious emphasis on the role of visual information about the hand and the target during the movement. Evidence in favor of this view comes from observations that movements made in the dark, or in other such situations where the hand is not visible, are less accurate than when the hand can be seen. While the speed/accuracy tradeoff tells us that accuracy decreases as speed increases, the effect is more pronounced when the hand is visible (Keele & Posner, 1968; Woodworth, 1899). Without lighting, movement accuracy is not affected by the speed of the movement because feedback-driven corrections cannot be used, and there is no "point of no return" for the faster movements. In one study (Keele & Posner, 1968), the point at which performance in a "lights-on" condition matched a "lights-off" condition was when movement duration

BOX 8.1:   MOVING IN VIRTUAL REALITY

The delay in processing visual feedback for accurate movements highlights a particular problem in one of the newest technologies: virtual reality. Most people may think of virtual reality (VR) as just a realistic three-dimensional world experienced while wearing a head-mounted display (HMD), and while that is an important part of VR, it is only just a part. Part of what makes the experience compelling is the ability to move around in the virtual environment (VE). When wearing the HMD, the user can turn their head left and right, as well as look up and down. These head movements must be accurately tracked by motion tracking sensors, usually attached to the HMD, so that the display is updated accordingly. For example, if one views an interesting sight to one's left in the VE and turns left to look at it, the motion tracker must detect this head turn and then quickly redraw the virtual scene to account for the user's new perspective. This has to happen quickly, or the user will experience latency, or delay, between their movements and the associated visual consequences. Noticeable latencies can cause disorientation and motion sickness (and associated nausea) in VR (see also Chapter 4).

These latencies are also relevant to other ways of interacting with VEs, such as reaching for and manipulating virtual objects. It is desirable that users of VR have a virtual body, or avatar, in the VE, rather than just being a disembodied eye. In actual reality, when we see something interesting and reach out to touch it, we see our hands—this is not always the case in VR unless motion tracking sensors are attached to the user's hands, and virtual hands are displayed correctly in the HMD. **Critical Thinking Questions**: Given what we know about the time it takes to process visual feedback, what might be the effect of latency in the display of virtual hand movements in VR? How might it affect reaching accuracy? Might the speed/accuracy tradeoff be magnified in VR? Why? What are some ways researchers can manipulate information about hand movement and position in VR, and in what ways can this be used to study basic aspects of skilled movement?

was a very short 190 ms (i.e., 0.190 seconds), because the movement was so fast that there was no time to make use of the available visual feedback. The next slowest movement duration recorded in the experiment was 260 ms—thus, the researchers estimated that the time needed to process visual feedback was between 190 and 260 ms. Other researchers have provided varying estimates of the feedback delay (see Carlton, 1992, for a review), but these results are generally similar to the above-mentioned values.

This work has obvious implications for the design of control panels and human–machine interfaces. In situations that require accurate movements to be performed by the user, the relevant controls should be located where the hand can remain visible. Further, the presence of occluding surfaces that may block the view of the hand during the operation of certain controls should be avoided.

8.4.2 Open-loop control

A second class of movement theories suggests that fast movements are guided by an **open-loop** control process, in which motor commands are issued to the relevant limb, and the effects of these commands are then executed without feedback corrections. Inaccuracy in the movement could thus arise due to several factors. Some authors have argued that the motor system contains some inherent noise, or variability, and in open-loop movements, such variability can lead to errors (Schmidt et al., 1979). Further, this noise is amplified when movement speed increases, making it more difficult to point to small targets, accounting for the speed–accuracy tradeoff.

Another source of inaccuracy in open-loop movements may be due to errors in perception. According to the principle of open-loop control, one must visually perceive the location of the target and, then using either vision or proprioception (or both), perceive the location of one's own hand. The information about the initial position of the hand relative to the target is then used to plan a movement trajectory that would transport the hand to the perceived target location. Any errors in the perceptual localization of the hand or target could lead to errors in movement planning. If closed-loop control is not employed, such errors could not be corrected once initiated and the movement would potentially be off target. Knowledge of the visual (see Chapter 3) and proprioceptive (see Box 8.2) capabilities of users is thus important for the consideration of human movement and the operation of controls.

BOX 8.2:   THE SIXTH SENSE: PROPRIOCEPTION

We perceive, on a near constant basis, the orientation of our body and limbs in space. This sense is called proprioception, after the Latin "proprio," which means "self," and could be translated as meaning "self-perception." This involves receptors embedded in the muscles and tendons that respond to the position and movements of the body and limbs. The physiologist Sir Charles Sherrington noted the significance of proprioception in motor skills. He reported that in animals in which proprioceptive feedback was removed surgically by severing the relevant nerves, their movements were jerky and uncoordinated. He pondered what it would be like for a human to be devoid of proprioception. Surely, it would be difficult to imagine having no sense of the presence of your body and what this would be like.

In 1991, the physician Jonathan Cole published an account of just such an instance in a man who had suffered sensory neuropathy (disease of the sensory nerves), presumably as the result of a viral infection. The patient had lost all cutaneous (skin) touch and proprioception below the neck. One could scarcely imagine this man's plight as he found himself, as the illness took hold, lying in a hospital bed feeling like a disembodied head. With no sense of where his body was or what it was doing, he found it almost impossible to move, even though his motor nerves were completely intact! This highlights the importance of proprioception in motor control: while the brain may be able to send motor commands to the limbs, without the sensory feedback that normally accompanies such movements and tells us where our limbs went as a result of these commands, we become uncoordinated and incapable of even the most basic activities. The patient reported that, while lying in bed those first few weeks of his illness, his arms would tend to flail about uncontrollably. Eventually, and with great effort through continuous visual concentration on his body, the patient was able to regain some motor skill. **Critical Thinking Questions**: How do you think this patient was able to regain his ability to move without any touch or proprioception? What would it be like to put on a turtleneck sweater without proprioception? What might you need to consider in control design to account for users with impaired touch or proprioceptive ability? What other situations might impair someone's sense of touch (hint: think about gloves, bulky clothing, etc.)?

8.5 CONTROLS

Most controls are operated by hand, but many are foot operated. Recent technological advances have allowed control by eye movements, head movements, by mouth (some devices for the disabled employ a joystick held between the teeth), or even by thought! We will focus primarily on hand and foot controls.

8.5.1 Types of controls

In kinesiology, the study of movement, an **effector** is a body part used to perform an action, usually an interaction with an object. The design of a control is influenced by the effector with which it is meant to be used. Some of the more common hand-controlled devices include buttons, joysticks, steering wheels, computer mice and keyboards, triggers, knobs, levers, switches, and cranks. Foot controls are generally limited to pedals, although buttons can be activated by a foot as well. If you consider the differences in the dexterity and anatomy of the hand and foot, the reasons for these differences become obvious. Such differences in the variety and types of movements of these two effectors are described in terms of their **degrees of freedom**, which refers to the number of movements that can be performed by that effector. With long fingers and an opposable thumb, the hands can grasp securely. The dexterity of the fingers, along with their tactile (touch) sensitivity, supports the performance of very small, precise movements, known as fine motor control. While the toes are generally constrained to move all together, the fingers can move individually, allowing the performance of a variety of sequential tapping-like movements (e.g., playing the piano). Further, the wrist can rotate in more ways than can the ankle, and the arm can move in more ways than the leg.

Figure 8.4 Discrete controls can be set into a finite number of states. For example, (a) buttons can be pressed or released, (b) toggle switches can be set into 1 of 2 positions, and (c) a rotary selector knob. Hybrid controls contain both discrete and continuous controls, such as (d) a game controller with multiple buttons plus two thumb-sticks and a directional pad that can take on numerous states along a continuum, (e) a virtual reality controller that can track the user's hand movements in three-dimensional space combined with multi-function buttons, (f) a video game steering wheel that rotates and includes multi-function buttons, (g) a standard 2-button computer mouse that tracks the user's hand movements in two-dimensional space, and (h) a fully continuous control featuring two foot pedals. (© Daniel S. McConnell.)

Controls can also be described based on the nature of the information that they input into the system they control (Figure 8.4). **Discrete controls** are those that allow information to be encoded in single, isolated chunks. Pressing a button or flipping a switch is a discrete response that sends a single command. A light switch is a typical example. **Continuous controls** allow the user to specify commands along a continuum, as the control can be set in multiple positions. The accelerator pedal in a car is a typical example. Other examples of continuous controls include steering wheels, computer mice, joysticks, rotary knobs, and cranks. Buttons also can be used in a continuous fashion, if the duration of the button press is meaningful, as might be the case with a doorbell.

8.5.2 Coding of controls

A key aspect of designing any control is ensuring that the user knows what it does and is able to select the desired control when multiple controls are present. When the user must select among multiple controls, this represents a problem of **control discrimination**. To increase the ease in recognition and discrimination of controls, we use **control coding**, the process of designing controls to be distinguished based on their shape, texture, color, size, location, mode of operation, or by label. Many technologies, from cars to airplanes, elevators to nuclear power plant control rooms, can present users with numerous choices between controls, and designers should take care to use good control coding principles to reduce the likelihood of errors.

8.5.2.1 Shape coding

Distinguishing controls by shape is known as **shape coding** and requires several considerations. First, users ought to be able to distinguish the shapes by touch alone (see Figure 8.5) because users may need to keep their eyes on important displays. For this reason, shape coding is appropriate for hand-operated controls, but not for foot-operated controls. Second, the shape of the control ought, if possible, to communicate to the user its function. Steering wheels are round, just like the wheels on the car that they control. Also, the U.S. Air Force developed aircraft controls that are shaped like the systems they operate. For example, the handle of the flap control is shaped like a wing, and the handle of the landing gear activation is shaped like a wheel. A factory worker may benefit from an emergency kill switch that, by its shape, signifies stopping. One possibility is an octagonal button; stop signs on roadways in many (but not all) countries are octagons, and so experienced drivers may associate this shape with the notion of stopping. This example also highlights the importance of knowing the user, as octagonal stop buttons may not carry the same significance across all cultures.

Figure 8.5 Top view of five separate rotary control knobs that can be discriminated by touch alone, similar to those tested by Hunt (1953). (© Daniel S. McConnell.)

8.5.2.2 Texture coding

The surface texture of controls can also be used as a means for users to distinguish controls and discern their function. Bradley (1967) investigated texture coding with rotary knobs. Through their sense of touch, participants were able to distinguish between smooth, fluted (cylindrical troughs notched into the surface), and knurled (straight or crisscrossing grooves etched into the surface) textures. Texture coding may be less desirable if users may be wearing gloves.

8.5.2.3 Color coding

Distinguishing controls by color requires consideration of the visual abilities of the user, including the ability to distinguish shades of color, especially under various conditions of illumination or varying levels of visual ability (e.g., color blindness). This topic and the limitations of using color are addressed more thoroughly with regard to visual displays in Chapter 3 and the design of the environment in Chapter 10. One useful property of color with respect to controls is its ability to communicate function. As the color red is generally related to danger or emergencies, large red buttons on machinery are generally understood to be emergency "off" buttons. Green buttons, sometimes located near the red ones, may be correctly understood as functioning to turn the machines back on. Clearly, workers may benefit if their octagonal emergency stop buttons are also red. This is an example of **redundant coding,** in which two different features of the button both signify its function.

8.5.2.4 Size coding

Coding by size may not always be the best method for distinguishing controls. The user must be able to discriminate the sizes by touch, and the human touch system is not well equipped to make fine distinctions of this dimension of perception. When increasing the size of a knob on a handle, it must change by a factor of at least 20% in order to be noticed (Chapanis & Kincade, 1972). Further, in many workplaces, space is limited, and large controls may not be practical.

Size coding can be useful, however, in signifying what a control does. Imagine a nuclear power plant equipped with a cooling system that can be manually operated with a set of circular control dials. Each dial controls a valve that regulates the flow of coolant through separate sub-systems. Some of these sub-systems may be more critical than others, and these dials that control cooling can be made larger than the dials controlling the less important systems. In this case, control size indicates the importance of the system to be controlled. If emergency stops are an important function, then those red octagon buttons should be the largest buttons in a workspace, which is conveniently consistent with the lesson learned from Fitts' Law.

8.5.2.5 Location coding

Controls can be distinguished based on where they are in the workspace or environment. Location can be used to indicate function, to discriminate between otherwise similar looking controls, or sometimes for convenience (light switches are placed near doorways) or safety (emergency shut-off controls are within easy reach of the user; conversely, in a spy car, one would not want to place the button for the rocket launcher or ejector seat in an overly conspicuous location where it might accidentally be activated). Sometimes the location of controls may be somewhat arbitrary, but based on user expectation cannot be rearranged. These user expectations are known as **population stereotypes,** based on the idea that different beliefs or expectations arise within a given population of users. In presumably all cars in the United States, the accelerator pedal is on the right and the brake pedal is on the left. After many years of driving such a car, it would be confusing and disturbing if you bought a new car in which the pedals had been swapped!

Another population stereotype related to location coding involves sinks or tubs with separate knobs for controlling the flow of hot and cold water. It is very common that the hot water control is on the left. In many cases, for aesthetic purposes, the knobs are indistinguishable by appearance; they look exactly alike and can only be discriminated by location. If you have ever visited a hotel where the pipes were installed differently, you may have experienced the unpleasant consequences of the violation of a population stereotype. Even in water faucets that have only one knob, we still expect that the farther left (counterclockwise for a rotary knob) we turn the knob, the hotter the water will get.

8.5.2.6 Mode of operation coding

Controls can also be distinguished by how they are operated or moved. It is generally apparent that buttons are meant to be pressed, that rotary knobs, cranks, and wheels are to be turned, and that switches should be flipped. Nevertheless, it is possible to design these controls to be used in different ways, but doing so runs the risk of confusing users, leading to errors and annoyance.

An important consideration in mode of operation involves the distinction between discrete and continuous controls described above. A light switch is a classic example of a discrete control, and it is easily operated because it can only be flipped into one of two positions. But which position is the appropriate setting for the desired state of the lights? Most light switches are mounted vertically, so that they can take the positions of either "up" or "down." Yet another population stereotype, at least in the United States, is that "up" means "lights on" and "down" means "lights off." Have you ever seen a light switch mounted horizontally? They are uncommon but do exist, and in such cases, there is no population stereotype that signifies "lights on" versus "lights off."

A related principle is that the operation of the control should be compatible with whatever aspect of the system it is controlling. If the system state is discrete, i.e., on or off, then a discrete, rather than continuous, control should be used to control this state. Hence, one would use a button or switch, and not a wheel or pedal, to turn lights off and on. Older model cars used a continuous control, a crank, for rolling up and down the windows. Newer cars have replaced cranks with multi-function buttons that can be operated discretely (a hard press rolls the window all the way down) or continuously (continuous press and release to set the window at an intermediate position). This issue of compatibility between a control and its function is explored more fully in Section 8.6.

8.5.2.7 Label coding

Coding with labels specifically refers to the use of text or other symbols to designate the function of a control. While labels can be useful in specifying the exact function of a control and the correct means of operation, they have several drawbacks. For one, labels can only be read when lighting conditions are sufficient. When multiple controls are present, the labels may be very small and difficult to read. Also, it may take time to read some labels, and sometimes responses need to be performed quickly. Lastly, the placement of labels near controls means that while they are being operated, the user's hand may partially or totally occlude view of the label. Labels can thus be useful, especially for someone who is just learning how to operate a particular piece of machinery, but they should not be the primary means for distinguishing controls. They should be redundant with other coding features instead of being the only means for the user to recognize the control.

Perhaps the all-time classic example of poor control coding and associated resorting to labels is illustrated by just about any door found in any office building, hospital, university building, or hotel. Often the doors are designed for aesthetics, that is, handles may be disguised or hidden in an effort to make the doors look nice and fit with the décor and design of the building. But when approaching such doors, how do you pass through? Are you supposed to pull or push? Or maybe you should slide the door sidewise? Which side of the door do you push? Which direction does it slide? Good handle design can make it obvious; handplates and pushbars generally signify pushing, and handles imply you should pull. But often the design of the handle is ambiguous, and at worst, the mapping between handle shape and function is completely backward. The result is usually that someone must hang a sign or label on the door. Figure 8.6 shows some examples of good and bad handle design. A good rule of thumb is that if a control needs a label, then it has not otherwise been designed properly (Norman, 1988).

Figure 8.6 Some doors are easier than others to operate. (a and b) Feature push bars and hand-plates designed for pushing, while (c) is affixed with similar handles on both sides of the door—which is pushed and which is pulled? (d) depicts an aesthetically pleasing entrance to a ballroom, but is this to be pushed or pulled? (e) depicts doors with pull handles that invite pulling but also labels discouraging use. (© Daniel S. McConnell.)

8.5.3 Resistance in controls

Controls can be made easier or safer to use by the amount of force required to activate them. Resistance opposes the force applied by the control operator. Highly resistant controls feel stiff and difficult to activate, and this can be an important safety feature that prevents accidental activation of controls that have drastic consequences, such as the aforementioned ejector seat button in a spy car. The same may be true in our ongoing consideration of emergency kill switches. We have discussed designing them so that they can be operated quickly, but prominent location and size may lead to accidental and annoying activation. Some degree of resistance can prevent this—although too much resistance could also interfere with successful activation.

It is useful for controls to offer some resistance, which can provide feedback to the operator in terms of their activation. Frequently, one may feel a "click" when pressing a button, and this tactile feedback informs the user that their attempt to press the button was successful. Another example is elastic resistance or spring loading. Many joysticks offer **elastic resistance**, so that they return to a neutral, or home, position when released. In addition, this type of resistance increases in strength the further the control has been moved, allowing the user to feel the state of the control. **Viscous friction** resists rapid and irregular movements, which is useful for making precise control adjustments. For example, the fine focus knob on a microscope may need only a very small adjustment, but if the feel of the knob is too "loose," it may turn too easily resulting in too large an adjustment. Thus, achieving just the right focus may prove difficult. Viscous friction provides enough resistance to enable small and precise movements while also ensuring that the control remains in the desired state when released.

8.5.4 Computer controls

The advent of desktop computing has brought with it a wide range of controls for the operator to use. Keyboards are used for text and numerical input, and a variety of different cursor positioning devices are available to consumers. These devices control the position of the cursor on the screen for the purpose of selecting or highlighting images, text, or other screen elements. The standard computer mouse is the most widely used cursor positioning device, but touchpads (common on laptop computers), trackballs, joysticks, digitizing tablets, cursor keys, and touch screens have also been employed. The relative effectiveness of these devices has been compared numerous times over the last several decades, especially as new types of devices get invented. The earliest findings that the mouse is the best device for typical point-and-click tasks, in terms of both speed and accuracy (Card et al., 1978), have been corroborated several times. So far, no new cursor control device has improved on the mouse.

BOX 8.3:   WHAT IS THE BEST KEYBOARD LAYOUT?

If you have used a computer or typewriter, then you know what the QWERTY keyboard layout is, even if you did not know it was called such. The layout is named for the first six letters on the top row of keys, beginning on the left. You might be tempted to think that the QWERTY layout has become the universal standard for keyboards because it is the layout that results in the fastest and most accurate typing rate. Surprisingly, it was intentionally designed to slow down typing speed! Before computers, or even electronic typewriters, mechanical typewriters employed typebars that would swing up to strike the page whenever a key was depressed. Early mechanical typewriters used a key layout that was in alphabetical order, but when typing very fast, the

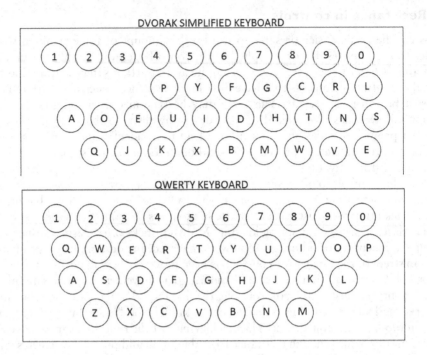

Figure 8.7 The Dvorak simplified keyboard (showing only the number and letter keys with punctuation keys omitted) and the standard QWERTY keyboard. (© Daniel S. McConnell.)

typebars would swing simultaneously and become jammed together. The QWERTY layout was invented to spread out letters that were commonly typed in sequences (such as the Q and U) so that the typebars approached the page from different directions, and to slow down typing rate, resulting in fewer jams. With computers, such concerns are obsolete, and it thus remains possible that there may be any number of keyboard layouts that are faster than QWERTY. One example is August Dvorak's (1943) Simplified Keyboard (Figure 8.7). This keyboard was designed based on four key principles: (1) Most users are right-handed, so the most commonly used keys should be in the right hand position, (2) the index and middle fingers are stronger and more skillful and should do more of the typing, (3) the most commonly used keys should be located on the home, or middle, row of keys, and (4) there should not be times when one hand is idle, leaving all the work to the other hand. The Dvorak Simplified Keyboard has found to be anywhere from 10% faster than QWERTY (Norman, 1983) to 40% faster (Nakic-Alfirevic & Durek, 2004). **Critical Thinking Questions**: Why do you think Dvorak's keyboard has not been more widely adopted? How would you design an experiment to test the superiority of various keyboard layouts?

8.6 IMPACT OF COMPATIBILITY ON PERFORMANCE

Another factor that influences the speed of responses to a stimulus is the compatibility of the response to the stimulus. That is, how natural is the mapping, or degree of correspondence, between the stimulus and the response to it? For example, a steering wheel is used to turn a car, and the wheel can only be operated via turning, hence there is a direct correspondence between the motion of the control and the effect it has on the controlled system. The mapping is intuitive and obvious, and cars are easy to steer.

8.6.1 Spatial compatibility

The nature of the stimulus-response mapping is often spatial in nature. **Spatial compatibility** refers to the relationship between the location of a control and the location of the stimulus object it controls. Consider an everyday example of the controls for the stovetop burners on a regular household kitchen range. If a pot begins to boil over or flare up with a flame, a speedy response to lower the heat is obviously desirable. Also, it is generally desirable that when turning on a burner, you would like to avoid accidentally turning on the wrong one. Consider the mapping of controls to burners shown in Figure 8.8. Is the spatial mapping shown ideal?

The effect of spatial compatibility is also relevant to the relationship between the direction a control moves and the direction of the object it controls. This is a common concern in aircraft flight controls. Pulling back on the control stick results in raising the nose of the plane and climbing altitude, while pushing forward on the stick drops the nose and results in descent. Is this the best mapping of control movement to airplane response? Such questions can be tested in an experiment, such as the classic performed by Fitts and Seeger (1953). In their study, a participant held a joystick that could be moved left or right, forward or backward, or diagonally in any direction. Participants operated this controller while viewing a circular display of eight lights mounted on a vertical board on a table before them.

In one condition, if the light at the top of the circle was lit, the participant was required to move the joystick forward as rapidly as possible; if the light at the bottom of the circle was lit, the participant was instructed to move the joystick backward as rapidly as possible. Similar stimulus (light) to response (joystick) mappings were created for all eight lights around the circle.

However, Fitts and Seeger (1953) also reversed the mapping such that, if the light at the top of the circle was lit, the participant would need to move the joystick backward. In this case, reaction times for the movements were slower, and more errors were made. It should be evident that the first condition involved a compatible stimulus-response mapping in terms of the spatial alignment of the lights in the array compared to the movements performed by the participants.

(a) (b)

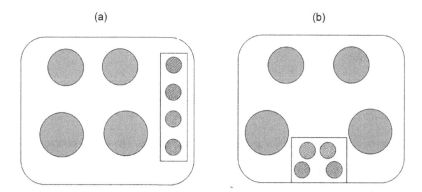

Figure 8.8 Two typical stovetop configurations of burners relative to their control knobs. (a) The spatial arrangement of the four control knobs is ambiguous with respect to which burner they operate. (b) The spatial mapping of the control knobs reflects the spatial arrangement of the four burners. (© Daniel S. McConnell.)

8.6.2 Movement compatibility

Just as spatial compatibility can affect reaction and movement times, another consideration in compatibility is the type of movement to be made relative to the type of movement displayed. For example, when tuning in a radio station on an old-fashioned analog display, the station indicator travels linearly across the display, but most radio knobs are dials that rotate. A good rule of thumb is that rotary dial controls are best for radial displays, while linear or sliding controls are best for linear displays. Usually, though, analog radio tuners are not difficult to operate because the arrangement of the control dial to the display helps make it obvious which direction of turning corresponds to a particular direction of the station indicator (see Figure 8.9). One example of this is the **Warrick principle** (Warrick, 1947), which states that the side of the rotary dial closest to the display should move in the same direction as the display. This is the situation shown in Figure 8.9.

Modern car radio tuners typically do not feature knobs and dials, however. Instead, buttons are used to tune to different frequencies, and the readout is no longer an analog display, but a digital LCD readout of the radio station's broadcast frequency. A new question arises, though, about how the tuning buttons should be arranged. Because broadcast frequency is an interval scale of numbers, and the user's typical goal is to scan up and down the range of numbers, arranging the buttons vertically makes sense. Unfortunately, design considerations sometimes prevent a vertical arrangement of tuning buttons, and thus a left/right arrangement is required. In this case, which button should correspond to a decrease in radio frequency? Typically, the left button is assigned this function, while the right-hand button

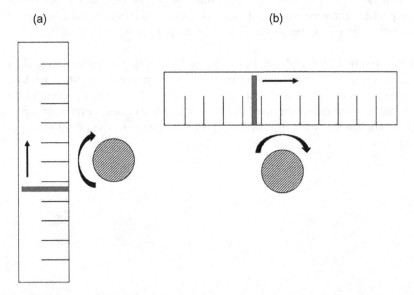

Figure 8.9 Stimulus-response compatibility using Warrick's principle. The round control knob moves the red indicator across the display. The side of the knob nearest the display moves in the same direction as the indicator, as shown by the arrows. (a) The display is oriented vertically, and a clockwise rotation of the knob moves the indicator up. If the knob was placed instead to the left of the display, a counterclockwise rotation would be expected to move the indicator up. (b) The display is oriented horizontally, and a clockwise rotation of the knob results in the top of the know moving rightward, corresponding to a rightward movement of the indicator. If the knob was instead located above the display, a counterclockwise movement would be expected to move the indicator to the right. (© Daniel S. McConnell.)

increases the frequency when pressed, and most users appear to have little trouble with this arrangement.

The above example represents a case in which standards and user expectations need to be considered in design. For many older users, they may recall the old-fashioned analog radio display, and when they want to increase the radio frequency, they envision the old analog pointer moving linearly to the right across the display (as in Figure 8.9b) and thus expect the right-hand button to accomplish this goal. But what about younger users who may have never seen the old-fashioned radio displays? Might they find this left/right mapping arbitrary and confusing? Perhaps, or possibly due to experience, they have simply grown accustomed to this arrangement, or maybe it makes sense because of the similarity of this scale to the typical number line taught in many elementary schools. This is another example of a population stereotype and represents the idea that compatibility is often determined by user expectation.

8.6.3 Control-display ratio

The examples above regarding using controls to move a pointer on a display highlights a special kind of movement task. While most often we move our hands to interact with controls directly, technology has provided several situations in which we use indirect control. In Figure 8.9, the hand directly operates the rotary dial and indirectly moves the pointer in the display. A more modern and much more common example is using a mouse to point to objects and icons on computer screens. We move the mouse with our hand, but we pay attention to the visual motion of the pointer traveling across the screen toward icons. Most computer screens are at least 16 in. (40.5 cm) wide, and many are even wider. Have you noticed that when you need to move the cursor from one side of the screen to the other, you do not actually have to move your hand (and mouse) the whole 16 in. across the desk?

The scaling between the movement of the controller (e.g., a computer mouse or joystick) and the displayed movement of a cursor or pointer determine the **control-display ratio**, the inverse of which is called **gain**. In most computer interfaces, the speed and distance of the display is larger than the actual movement produced by the hand. Fortunately, this difference does not decrease performance, as most computer users have no problem controlling a pointer that moves faster than the mouse.

Most computers use a nonlinear mapping between control and display. That is, the pointer speed is not simply a multiple of the speed of the mouse. The speed of the pointer increases as a function of the speed of the mouse, known as **velocity gain**. When the user moves the mouse slowly, the gain in velocity is low, so that the motion of the pointer closely matches that of the mouse. In contrast, when the user moves rapidly, the gain in velocity is high.

This dynamic combination of high and low gains in velocity is very useful. Imagine a situation in which the gain is always very high. If the user needed to mouse all the way across a large computer screen, they could do so very rapidly by making only a small displacement of the mouse. Imagine what would happen, though, once they have arrived in the general area of the desired target, possibly a very small screen element such as a small cell in a spreadsheet. In this case, the small movements of the mouse would be magnified into large displacements of the on-screen pointer, and it would be difficult to place the pointer in the cell and keep it there. The user would probably have to make several extremely small mouse movements to complete this task. On the other hand, if the gain was always low, such precise pointing tasks would become easy, but would come at a cost whereby traveling a large distance across the display would require long and inefficient movements. By using a combination of high and low gains, computer users obtain the benefits of both.

8.7 TRACKING

The situation of indirect pointing with a computer mouse represents a special case of a more general type of control called **tracking**. Whenever we operate system controls, the system responds in some fashion, and we usually want to monitor how the system responds so that we can appropriately modulate the controls. From a systems perspective, we usually define the operation of a control as an **input** to the system, and the response of the system is called the **output**. Consider a pilot who wishes to ascend to a certain altitude. They must initiate a degree of deflection on the control stick by pulling it back, and the plane will begin to climb in altitude. How does the pilot know when to release the stick and level off their flight? They must perform a basic tracking task: the control input (pulling back on the stick) is made, and the pilot monitors the system output (how much the altitude has changed) and then tracks this output until the altitude is at the desired state. This is called the **tracking loop**.

We start with having a desired state. Initially, the plane is too low—perhaps air traffic control has called in a weather warning and advised the pilot to climb to 35,000 ft to avoid turbulence. Because the current altitude of the plane may be 32,000 ft, this is defined as an error state. The goal is to reduce the error by pulling back on the control stick until the error goes to zero.

To reduce the error, the pilot must apply a force to the control, which is done by applying pressure on the lever. This action results in a deflection of the lever away from its initial position by some amount, which is the result of the amount of force applied by the pilot and the resistance to movement of the stick itself (see Section 8.6.3). Changes in the control position then influence the system, which in this case would be the plane's ascent. This is the system output.

When engaging in this tracking task, the pilot is comparing the system output to the desired state. Since the goal is to reduce error by setting the system state to be the same as the desired state, the pilot will continue to apply control inputs until this state is achieved and the error state is null.

Tracking tasks can be discrete, requiring a single movement to achieve the desire state, as in computer pointing. Other tasks involve continuous tracking of system variables and adjustment of control commands. A familiar example is steering a car. The desired state would be defined as maintaining lane position on the roadway (i.e., remaining centered in one's lane). Any number of factors might cause the vehicle to deviate from this desired state, such as wind gusts or uneven road surfaces. The resultant swerving out of the lane, say to the left, creates an error state that a driver detects visually by comparing the car's current position relative to the lane markers on the road. The driver attempts to reduce error in this case by applying a force (to the right) to the steering wheel, which turns by some degree and then moves the car back to the right by some amount.

Note that control inputs are designed to oppose the nature of the error. When the plane is too low, a pilot performs control input to increase altitude; conversely, when the plane is too high, control inputs are made to decrease altitude. When a car swerves to the left, the driver steers back to the right. These are examples of **negative feedback loops**. A feedback loop is any situation in which the control inputs lead to the system's response and is fed back to the operator, who would then determine whether additional control adjustments are required. It is considered negative feedback whenever control inputs oppose, or negate, the error detected in the system.

The challenge of a continuous tracking loop is illustrated by the observation that single inputs rarely eliminate the error exactly. In the case of driving and lane maintenance, when the car swerves to the left and the driver steers back to the right, it is possible that the driver may over-steer, which will result in a new error state, this time to the right of the desired lane position. Control adjustments are now used to steer back to the left. All drivers perform this tracking loop while driving because cars never stay perfectly in the center of a lane, but rather continuously drift a little to the right or left, and drivers then are continuously detecting these slight lane shifts and make small steering adjustments to stay close to the center of the lane.

8.7.1 Control dynamics

The concept of **control dynamics** defines the relationship between the command input and the system response to it. The **order of control** defines the complexity of this relationship. Some commands change the position, or state, of a system, which is known as position control. Other commands affect the rate of change in the system state, which is called velocity control. Yet other commands can affect the rate of the rate of change in the system state, which is called acceleration control.

8.7.1.1 Position control

Also known as zero-order control, **position control** is the simplest control order, and it involves a simple, linear relationship between command input and system output. It is called position control because changing the position of the control changes the *position* of the system. Using a computer mouse to point to on-screen targets is an example: changing the position of the mouse changes the position of the pointer. When the mouse is not moving, the pointer is not moving. Further, the magnitude of change in position of the mouse, i.e., how far the mouse is moved on the mouse pad, is proportional to the distance traveled by the pointer on the screen.

8.7.1.2 Velocity control

Also known as first-order control, **velocity control** involves a slightly more complex control dynamic. In this case, changing the position of the control does not affect the position of the system; rather it affects the *velocity* of the system. Using a joystick, rather than a mouse, to control the pointer on a computer screen is an example. Moving the joystick affects the velocity of the pointer, where the velocity is proportional to the angle of deflection. A slight deflection of the stick results in a low velocity movement of the pointer, and a large deflection results in a high velocity movement. A constant position of the joystick results in a constant velocity movement of the pointer. When the joystick is in the neutral position, velocity is zero, but holding the joystick at some angle away from neutral results in continuous movement of the pointer at a constant velocity proportional to the angle of deflection. Speed control of a car is also first-order control, because the angle of deflection of the gas pedal is proportional to the speed of the vehicle. Consider how speed control would work if the pedal worked according to zero-order control. In this case, change in pedal position would affect the position of the car. Pushing the pedal a small amount would cause the car to travel a small distance, while pushing the pedal more would cause the car to travel a longer distance. Obviously, this is not an efficient way to propel a vehicle!

8.7.1.3 Acceleration control

Also known as second-order control, **acceleration control** represents the most complex control dynamic commonly used in tracking loops. Per this dynamic, the state of the control position determines the *acceleration* of the system. Such higher ordered control systems are usually difficult to control. One example is steering a car. The position of the steering wheel determines the angular position of the wheels, which in turn determine the rate at which the car moves laterally. Thus, a given rotation of the steering wheel determines the lateral acceleration of the car. While steering the car may not seem difficult for experienced drivers, many novice drivers exhibit under-steering or over-steering mistakes until they get a feel for how the car responds to control inputs.

8.7.2 Lag

One important aspect of control dynamics is the system **lag,** or time-delay in responding to control input. In steering a car, for example, it takes time for a change in the steering wheel angle to be transmitted through the linkages in the steering column and to the axle and wheels and hence there is a delay, however slight, between the moment the driver turns the wheel and the time the car actually responds by changing direction. Imagine an unrealistic situation in which 5 seconds elapsed between the time the steering wheel was turned and the time at which the car changed heading. Obviously, this would be disconcerting to the driver because it would take 5 seconds before the driver knows whether the amount of steering that was applied was the appropriate amount to correct for any deviation in their desired lane position. Such a car would be almost impossible to drive in a stable fashion; it would probably end up swerving all over the road.

While long lag times are intuitively undesirable and result in instability in control dynamics, they are not always avoidable. This is especially true for second-order control dynamics, because such systems tend to be sluggish in their response. Another factor that affects lag is the inertia of the system. **Inertia** is a system's resistance to movement. More force is required to move a heavy object than a light one, and it takes longer to get heavy objects moving. Imagine driving a pick-up truck under two different conditions. In one condition, the bed of the truck is empty, and in the other, it is loaded down with 2,000 lbs of bricks. In the latter case, pressing on the accelerator does not have an immediate effect, so the driver may end up pressing just a little more to get the truck moving. However, that degree of acceleration may end up being more than was desired, and the truck ends up going too fast. The driver must then let off the pedal, but the inertia of the moving truck keeps it traveling at a high speed for a while, so the driver keeps his foot off the pedal too long, and then the truck eventually slows down more than was desired. This constant switching between too much acceleration and too little is the kind of instability associated with laggy control dynamics.

If one can anticipate, or predict, the system's response to the controls, one can reduce the instability. Usually, this kind of prediction is achieved through practice and experience; an experienced driver of heavy trucks learns just how much to press on the accelerator to obtain a desired speed. Similar problems are faced in piloting airplanes or steering large ships. In both cases, the inertia of these vehicles makes them slow to respond to steering inputs, and thus over- or under-steering can be common when learning to operate them. Aside from user expertise, smart technologies can often predict future systems for the user and display that information back in the form of predictive displays, which depict future states so that users can better anticipate the effects of control inputs.

8.7.3 Tracking displays

In the previous section on tracking loops, we focused on the relationship between control inputs to a system and the nature of the system's output, or response. The operator must monitor the system response, but we have not paid much attention to how that information is provided to the operator. A key component to a tracking task is the nature of the display that provides information to the operator about the system state and the effect of his or her commands. As seen in previous examples, drivers rely on speedometers to provide feedback about a car's response to acceleration, as well as seeing the front end of the car and the lines on the road through the windshield in order to get information about a car's response to steering. Computer users rely on seeing a pointer move on the screen to provide information about the movement of the mouse. Further, pilots must rely on their instruments for information such as altitude, heading, pitch, and airspeed to know how the airplane responds to the controls.

There are two kinds of tracking displays: pursuit and compensatory. **Pursuit tracking** displays show the system state and desired state separately, the difference between which is the error state. The operator's goal is to reduce error by getting the system state in the display to match the desired state. In a sense, the operator makes the system state *pursue* the desired state in the display. Seeing the front of the car and the road lines through the windshield while steering is an example of a pursuit task. The front of the car is the system and the lane markers are the target, and the steering achieves the goal of reducing error by keeping the visible portion of the car between the lane markers.

Compensatory tracking displays show only the error state. No information is given to the operator as to whether the error state is due to changes in the system state (e.g., the car has moved away from the desired state) or control state (e.g., the steering wheel has moved away from the desired state). Nevertheless, the operator can use information about the existence of the error state to reduce it. Speedometers are compensatory displays because the desired state cannot be displayed, as the desired state will change based on driving conditions but is nevertheless specified as a fixed point in the display. For example, if the goal is to maintain speed at 70 mph, then there is a fixed point on the display denoting this state. If the speedometer indicates a higher rate of speed than the desired state, the driver modifies the system input by letting up on the accelerator pedal and does not reapply pressure until the speedometer needle falls back to the desired state. From above, this is another example of a negative feedback loop.

As might be expected, pursuit displays tend to result in better performance (e.g., Briggs, 1962; Briggs & Rockway, 1966), probably because the operator has more information about the source of the error state (i.e., whether it is due to unexpected changes in the system or controls; Adams, 1989). Examples of pursuit and compensatory displays are shown in Figure 8.10.

8.8 APPLICATION TO SPECIAL POPULATIONS

We know that a central tenet in human factors design is consideration of the characteristics of the user. Further, we know that not all users are the same. In Chapter 9, we will see that they can come in all shapes and sizes, and anthropometric data are useful in designing the size and configuration of controls and other objects. Another important factor to be weighed in human factors design is the age of the user, especially when it comes to the consideration of motor skills.

(a)   (b)

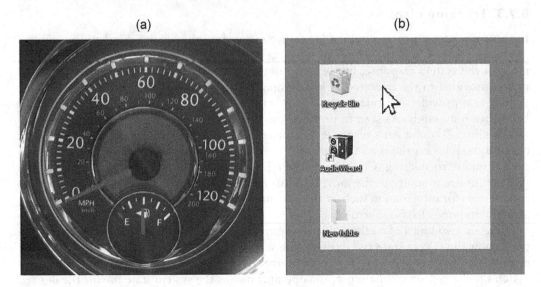

Figure 8.10 Compensatory and pursuit displays. (a) A typical car speedometer is a compensatory display. Error is defined as the distance between a desired speed and the car's speed as indicated by the red needle. Speed adjustments are made using the accelerator pedal to speed up or slow down accordingly, thus compensating for deviations from the desired state. (b) Using a mouse to control the on-screen cursor is a pursuit display. The to-be-selected icon defines the target or goal state, and the arrow defines the current cursor location. Error state is defined as the distance between the current and target locations. Mouse movements are performed until the error state is reduced to zero. (© Daniel S. McConnell.)

Young to middle adulthood typically represents the peak of human motor skills, corresponding roughly to an age range of 18–50 years. As infants, we gradually learn to control our bodies, learn to walk, reach and grasp for objects, and hold things in both hands. During early childhood, we refine these skills, adding strength and additional fine motor skills, such as the ability to grasp and manipulate small objects, as well as improving balance (McGraw, 1943). In the teen years, strength and coordination continue to improve until they reach approximate peak adulthood levels. Then, as older adults, some motor skills begin to deteriorate. We may lose some fine motor skills, our muscles may lose strength, and we generally tend to move slower (Goggin & Stelmach, 1990).

Understanding these age-related differences is key to designing technology that is appropriate to the user. For example, toys are commonly labeled with an appropriate age. Children younger than the prescribed age may not be able to perform some of the necessary motor behaviors to operate the toy. Infant and toddler toys should be designed with large, easy-to-move interactive controls. It is common to employ buttons on some toys that, when pressed, trigger music to be played. Such buttons should be big enough that the child can activate them easily, using the whole hand, as infants typically do not have the ability to perform precise movements using only their fingertip. Further, the buttons should offer little resistance to pressure, so that the typical child is strong enough to depress them.

The design of safety devices in "child-proofing" a home also benefits from knowledge of children's motor skills. In this case, items are designed intentionally to exceed children's abilities. The operation of these devices typically entails dexterous two-handed manipulation of parts and requires some hand strength, especially a strong pinch-grip, i.e., the ability to exert pressure between the thumb and index finger of the same hand. Some safety latches for kitchen cabinets may require inserting two parts simultaneously through an aperture small enough so that they will only fit when pinched tightly together. Medicine bottles typically require the ability to press down firmly with the palm, while simultaneously turning the wrist. The ability to perform these types of actions does not develop until mid- to late-childhood, thereby preventing infants and toddlers from defeating these safety features. In an unfortunate twist, these same features that make it difficult for children to open medicine bottles may also make it difficult for older adults who experience declines in fine motor control and hand strength.

As noted above, we tend to lose some motor abilities as we progress into late adulthood, ranging from some very basic motor skills to more complex manual performance. Older adults tend to move slower. This affects reaction time and total movement duration. However, while older adults fare less well in tasks that require rapid responses, their movement accuracy is not significantly impaired, and they can thus perform many skilled actions if rapidity is not a major requirement. Some research has demonstrated that these deficits can partially be overcome if the movements are predictable and well-learned (Stelmach et al., 1987), hinting at the possibility that older adults who perform a particular job task over many years may not show a significant drop in performance. Recent studies have found that age-related slowing of reaching movements is attributable to declines in processing the sensory feedback used to control such movements (Van Halewyck et al., 2015a, b). Research also suggests that older adults who maintain high levels of physical activity exhibit less of a decline in sensory feedback (Adamo et al., 2009) and consequently show less of a decline in reaching performance (Van Halewyck et al., 2014, 2015a, b).

Another issue for older adults involves strength. Manual grip strength, including pinch-grip strength, declines by over 25% in older adults (Ranganathan et al., 2001), although exercise and continued physical activity may offset these declines (Hughes et al., 2001).

Lastly, certain complex manual tasks can be affected by age. Many tasks require that the two hands do two different things at once. For example, when Antonia plays her balloon popping game on a tablet, one hand is likely used to hold the tablet securely, while the other performs the various tapping tasks within the game. Subtle movements of the supporting hand will change the orientation of the tablet screen and can affect her ability to correctly tap on the moving targets. Antonia needs to be able to coordinate these motions so that a shift in the tablet's position with one hand is immediately compensated for in her aiming hand. One study found that when older adults were asked to move their hands back and forth in various rhythmic patterns, they exhibited significantly more movement variability than did young adults (Serrien et al., 2000), which implied less two-handed coordination among the older subjects.

8.9 SUMMARY

The study of human movement is crucial to understanding how people interact with technology. Many aspects of the design of controls and displays can lead to errors and annoyance of the user because of mistakes in control use and activation. This chapter covered factors that affect the speed and accuracy of movements, along with a discussion of theories of how movements are controlled. This chapter also discussed how controls are designed and how their operation should correspond to the system they are controlling—as well as the various ways a control response can affect a system state.

LIST OF KEY TERMS

Closed-loop control	Lag
Compensatory tracking	Movement compatibility
Continuous control	Movement time
Control coding	Negative feedback loop
Color coding	Open-loop control
Mode of operation coding	Order of control
Redundant coding	Acceleration control
Shape coding	Position control
Size coding	Velocity control
Label coding	Output
Location coding	Population stereotype
Texture coding	Pursuit tracking
Control discrimination	Reaction time
Control dynamics	Decision time
Control-display ratio	Response selection
Degrees of freedom	Simple reaction time
Discrete control	Spatial compatibility
Effector	Speed/accuracy tradeoff
Elastic resistance	Tracking
Fitts' law	Tracking loop
Hick–Hyman law	Velocity gain
Index of difficulty	Viscous friction
Inertia	Warrick principle
Input	

SUGGESTED READING

Cole, J. (1991). *Pride and a Daily Marathon*. Cambridge, MA: The MIT Press.
   This book tells the compelling story of the patient without proprioception, mixing both the neuroscience and medical perspective with the human perspective of how the patient was able to overcome his illness.

Elliott, D., Helsen, W. F., & Chua, R. (2001). A century later: Woodworth's (1899) two-component model of goal-directed aiming. *Psychological Bulletin*, 127(3), 342–357.
   This paper reviews nearly all of the research on motor skill pertaining to the speed/accuracy tradeoff through the entire 20th century.

CHAPTER EXERCISES

1. Describe the difference between discrete and continuous controls.
2. Describe the seven types of control coding described in the text and identify a real-world example of each type.
3. Define elastic and viscous friction and the advantages of using them as a form of resistance in a control.
4. According to research presented in the text, what is the relationship between motor skill and aging, and what effect does experience and expertise have on this relationship?
5. List and describe three factors that can affect reaction time.
6. What are the task parameters that make up the Index of Difficulty in Fitts' law? How do these parameters individually affect movement time?
7. Compare and contrast open-loop and closed-loop control.

8. Explain the differences between position, velocity, and acceleration control.
9. Define lag and the effect it has on tracking performance.
10. Compare and contrast pursuit and compensatory tracking displays.

REFERENCES

Adamo, D. E., Alexander, N. B., & Brown, S. H. (2009). The influence of age and physical activity on upper limb proprioceptive ability. *Journal of Aging and Physical Activity*, *17*(3), 272–293.

Adams, J. A. (1989). *Human Factors Engineering*. New York: Macmillan.

Air Force System Command (1980, June). *Design Handbook 1–3, Human Factors Engineering* (3rd ed). Dayton, OH: Wright-Patterson Air Force Base, U.S. Air Force.

Bradley, J. V. (1967). Tactual coding of cylindrical knobs. *Human Factors*, *9*(5), 483–496.

Briggs, G. E. (1962). *Pursuit and Compensatory Modes of Information Display: A Review*. Wright-Patterson Air Force Base, OH: 6570th Aerospace Medical Research Laboratories. Technical Documentary Report AMRL-TDR-62–93, August.

Briggs, G. E., & Rockway, M. R. (1966). Learning and performance as a function of the percentage of pursuit component in a tracking display. *Journal of Experimental Psychology*, *71*, 165–169.

Card, S. K., English, W. K., & Burr, B. (1978). Evaluation of mouse, rate-controlled isometric joystick, step keys, and text keys for text selection on a CRT. *Ergonomics*, *21*, 601–613.

Carlton, L. G. (1992). Visual processing time and the control of movement. In L. Proteau & D. Elliott (Eds.) *Vision and Motor Control*, pp. 3–31. Amsterdam: North-Holland.

Chapanis, A., & Kincade, R. G. (1972). Design of controls. In H. P. Van Cott & R. G. Kincade (Eds.). *Human Engineering Guide to Equipment Design*. Washington, DC: U.S. Government Printing Office.

Donders, F. C. (1869/1969). On the speed of mental processes. *Acta Psychologica*, *30*, 412–431.

Dvorak, A. (1943). There is a better typewriter keyboard. *National Business Education Quarterly*, *12*, 51–58.

Fitts, P. M. (1954). The information capacity of the human motor system in controlling the amplitude of a movement. *Journal of Experimental Psychology*, *47*(6), 381–391.

Fitts, P. M., & Seeger, C. M. (1953). S-R compatibility: spatial characteristics of stimulus and response codes. *Journal of Experimental Psychology*, *46*, 199–210.

Goggin, N. L., & Stelmach, G. E. (1990). Age-related deficits in cognitive-motor skills. In E. A. Lovelace (Ed.). *Aging and Cognition: Metal Processes, Self Awareness and Interventions*, pp. 135–155. Amsterdam: Elsevier Science Publishers.

Hick, W. E. (1952). On the rate of gain of information. *Quarterly Journal of Experimental Psychology*, *4*, 11–26.

Hughes, V. A., Frontera, W. R., Wood, M., Evans, W. J., Dallal, G. E., Roubenoff, R., & Fiatorone-Singh, M. A. (2001). Longitudinal muscle strength changes in older adults: Influence of muscle mass, physical activity, and health. *The Journals of Gerontology: Series A*, *56*(5), B209–B217.

Hunt, D. P. (1953). *The Coding of Aircraft Controls*. Tech. Report 53–221. Dayton, OH: U.S. Air Force, Wright Air Development Center.

Hyman, R. (1953). Stimulus information as a determinant of reaction time. *Journal of Experimental Psychology*, *45*, 423–432.

Keele, S. W. (1986). Motor control. In K. R. Boff, L. Kaufman, & J. P. Thomas (Eds.). *Handbook of Perception and Human Performance, Vol. II: Cognitive Processes and Performance*, pp. 1–60. New York: Wiley Interscience.

Keele, S. W., & Posner, M. I. (1968). Processing of visual feedback in rapid movements. *Journal of Experimental Psychology*, *77*, 155–158.

McGraw, M. B. (1943). *The Neuromuscular Maturation of the Human Infant*. New York: Columbia University Press.

Nakic-Alfirevic, T., & Durek, M. (2004). The Dvorak keyboard layout and possibilities of its regional adaptation. In 26th Annual Conference on Information Technology Interfaces, Cavtat, Croatia.

Norman, D. (1983). The Dvorak revival: is it really worth the cost? *Consumer Products Tech Group: The Human Factors Society*, 8(3), 5–7.

Norman, D. (1988). *The Design of Everyday Things*. New York: Doubleday.

Ranganathan, V. K., Siemionow, V., Sahgal, V., & Yue, G. H. (2001). Effects of aging on hand function. *Journal of the American Geriatrics Society*, 49(11), 1478–1484.

Schmidt, R. A., Zelaznik, H. N., Hawkins, B., Frank, J. S., & Quinn, J. T., Jr. (1979). Motor output variability: a theory for the accuracy of rapid motor acts. *Psychological Review*, 86, 415–451.

Serrien, D. J., Swinnen, S. P., & Stelmach, G. E. (2000). Age-related deterioration of coordinated inter-limb behavior. *Journal of Gerontology: Series B*, 55(5), P292–P303.

Stelmach, G. E., Goggin, N. L., & Garcia-Colera, A. (1987). Movement specification time with age. *Experimental Aging Research*, 13(1), 39–46.

Van Halewyck, F., Lavrysen, A., Levin, O., Boisgontier, M. P., Elliott, D., & Helsen, W. F. (2014). Both age and physical activity level impact on eye-hand coordination. *Human Movement Science*, 36, 80–96.

Van Halewyck, F., Lavrysen, A., Levin, O., Boisgontier, M. P., Elliott, D., & Helsen, W. F. (2015a). Factors underlying age-related changes in discrete aiming. *Experimental Brain Research*, 233(6), 1733–1744.

Van Halewyck, F., Lavrysen, A., Levin, O., Elliott, D., & Helsen, W. F. (2015b). The impact of age and physical activity level on manual aiming performance. *Journal of Aging and Physical Activity*, 23, 169–179.

Warrick, M. J. (1947). *Direction of movement in the use of control knobs to position visual indicators*, USAF AMC Report, no. 694-4C. Dayton, OH: Wright-Patterson Air Force Base, U.S. Air Force.

Woodworth R. S. (1899). The accuracy of voluntary movement. *Psychological Review*, 3, 1–119.

Chapter 9

Anthropometry and biomechanics

Chapter Vignette

For as long as he can remember, Jeff has wanted to be a Navy pilot. He is an athlete, is an excellent student, is in outstanding health, and has 20/15 vision. Now that he is old enough, he applied for pilot training school and just received a letter while at home during the summer break. He walks into the kitchen where his parents are cleaning up after dinner while holding the open letter. "I can't believe it, I've been denied!" Incredulously, his mom asks, "But, why?" "Because I am too tall," Jeff groans. As someone who is 6′ 5″, Jeff would have a difficult time fitting into the cockpit of a fighter jet.

In somewhat of a state of shock, Jeff begins to help his mother put away clean dishes, but he is placing all the dishes on the top shelf. "Ah…, dear…if you place those up there, I won't be able to reach them. Even if I could reach them, I'm not sure I could lift the dishes so far over my head without dropping them." Given that his mother is 5′ 6″, she would have trouble reaching those dishes without a chair or stool. "Sorry, Mom, I'm just a bit upset, I suppose, and not thinking. I never thought that being tall would be a disadvantage." With that, Jeff leaves the kitchen and heads for the outdoors. While walking out, he strikes his head on the door jam. "Oh, *%&!%$#!"

These issues Jeff encountered are all related to his height. That is, the cockpit was too small; the cupboards' top shelves were a perfect height for him, but too high for his mother; and the door jamb was just a bit too low for his tall stature. Because Jeff is taller than most people, it is also likely that he will have trouble finding a comfortable workspace or automobile because most things are built for a range of users that includes the greatest number of people, not those who are at the extremes of the distribution of sizes. Office desks, student desks, automobile seats, and kitchen cupboards are not usually built to accommodate those individuals who are extremely tall or extremely short. Individuals who are extremely small or extremely big, or extremely short or extremely tall, will either need to have something custom made or make various adjustments to the environment such as using a step stool or an adjustable desk to accommodate them well.

Jeff's mom also faces the issue of having dishes placed by Jeff on a shelf too high for her. If she had to retrieve the dishes with her arms extended over her head, she would be in an unsafe posture reaching as far as possible straining her arms. We should not be working with our arms extended over our heads. Certain body postures, twists, turns, or extensions can cause strain on the body. Although the anthropometric aspects are important to ensure environments "fit" the user, we also need to make sure that the functions of the user, or biomechanics, are appropriate for the job. Therefore, in order to create environments that fit most people or environments that can accommodate people who are extremely tall or short, or have physical limitations such as being in a wheelchair, we need to understand anthropometry and biomechanics.

DOI: 10.1201/9781003515463-9

9.1 CHAPTER OBJECTIVES

After reading this chapter, you should be able to:

- Define what anthropometry is.
- Define various anthropometric terms.
- Describe the various design criteria.
- Explain what anthropometric tools exist and how they are used.
- Explain how anthropometric data are collected.
- Explain how anthropometric data are analyzed.
- Describe the need for and use of various design standards.
- Explain the essentials of proper seating, work surface height, and workspace design.
- Define various biomechanical terms.
- Identify and describe the injuries that can occur when anthropometry or biomechanics are not properly applied to a situation.
- Explain what manual materials handling is and how it relates to biomechanics.
- Determine the proper workspace design for various activities.

9.2 WHAT IS ANTHROPOMETRY?

Anthropometry originates from physical anthropology (Herzberg, 1960; Nowak, 1997), which documents body differences among the human races (Herzberg, 1960). Therefore, **anthropometry** is the measurement of the human body or, more specifically, of human body parts. Some examples of body dimensions include stature, leg length, forearm length, head width, sitting height, and hand width. Anthropometric data are used to design environments and equipment for human use. This ensures that the space and equipment fit or conform to the size and shape of humans. Anthropometric measures can also include measures of range of motion and muscle strength (O'Brien, 1996), highlighting the importance of biomechanics, which are discussed later in the chapter. In fact, for anthropometric data to be applied, one must consider the conditions or constraints set by the biomechanics involved (Pheasant, 1990).

9.2.1 Common measurement terms

When working with anthropometric data, it is important to be specific about what is being measured. For example, if we want to measure foot width, it is important to specify which part of the foot to measure. If we measure across the ball of the foot, it is likely to be wider than a measure across the heel of the foot. Similarly, when measuring waist height, anthropometrists use two different landmarks to measure this distance from the floor: the navel and the narrow part of the waist (Roebuck, 1995). Even if there is only one landmark, it is also specifically identified. For example, eye height is the "vertical distance from the floor to the inner canthus (corner) of the eye" (Pheasant, 1996, p. 31). Therefore, it is important to know exactly what is being measured, using common measurement terms.

As mentioned above, specific measures are based on landmarks on the body. **Landmarks** are specific spots on the body that are physically visible such as a specific bone end or joint on the knee or shoulder, but not always. When a landmark is not so easily identifiable, some type of marker is used to identify the landmark on the person. For complete illustrations of landmarks, see Roebuck (1995).

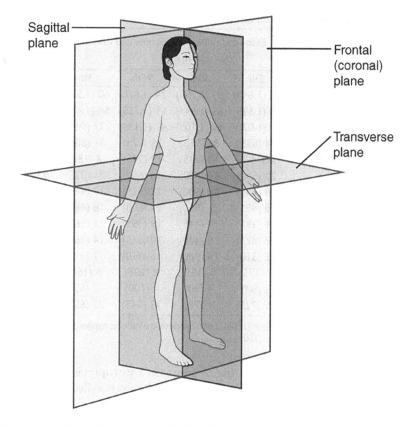

Figure 9.1 Orientation planes. (Connexions, CC BY 3.0 <https://creativecommons.org/licenses/by/3.0>, via Wikimedia Commons.)

When discussing measurements, it is helpful to define the human figure by a three-dimensional x–y–z graphing plane. The plane that traverses through the body *posterior* to the *anterior* (i.e., from the back to the front) area of the body is the **coronal (frontal) plane** (see Figure 9.1). The **sagittal plane** moves through the body from side to side to both *lateral* positions. The **transverse plane** travels from the feet or *inferior* location to the head, the *superior* location. The middle part of the individual is the *medial* position. Finally, regardless of direction, body parts that are farther from the core of the body are said to be **distal** compared to **proximal** body parts that are closer to the trunk of the body. For example, the elbow is proximal compared to the hand, which is distal.

Kroemer (1987) identified other common terms including **height**, a straight vertical (i.e., inferior–superior) measurement; **breadth**, a straight horizontal (i.e., lateral–lateral) measurement across the body; **depth**, a straight, horizontal measurement from the front of the body to the back (i.e., anterior–posterior); and **circumference**, usually a closed, but non-circular measurement, whereas **curvature** is usually an open non-circular measurement. Applying these terms, sitting height is measured from the seat to the top of the head (Pheasant, 1990) and shoulder breadth is measured between two shoulder landmarks. Similarly, sitting depth (or more specifically, the buttock-popliteal length) is measured from the chair back to the **popliteal**, or the back of the knees (see Table 9.1). Another commonly used measure is **stature**, which is standing height from the floor to the top of one's head (i.e., the vertex or crown; Pheasant, 1990, 1996). For an in-depth look at various anthropometric measures and illustrations identifying what is measured, see Pheasant and Haslegrave (2006)

Table 9.1 Common anthropometric measurements

| | Estimated values for U.S. men and women in inches (and millimeters) | | | | | |
| | Men percentiles | | | Women percentiles | | |
Measurement	5th	50th	95th	5th	50th	95th
Stature	65 (1,640)	69 (1,755)	74 (1,870)	60 (1,520)	64 (1,625)	68 (1,730)
Eye height	60 (1,529)	65 (1,644)	69 (1,759)	56 (1,416)	60 (1,519)	64 (1,622)
Elbow height	40 (1,020)	45 (1,105)	47 (1,190)	37 (945)	40 (1,020)	43 (1,095)
Sitting height	34 (855)	36 (915)	38 (975)	31 (800)	34 (860)	36 (920)
Sitting elbow height (elbow rest height)	8 (195)	10 (245)	12 (295)	7 (185)	9 (235)	11 (285)
Buttock-knee length	22 (550)	24 (600)	26 (650)	21 (525)	23 (575)	25 (625)
Buttock-popliteal length	18 (445)	20 (500)	22 (555)	17 (440)	19 (490)	21 (540)
Knee height	19 (495)	22 (550)	24 (605)	18 (460)	20 (505)	22 (550)
Popliteal height	16 (395)	18 (445)	19 (495)	14 (360)	16 (405)	18 (450)
Shoulder breadth (bideltoid)	17 (425)	19 (470)	20 (515)	14 (360)	16 (400)	17 (440)
Hip breadth	12 (310)	14 (360)	16 (410)	12 (310)	15 (375)	17 (440)
Hand length	7 (175)	8 (191)	8 (205)	6 (160)	7 (175)	7 (190)
Hand breadth	3 (80)	4 (90)	4 (100)	3 (65)	3 (75)	3 (85)
Forward grip reach	29 (725)	31 (785)	33 (845)	26 (655)	28 (710)	30 (765)

Source: Adapted from Pheasant and Haslegrave (2006). For complete definitions, applications, and corrections for each measure, see Pheasant and Haslegrave (2006).

or Roebuck (1995). Table 9.1 presents a list of common anthropometric measurements for U.S. men and women 19–65 years of age. More recent data sets reflect the changes in some measurements including increases in waist circumference. Therefore, these original data sets might not apply to the current population.

9.3 THE NEED FOR ANTHROPOMETRIC MEASURES

Anthropometric measurements are used for designing workspaces as well as many things a person uses such as controls, tools, and equipment (Kroemer, 1987). If you were going to start producing bicycle helmets, you would need accurate data about male and female head dimensions to know what size and shape to make the helmets! In particular, the helmet for a child would need to be much smaller than for most adults, but how much smaller? Similarly, the international space station must be designed for individuals who will use it under weightlessness. As the body expands under weightlessness, clothing must be designed to accommodate this slight expansion. In contrast, the opposite effect occurs when SCUBA diving. As a diver descends, weight belts become looser. It is important not to tighten the belts while diving or they will be too tight as one ascends to the water's surface and one's waist expands to its original size. Space suits and gloves are also designed with anthropometric data. In addition, tools must be designed so astronauts can grasp and effectively use the tools during space walks while wearing their protective, but bulky space suits and gloves. Imagine what it is like to wear gloves or mittens during the winter, making it difficult to grasp certain items.

Most of us, though, are not astronauts. Nevertheless, we may have been in a chair so big that it seemed to swallow us, similar to a child in an adult chair with legs sticking out. Or perhaps you sat in a row of chairs placed side-by-side and were uncomfortable because you could not sit back in the chair without touching the shoulders and arms of the persons on each side of you. The use of anthropometric data in design allows us to ensure that chairs and other things people use physically "fit" the user. Unfortunately, no single design usually accommodates all users. For example, if we design a checkout counter for standing individuals, this will impact the accessibility for someone in a wheelchair.

Although we can design for some adjustability, it may not always be possible to accommodate everyone. For example, the cockpit of a fighter aircraft is a tight compact area, which restricts the seat position. Only individuals within a certain range of leg length, sitting height, arm length, and body width can be accommodated, and, as Jeff discovered, restricts who qualifies to be fighter pilots. Therefore, consideration of anthropometric data is extremely important in the design of cockpits, passenger aircraft seats, and, of course, the desks and tables at your school, university, or home. As passenger airline seats are not likely to get wider (as fewer seats can be sold to passengers), many airlines require individuals who cannot fit within a single seat with the armrests down (in economy class the width is just over 17 in. or 43.2 cm) to purchase another seat.

Keep in mind that, while some designers claim to design for the "average" person, this person does not exist. A person of average height might have short legs and a long torso, or long legs and a short torso. The challenge is to design tools, equipment, and furniture to fit most people so they can use the items appropriately and effectively, reducing error and possible injury by reducing biomechanical stresses (Kroemer, 1987) and, hence, fatigue (Seidl & Bubb, 2006).

9.3.1 Design criteria

When designing workspaces or equipment, the intent of anthropometric data is to meet three main criteria: comfort, performance, and health and safety (Pheasant, 1990). These three criteria, though, are conceptual criteria that we must operationalize (see Chapter 2) to understand how these criteria are being met. For example, if we are designing a chair, the conceptual criterion of comfort may be divided into sub-criteria such as avoiding pressure hot spots and providing postural support. The lowest (most detailed) level of definition would include specification of the exact backrest angle, seat height, and other seating criteria for achieving postural support (Pheasant, 1990).

9.3.1.1 Clearance, reach, and limiting users

In order to achieve comfort, performance, and health and safety, other anthropometric criteria must be met, including clearance, reach, and the limiting user (Pheasant, 1990). **Clearance** involves determining at what point a space is too small (Pheasant, 1990). For example, passageways must be high enough and wide enough for people to pass through that area. As Jeff experienced, the doorway clearance was not high enough for him to avoid hitting his head without ducking. Similarly, your computer desk should have appropriate knee clearance and your car should give you ample elbow room. When considering hand controls, if you must reach into a space to turn a control, it is important that the opening to the control is large enough for your fingers or hand to fit when gripping and turning the control.

This issue of clearance is considered a *one-tailed* or **one-way constraint** because the space can be too small, but the space cannot be too big (Pheasant, 1990). Therefore, clearance is normally defined for the largest user.

Even though clearance cannot generally be too big for a particular task, it can be too big for other reasons. For instance, what size gap between rows of seats in an airplane is needed for safe evacuation in case of an emergency? If the clearance is large, then passengers can evacuate more easily, but this restricts the number of seats for passengers. Fewer seats means fewer passengers, which translates into lost company revenue. Research has determined the minimum distance possible between rows of airplane seats that allow for safe evacuation. Some airlines include a few rows with more leg room, but you pay more for those seats.

In contrast to clearance, **reach** involves determining when the distance between two points is too great for effective functioning (Pheasant, 1990). As in the opening scenario, the top kitchen shelf was too high for Jeff's mom, as her reach was not as great as his reach. Therefore, the location of anything the user may need to access or use, such as various controls, should be within reach. When driving, it is best if windshield wiper, stereo volume, and light controls are all within easy reach of the driver. Otherwise, the driver will stretch to use the controls, which could negatively impact driving performance. Similarly, if we cannot comfortably reach the gas and brake pedals, we cannot safely drive the vehicle.

Reach is also a one-tailed or one-way constraint in that the distance can be too great, but it cannot be too small. To accommodate most people, we should design for the users with the smallest reach.

Based on what we know about clearance and reach, it is possible to design things that have too little clearance or require too large a reach. Therefore, we often consider the limiting user. The **limiting user** is that potential person from the population who places the greatest number of constraints on the design because of certain individual characteristics (Pheasant, 1996). A limiting user in a clearance problem could be someone who is extremely tall or extremely wide (i.e., obese). In contrast, the limiting user for reach would be an extremely small or short person or possibly someone with an injured or disabled limb.

9.3.1.2 Maintaining appropriate posture

In order to achieve comfort, performance, and health and safety, we also should have appropriate posture. In contrast to clearance and reach, which are one-way constraints, posture is a *two-tailed* or **two-way constraint** (Pheasant, 1996). When designing a workstation, it is possible for the desk height to be too low or too high. For appropriate posture, it is important that the working surface height fits the user and the task. For example, when a desk is too low, users often push their chair back and lean or sit on the edge of the chair, which eliminates any support from the chair's backrest (Pheasant, 1990).

Therefore, the use of anthropometric data in design can help ensure that users can maintain appropriate posture. Consider the young child working at a school desk, a data entry person working at a computer workstation, or an individual sitting on the floor working on a laptop as demonstrated in Figure 9.2. Neck or shoulder tension or lower back pain can result if chairs are not designed correctly such as when the desk or workstation is too low or too high. These problems arise for two reasons: (1) the body is held in one position for a long period of time and (2) the body posture is not appropriate.

A healthy body is a body that moves. Unfortunately, those working in office jobs often sit for long periods of time at a desk, table, or computer station. If we are going to sit in one position for a long period of time, we should at least select appropriate furniture and arrange the environment to help us maintain proper posture. Improper postures often cause too much load on the back, arms, or joints. **Load**, or loading, is the amount of pressure placed on various muscles or joints, or the back. Excessive or long-term loading can lead to injury and cumulative trauma disorders. Most commonly, people report "pains" in the neck, shoulders, and back generally due to improper sitting or standing posture, leading to muscle fatigue in the back, shoulders, and arms (Kroemer, 1987).

One way to deal with this problem of sitting in one position too long is to drink a lot of water. Sooner or later you will need to move! Specific discussions on how to use anthropometric

Figure 9.2 Improper posture. (From Pixabay, https://pixabay.com/en/computer-female-girl-isolated-15812/.)

data to help individuals maintain more appropriate posture and reduce loading and the likelihood of injury follow in later sections of this chapter. To know more about how to evaluate working postures, see Delleman et al. (2004) and Delleman and Dul (2006a, 2006b).

9.3.1.3 Strength

When designing a workspace to maintain appropriate posture, a fourth criterion to consider besides clearance, reach, and the limiting user is strength. When operating controls, it is important to determine how much force a person can apply when reaching for and manipulating a control. Reaching for and turning a knob or handle could be more taxing than reaching for and flipping a switch, especially if one must reach far. Pushing and pulling strength is also influenced by one's biomechanics (discussed later in this chapter), which is influenced by the positions of the shoulder and elbow when in a seated position (Kroemer, 1987). Strength would usually be considered a one-way constraint when designing for the weakest limiting person, unless it is possible to apply too much force (Pheasant, 1996).

9.4 ANTHROPOMETRIC TOOLS AND MEASUREMENT

9.4.1 Anthropometric tools

Common tools for measuring height, breadth, depth, and circumference include anthropometers, spreading and sliding calipers, tape measures, and cones. Scales are used for measuring weight. An **anthropometer** is used to measure height (Roebuck, 1995). **Spreading calipers** are used for measuring the head or any place where it is necessary to reach around obstacles, as the prongs of the jaw are deeper and curved. A **sliding caliper** with shorter and straight prongs (see Figure 9.3a) is used for measuring hand breadth. To measure hip or chest circumference, a **tape measure** (see Figure 9.3b) is used unless it is grip circumference, then **cones** are used whereby an individual wraps one's hand around a cone-shaped object (Roebuck, 1995). Tools are evolving with newer technology; however, these traditional tools have been used for over 80 years and are likely to remain essential (Seidl & Bubb, 2006).

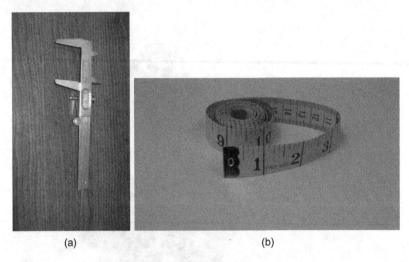

(a)   (b)

Figure 9.3 Common anthropometric tools. (a) A caliper (© Nancy J. Stone) can be used to measure inside and outside distances and (b) a tape measure (© Nancy J. Stone) is best used for measuring circumferences. Sometimes, the caliper has curved edges to make it easier to place it around the object being measured.

The newer digital methods of human modeling, or human modeling systems (HMS), can produce three-dimensional (3D) manikins (Oudenhuijzen et al., 2002). This technology was originally used in the car industry for the design of seating. Now, 3D scanning assists in the design of things such as what people wear on their heads (e.g., helmets; Bradtmiller, 2023). Even with these digital methods, the manual measurements serve as the basis for developing digital manikins or scans of humans (Bradtmiller, 2023; Oudenhuijzen et al., 2002).

9.4.2 Anthropometric data collection methods

With these traditional and newer digital tools, it is easier to collect the appropriate data. Before collecting the data, we first determine the population of interest (i.e., the beneficiaries of these data; Oudenhuijzen et al., 2002; Seidl & Bubb, 2006). The population you want to study will be different if you are collecting data to design school desks for elementary aged children or entrance ramps for individuals in wheelchairs. In the first case, the population you want is children between the ages of 5 and 12. Given the potential for rapid and differential growth rates over the age range, it may be desirable to define smaller age ranges such as ages 5–7, 8–9, and 10–12 years. Similarly, important body changes occur in adults as they age, and it is important to characterize these changes (Seidl & Bubb, 2006).

Besides age, other important considerations for defining the population include sex, ethnicity, and, for adults, occupation (Pheasant, 1990). The sex of the user is important because males tend to be bigger than females, except in hip breadth (Kroemer & Grandjean, 1997). There are also racial or ethnic differences among anthropometric measures. For example, Asians tend to be shorter than other populations (Seidl & Bubb, 2006). Once the population of interest has been identified, one should randomly sample from that population (see Chapter 2).

**BOX 9.1:   DOES THE SCHOOL FURNITURE FIT OUR
CHILDREN OF DIFFERENT ETHNICITIES?**

Although children's furniture should be smaller than adult furniture, we need to consider how the dimensions might differ for children as well as what differences might exist among children. It appears as though size differences and different growth rates of children from different ethnic groups impact the "fit" of school furniture. Lance Cotton, Dennis O'Connell, Phillip Palmer, and Marsha Rutland (2002) investigated the mismatch of school furniture for children while considering ethnicity and age. For 6th, 7th, and 8th grade boys and girls in Abilene, Texas, it was found that 99% of the students from all grade levels did not fit the seat height and seat depth of their chairs! In addition, when considering just seat height, fewer than half of all students could find a desk with a seat height low enough for the students; however, the 6th graders had the best fit when using the smallest chairs.

When considering ethnicity, African American students tended to have a better fit with the chairs, probably because their legs tended to be longer at these ages than the legs of Caucasian American and Mexican American children especially at the younger age (6th grade). In fact, Cotton et al. (2002) found that knee height, popliteal height, and buttock-popliteal length were all significantly greater for the African American children than either the Caucasian American or Mexican American students.

For all students, the desks tended to be too high as the surface level tended to be higher than the children's functional elbow height. Fortunately, the tall desks provided plenty of knee clearance. Cotton et al. (2002) also found sex differences appearing in the eighth grade. Given all these potential differences, as well as the potential negative impact of improper posture, it would seem appropriate to design better seating for our children. **Critical Thinking Questions**: How might you design the school furniture for children to accommodate these differences? If you could not change the type of furniture, how might you modify the furniture or environment to help the children assume the correct seating posture?

The next step is to determine what measures are needed (see Table 9.1). If you are designing student desks for individuals in wheelchairs, it is important to know the wheelchair width and sitting height, and possibly the turning radius, to determine how much room is needed for turning in order to place one's legs under the desk.

9.4.2.1 Static versus dynamic measures

We often collect both static and dynamic measures. **Static measures** are collected when an individual is standing or sitting in a still and somewhat rigid position. Static measures are not realistic of how someone usually interacts with the environment, but they do give us information for making a number of initial design decisions regarding clearance and reach. It is also important to collect data when the individual or computerized manikin is moving, giving us **dynamic measures**. Dynamic measures help us understand if the design is appropriate for our biomechanics (O'Brien, 1996). In particular, the nascent technology of 4D scanning includes the measurement of movement (Alemany et al., 2023).

9.4.2.2 Fitting trials

Besides using the traditional and new digital tools to collect the basic measurements (e.g., stature or thigh thickness), it is important to determine if a design feels right to a user and if most users consider the design to be the right fit. This process is known as a fitting trial. A **fitting trial** is an experiment whereby mock-ups of the space to be evaluated are adjusted, and the users rate the situation's "feel" or comfort along continua such as too high to too low or easy to use to difficult to use, with the middle rating labeled "just right" (Pheasant, 1996).

If we want to determine the desired height and width of a door for a particular application, say for firefighters wearing their safety gear, we can adjust the dimensions of height and width. Actually, we could ask the participants to adjust the door height and width to their lowest (narrowest) and highest (widest) acceptable measures, but it is more systematic if we control these measures. After selecting an appropriate representative sample, we would create a mock-up with a height requiring everyone to duck before walking through or one too narrow to walk through without contacting the sides of the door frames. Then, we would gradually increase the height or width and obtain the users' assessment of whether the height is too low or the width is too narrow. This is an example of a one-way constraint and our graph should be a decreasing line similar to the "Too low" line displayed on the left in Figure 9.4. Instead of asking if the door is too high or too wide, we would ask the participants to rate whether one "can pass through without ducking" to assess the fit. As the height increases, the percent of individuals who respond that they can pass through without ducking would increase, as represented by the dotted satisfaction line on the left in Figure 9.4.

In contrast, determining the appropriate height for a student desk represents a two-way constraint, as there are limiting users at both ends of the spectrum. We would start with a desk that is too short for all students. As we gradually increase the height, the number of individuals who indicate that it is "too low" will decrease, while the number of individuals who indicate it is "satisfactory" or "just right" will increase to a point and then it will begin to decrease, as the desk height will become too high for many students (see Figure 9.4). Soon we will reach a height that is too high for all students. As indicated in the right drawing in Figure 9.4, we tend to get three different curves from our data representing "too low," "too high," and "satisfactory" (Pheasant, 1990, 1996). To satisfy most people, we would want a desk size that falls within the "satisfactory" curve. If we were to graph a curve that represented students' perceptions of a "just right" height, this graph would be similar to the

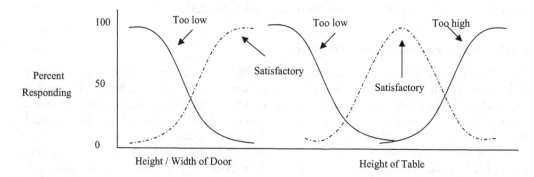

Figure 9.4 Examples of curves for fitting trials. The curves on the left deal with the height or width of a door, which is a one-way assessment (it generally cannot be too high). The curves on the right for the height of a table is a two-way assessment in which the table can be too low for some or too high for others; therefore, we wish to select a height that is satisfactory for most. (Adapted from Pheasant, 1988.)

"satisfactory" curve, but with a smaller distribution. Therefore, we try to fit most people within the satisfactory range, which would generally exclude those individuals who are extremely small or extremely large.

In the example above, we started with the shortest desk and increased the height, which represents an **ascending trial**. A complete fitting trial includes both ascending and **descending trials** (Pheasant, 1990), where the height or size is decreased to see if the results are similar to the ascending trial (and the results often are different).

To summarize the process of a fitting trial, a participant comes in and "experiences" the mock-up. The first experience is set to be outside an expected range (e.g., too high or too low). Then, the desk is raised or lowered, and the participant rates all levels of the ascending and descending trials. These ratings could be collected by surveys or interviews, as discussed in Chapter 5.

As you might imagine, fitting trials can be extremely time-consuming and costly. Computer technology including virtual environments (VE), environments that allow individuals to "experience" a 3D view of new designs by wearing goggles, can reduce the cost of evaluating new designs (Oudenhuijzen et al., 2002). Similarly, it is possible to use two-dimensional manikins and computer-generated models to conduct fitting trials and acquire data. The need for fitting tests with humans could be reduced as 4D scanning technology improves (Alemany et al., 2023). For a more thorough discussion on fitting trials, see Pheasant (1990).

9.4.3 Anthropometric data tables

Because the collection of anthropometric data can be extremely time-consuming, it is advantageous to use existing data when possible. Anthropometric data tables should include information about the population sampled and the types of measures collected (e.g., static versus dynamic, stature). Many of these tables, though, are based on military samples representing mostly men, which do not always generalize well to a civilian population (Albin & Molenbroek, 2023). NASA's *Anthropometric Source Book* (NASA, 1987) is one of the most comprehensive collections of anthropometric data (Kroemer, 1987; Pheasant, 1990; Roebuck, 1995), but dated. Anthropometric tables for U.S. adults, adults from other countries, as well as infants and British children can be found in Pheasant and Haslegrave (2006). For a list of anthropometric data sources, see Roebuck (1995). Table 9.1 presents some common measures.

Some more recent data sets are publicly available including ANSUR II (https://www.openlab.psu.edu/ansur2/) and DINED—anthropometric database (Molenbroek, 2018). ANSUR II data are based on a survey of 6,000 U.S. Army personnel. One can purchase access to more than 68 data sets including ANSUR and CAESAR from the World Engineering Anthropometry Resources (WEAR). WEAR data include one- and three-dimensional scans from a more diverse population than a military population (Albin & Molenbroek, 2023). There are computational procedures that can be used to determine the missing data from an existing data set (see Kroemer, Kroemer, & Kroemer-Elbert, 1997; Pheasant, 1996).

Finally, there are extremely limited data, if any, for special populations such as the elderly or people in wheelchairs. This is why researchers often collect their own data and use fitting trials for a new design. To effectively use the fitting trial data specific to our situation and population of interest, we need to understand the measures used in analyzing anthropometric data.

9.5 COMMON MEASURES USED IN ANTHROPOMETRIC ANALYSES

Once anthropometric data have been collected, what do we do with the data? After measuring the stature of ten males and ten females, you need to put those data into an understandable and useful form. The most common measures used with anthropometric data are the mean, standard deviation, range, and percentile.

9.5.1 Range, mean, and standard deviation

The range of the data set is important because we want to accommodate most people. Knowing the smallest and largest or shortest and tallest measures can be helpful. Although the mean and standard deviation for the data (see Appendix of Chapter 2) are valuable, they are more valuable when used to calculate percentile points, which are commonly used with anthropometric data.

9.5.2 Percentiles

Percentiles specify how many individuals are included at or below a particular value in our data set. For example, you probably took either the ACT or SAT to get into college. If you scored at the 78th percentile in math, this means that 78% of the people taking the test did worse than you, but 22% performed better. Similarly, if your English or science scores were at the 92nd percentile, this means that you performed better than 92% of the people, but 8% performed better than you did.

If you have large data sets (anything greater than a sample of 30 according to the law of large numbers), the data should be normally or close to normally distributed. Given that many data sets were created using military samples that number is in the hundreds if not thousands, you can be assured that the data set is normally distributed, at least for that population being sampled. Furthermore, body dimensions tend to fit a normal distribution (Pheasant, 1990). A **normal distribution** is a bell-shaped curve (see Figure 9.5), and the 50th percentile would be the mean.

Note that in Figure 9.5, the mean is zero, represented by a z-score of zero. A **z-score** is a standardized value that tells us how many standard deviations above or below the mean value our score falls and z-scores have a distribution with a mean of 0 and a standard deviation of 1. If you have a z-score of +1, then you are one standard deviation above the mean

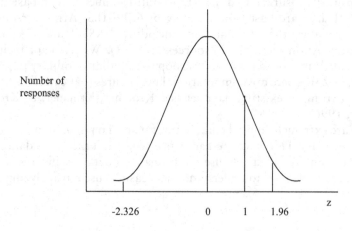

Figure 9.5 The normal distribution and z-scores

Table 9.2 z-Scores for calculating common percentile values

z-Score	Percentile
−2.326	1st
−1.645	5th
−1.28	10th
0	50th
1.28	90th
1.645	95th
2.326	99th

(see Figure 9.5). The standardization of this distribution gives us information about the percent of people who fall below that score, giving us percentiles. Percentiles less than 50% have a corresponding z-score that is negative (see Table 9.2 and Figure 9.5). Table 9.2 lists z-scores that correspond to the more commonly used percentiles.

When we know what z-scores correspond to which percentiles, it is possible to calculate the measurement size that should accommodate a certain percentile for your population. Because we normally want to design from the smallest measurement (e.g., 5th percentile female reach) to the largest person (e.g., 95th percentile male stature), we need to determine what value matches the 5th percentile reach for woman or the 95th percentile stature for men. Thus, we do not calculate the percentile, but rather the value, measure, or score that gives us that percentile, or the **percentile point**. To do this, we need to know the mean and standard deviation of the distribution as well as the relevant z-score.

To calculate a percentile point, we add to the mean the multiplicative value of the z-score times the standard deviation:

Percentile point $= \bar{x} + zs$

If our data have a mean of 50 and standard deviation of 10, the 5th percentile (the smaller person's dimension) would be $50 + (-1.645)(10) = 33.55$. Anthropometric data are often rounded to a convenient number. Thus, this number might be rounded to 35. To calculate the 95th percentile, we would have $50 + (1.645)(10) = 66.45$, which might be rounded to 65. Of course, this rounding would decrease the number of people accommodated by this design. To accommodate more people, someone might prefer to round to 30 and 70, depending on the circumstances.

It is also possible to graph the cumulative percentiles as opposed to the normal distribution, as depicted in Figure 9.6. Using the cumulative graph, it is easy to determine what percent of the population is affected or accommodated by a certain height or size (Pheasant, 1990, 1996).

Finally, a word of caution using percentiles. We should not refer to someone as a percentile (e.g., 5th percentile woman), as that person is unlikely to have all anthropometric dimensions at the 5th percentile. In addition, when we are trying to address multiple dimensions, the use of percentiles becomes complex. For example, if we want to know the 95th percentile eye height for seated individuals, we cannot just add the 95th percentile popliteal height to the 95th percentile seat to eye height (see Albin & Molenbroek, 2023).

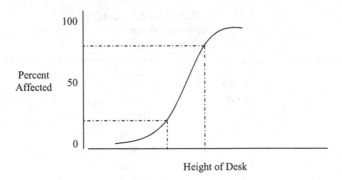

Figure 9.6 A cumulative percentile curve

9.5.3 Method of limits

Once you have anthropometric data on a particular population and have calculated means, standard deviations, and percentiles, it is possible to determine the "just right" fit without the use of a fitting trial. The **method of limits** involves calculating the specified dimension (e.g., height or width) based on the given anthropometric data (e.g., Pheasant & Haslegrave, 2006). For example, the best height for a standing work surface is between 50 and 100 mm (about 2–4 in.) below one's elbow height (Kroemer & Grandjean, 1997). The mean elbow height for European and North American men is about 1,070 mm (~42 in.) and about 1,000 mm (~39 in.) for women (Kroemer & Grandjean, 1997). Using the method of limits, we can estimate that a range of 970–1,020 mm (~38–40 in.) would be appropriate for men and a range of 900–950 mm (~35–37 in.) for women, but this would only be for "average" men and women for this one dimension and would not meet the need of extremely short or tall users. If the workstation can be adjusted, then we need the standard deviation to calculate and apply the 5th and 95th percentiles to establish the range of the adjustable workstation.

As mentioned before, because many of the data sets include military samples, they tend to include a large sample of men, but the sample of women is relatively small. Therefore, when using data tables, you must be sure that the sample is representative of the population you want to fit. Recall from Chapter 2 that the sample should be a representative subset of the population you are trying to study. Some believe that a military population is representative of the general population, or at least of the working population, whereas others question how well it represents the population. This is a decision you would have to make based on the intended use of the data.

9.5.4 Corrections

Regardless of how you get your data, corrections are often necessary. Anthropometric data are usually collected on nudes (Seidl & Bubb, 2006), whereby the measurements do not account for the thickness of clothing. In addition, people do not normally sit or stand in the stiff, rigid positions used for collecting anthropometric data. Therefore, it is necessary to make adjustments for various items such as clothing or shoe heels as well as for the sitting slump (Pheasant, 1996). Pheasant (1996) and Pheasant and Haslegrave (2006) provide recommended corrections for each type of measure (e.g., 25 mm for men's shoes and 35 mm for helmets worn for protection).

9.5.4.1 Clothing

If individuals will wear heavy or thick clothing in the designed environment, it is best to test the fit of the environment when participants are wearing the clothes for extreme weather such as cold weather boots or winter jackets (O'Brien, 1996). The clothing will have an impact not only on fit, but also one's biomechanics, affecting reach and one's ability to manipulate controls and tools (see Section 9.10).

9.5.4.2 Secular growth

Another correction that is often considered is the size of the next generation of people who will use the designed equipment or environment. Young people appear to be getting taller (about 1–2 cm to 0.4–0.8 in.— every 10 years; Seidl & Bubb, 2006), which is known as **secular growth**. The stature growth trend may be slowing for the British and North Americans (Pheasant, 1990). In addition, body weight continues to increase as well as shoulder breadth and chest circumference (Kroemer, 1997). Thus, these potential changes should be considered in design.

9.6 DESIGN FOR ADJUSTABILITY

Because there is so much human variability, it is preferable to make the environment and equipment adjustable, if possible, to fit people between the 5th and 95th percentile for various dimensions. That is, we hope to accommodate 90% of the population. If it is not possible to design for adjustability, one usually designs for one extreme person of the population. For example, if a work surface is stationary, then it is better to design for the tallest person because shorter individuals can be accommodated with a platform (Kroemer & Grandjean, 1997). Because there is a great deal of knowledge about what designs are best under different circumstances, various standards have been developed to help with one's design.

9.7 STANDARDS

Standards or guidelines help ensure that there is uniformity across designs, which allows for interchangeability (Sherehiy et al., 2006). Although there are often regional, national, and international standards, these standards are not usually law; however, they can be the impetus to make a law if it is deemed appropriate.

The International Organization for Standardization (ISO) oversees the development of standards and you can find all ISO standards online at www.iso.org. The ISO has subcommittees (SC) and one in particular, SC3, deals primarily with Anthropometry and Biomechanics. Each subcommittee has working groups (WGs). The SC3, WG1 covers anthropometry. Some examples of relevant anthropometric standards are listed in Table 9.3.

Other available standards come from the military (e.g., MIL-STD-1472), NASA (e.g., NASA 1987), and the FAA. If you are in need of standards but do not know where to look, Carol Stuart-Buttle (2006) presents an overview of where to find standards and guidelines for various projects. You also can search the ISO website (www.iso.org) for "anthropometric" standards in the ISO store (yes, you have to order and pay for these). Another source for standards is the American National Standards Institute (ANSI—www.ansi.org).

Table 9.3 Examples of relevant anthropometric ISO standards

ISO standard	Title
ISO 6385:2016	Ergonomic principles in the design of work systems
ISO 7250-1:2017	Basic human body measurements for technological design—Part 1: Body measurement definitions and landmarks
ISO 8559:1989	Garment construction and anthropometric surveys—Body dimensions
ISO 9241-1, 1997	Ergonomic requirements for office work with visual display terminals (VDTs) —Part 1: General introduction
ISO 11226:2000	Ergonomics: Evaluation of static working postures
ISO 11228-1:2021	Ergonomics—Manual handling—Part 1: Lifting, lowering, and carrying
ISO 14738:2002	Safety of machinery—Anthropometric requirements for the design of workstations at machinery
ISO 15535:2012	General requirements for establishing anthropometric databases
ISO 15537:2022	Principles for selecting and using test persons for testing anthropometric aspects of industrial products and designs
ISO 16840-1:2006	Wheelchair seating—Part 1: Vocabulary, reference axis convention, and measures for body segments, posture, and postural support surfaces
ISO 20685-1:2018	3D scanning methodologies for internationally compatible anthropometric databases=Part 1: Evaluation protocol for body dimensions extracted from 3D body scans
ISO 24563:2023	Accessible design—Ease of operation

9.8 SEATING

Because many of us spend a lot of time sitting, let's look at specifics for appropriate seat design. One use of anthropometric data is for proper seat design to help us maintain appropriate posture. When viewed from the side, the human spine is naturally curved. When we sit, especially on hard surfaces, we lose the natural lumbar lordosis, or the natural curvature of the spine (Kroemer et al., 2001). The loss of the natural lumbar lordosis increases the loading on the back, or more specifically, the disks in the spinal cord. In addition, the back muscles must work harder to maintain the proper posture, which will fatigue over time. The muscles also can cause compression in the disks of the spine (Kroemer et al., 2001). This is why proper seating must provide lumbar support to help us maintain the appropriate curvature in the lumbar area.

Other criteria to consider in seat design include seat height, seat depth, seat width, backrest dimensions, backrest angle or "rake," seat angle or "tilt," armrests, legroom, and seat surface. Remember, maintaining a single body posture is not healthy. If we must sit for long periods of time (e.g., sedentary desk jobs), these criteria can help reduce loading, fatigue, and possible injury, but we should still move occasionally.

9.8.1 General criteria for good seating design

For seat height, we want to allow for the natural body posture. That is, we do not want the legs bent up too high or too low, which changes the curve of the lumbar region (see Figure 9.7a). A seated person should be able to place one's feet flat on the floor, keeping the thighs horizontal (Pheasant, 1996; Kroemer et al., 2001). To determine the appropriate **seat height**, we should consider the popliteal height of the users, which is the height from the floor to the crease in the back of the knee. A low popliteal height is preferable to having it too high. If the seat is too high, the front edge might cut off circulation in the back of the legs. Hence, we often design for the 5th percentile female seat height (Pheasant, 1996). Of course, seat height must be coordinated with the desk or working surface height so the arms and shoulders are in as close to a normal resting posture as possible, as demonstrated in Figure 9.7b.

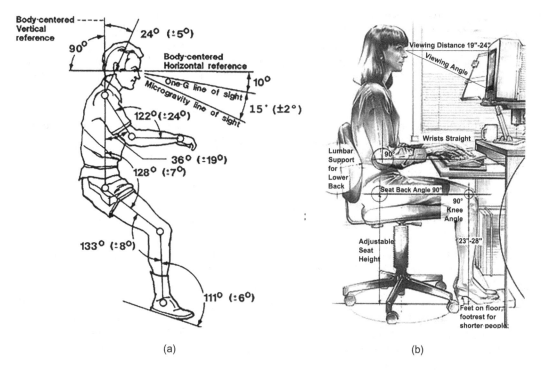

Figure 9.7 (a) Sitting in a natural relaxed posture in space (Adapted NASA, Public domain, via Wikimedia Commons.) (b) Sitting at a workstation. (Yamavu, CC0, via Wikimedia Commons.)

The **seat depth**, the distance from the front of the chair to the back rest, needs to be no greater than the 5th percentile dimension for woman so individuals can reach and use the backrest. According to Stephen Pheasant (1996), the seat depth can be quite small (300 mm; ~12 in.) and still give appropriate support to the **ischial tuberosities** (bones of the buttocks), although tall people might complain. The user's weight also should be distributed along the ischial tuberosities so people do not get sore when sitting and there is not too much pressure on the underside of the thighs (Pheasant, 1990).

The **seat width**, or breadth, the lateral distance of the seat pan, needs to support the width of the widest woman (95th percentile) because women generally are bigger than men across the hips (Kroemer & Grandjean, 1997). Again, though, the width can be small (25 mm less, ~1 in., on each side) and still provide sufficient support (Pheasant, 1996). It may be more important to consider the 95th percentile elbow width of a clothed man (Pheasant, 1996) in order to provide enough shoulder and arm clearance for the users.

Next, we should consider the backrest. Remember, it is important to provide appropriate lumbar support. A 40 mm (~1.5 in.) lumbar support placed at the midpoint of the curve of the lumbar region is sufficient to keep the back in its natural position (Andersson & Ortengren, 1974). Besides lumbar support, we also must consider the height of the backrest. The higher the back rest, the more support for the torso (Pheasant, 1996). Because the backrest is supposed to support the torso and the head, it is best to make the backrest as large as possible (at least 85 cm, ~3 ft high and 30 cm, ~ 1 ft wide) given the limitations of the workspace (Kroemer et al., 2001). In some work conditions, though, it is necessary to allow room for the shoulders to move.

Besides height, the backrest angle or **"rake"** affects how much of the trunk weight is supported. The farther one leans back from an upright position, the higher the backrest must be in order to support the torso and head (Pheasant, 1996). Leaning backward helps increase the angle between the torso and the thighs to be in a more natural position (see Figure 9.7a).

Pheasant (1996) recommends an optimal angle of 100°–110° from a perpendicular position. In contrast, Kroemer et al. (2001) suggest the angle be adjustable from 95° to 30° back (120° from the floor horizon). This discrepancy could be due to differences in the specific tasks of the users.

If the job or task allows, a semi-reclined position with some lumbar support reduces the load on the back and increases comfort (Pheasant, 1996). Of course, the backrest and lumbar support should be adjustable to allow for different types of work. Obviously, a reclining position is problematic when using a flat surface such as when writing. Therefore, it is also important to consider the user's performance, as well as comfort, preferences, and interests when designing chairs and environments (Pheasant, 1990).

Armrests also can be useful to help support the arm and keep the hands in proper positions, but must be positioned carefully so as to not impede movement. It is also possible to reduce the load to the shoulders and neck with wrist supports (Pheasant, 1996).

Seat angle or "tilt" can help the user maintain contact with the backrest without too much bend at the hips or sliding out of the seat. Ease of getting in and out of the chair also should be considered.

In the case of desks or tables, we need to consider vertical and forward leg room. For vertical leg room, we consider the knee height or, more importantly, men's 95th percentile thigh clearance (Pheasant, 1996). For forward clearance, although people are likely to sit farther back from the working surface, we assume that the abdomen touches the desk. For someone seated upright at a desk or table, we would subtract the abdominal depth from the buttock-knee length and then add the distance for the extended legs (about 150 mm, ~6 in.; Pheasant, 1996) to determine the depth needed for extended legs. If the user is in a seat lower to the ground, the person is more likely to stretch out one's legs and would need additional extended legroom.

Seat surface, as well as the seat depth and width, is used to distribute the weight along the ischial tuberosities. Pheasant (1996) suggests a simpler surface compared to a shaped surface, along with a rounded front to protect the back of the thighs and a material that allows for ventilation.

BOX 9.2:   SEATING FOR STUDENTS

As you know well, children spend a great deal of time sitting in chairs at desks in school for much of their young lives. One question arises as to whether the seating and desk design is appropriate for learning as well as growing bodies. Grenville Knight and Jan Noyes evaluated the posture of 9- and 10-year-old children in Bristol, England, when working in different types of chairs. Knight and Noyes (1999) sought to identify a chair that allowed students to sit in a relaxed position (see Figure 9.7a, b). As students spend up to a third of their school day leaning forward to write, in a traditional chair the students tend to sit inappropriately, adding load to the lumbar and hip areas. Chair 2000 is lower than a traditional student chair and has slightly more curving at the front of the chair to allow less hip flexion and less strain on the lumbar region. The back of the seat pan slants slightly backward, to place the student in a better seated position when they are leaning against the backrest. Knight and Noyes reported that students in traditional chairs spent more time in inappropriate postures than children in the Chair 2000. The inappropriate posture was suspected to be caused by chairs that were too tall; the chair's seat height was higher than the student's seat height. Given that inappropriate posture can lead to long-term musculoskeletal problems, as well as possibly distract students, it is probably a good idea to ensure that their workspace is designed appropriately. **Critical Thinking Questions**: How would you begin to evaluate this situation? What appear to be two of the main problems in this situation? How would you redesign the student chairs to ensure proper posture when students are working on computers?

9.8.2 Design for adjustable seating

Designing adjustable seating allows us to accommodate anyone on any dimension between the 5th percentile for women to the 95th percentile for men. If the seating is not adjustable, then a footrest can be used to support the users' feet, but the surface area should be large enough to allow users to stretch out their legs.

9.9 WORK SURFACE HEIGHT

Besides seating, work surface height is another important design issue. When determining the work surface height, we again want to keep the body in a fairly natural position. The arms should not be bent too much at the elbows depending on the type of work, and we want to avoid having the shoulders raised. If the surface is too high, we tend to raise our shoulders possibly causing neck and shoulder cramps. If the surface is too low, we often slump over potentially causing back problems (Kroemer & Grandjean, 1997). Again, the design needs to be "just right!" There are differences in what is the "right" height, though, based on whether the individual is standing or seated.

9.9.1 Work surface heights when standing

The work surface height is greatly influenced by the type of work performed. According to Kroemer and Grandjean (1997), when standing and performing handiwork, the best surface height is generally 50–100 mm (~2–4 in.) below elbow level. For different types of tasks, though, they recommend different surface heights. For example, when the work is more detailed or delicate such as drawing, they recommend a surface height 50–100 mm (~2–4 in.) above elbow height. When tools are used or there is a need for containers or various types of materials, then the surface height should be 100–150 mm (~4–6 in.) below elbow height. Finally, if the task requires some force or pressure (e.g., sanding), the surface height should be 150–400 mm (~6–16 in.) below elbow height (Kroemer & Grandjean, 1997).

9.9.2 Work surface heights when seated

Obviously, the appropriate surface height when seated also will depend on the task. When writing, the desktop or table surface should be just above the user's natural elbow height (Pheasant, 1996). Similarly, office work that requires the individual to process a great deal of paperwork should have the work surface 75 mm (~3 in.) above the sitting elbow height (Pheasant, 1996). In both cases, it is important that the shoulders remain relaxed as opposed to being lifted as if you are in a permanent shrug.

Given the prevalence of computers, many work surfaces used when seated should be designed for some form of data entry such as creating your paper for human factors! Again, the desire is to keep the shoulders relaxed as if hanging freely (see Figure 9.7b). The forearms should be close to horizontal (Pheasant, 1996) and, ideally, the keyboard will be at the level of your lap. Essentially, the home row of a keyboard (ASDFG), where you position your fingers for typing, should be close to sitting elbow height not above it (Pheasant, 1996). All of these positions help to reduce fatigue in the neck, shoulders, and wrists.

9.10 CONTROLS AND HAND ANTHROPOMETRY

In Chapter 8, you were introduced to the notion that controls should be designed so that they are compatible with the movement of dials, displays, or controlled movement. In this chapter, we are concerned with the dimension of the hands because the controls have to be of the "right" size so individuals can grasp them appropriately in order to create enough force to turn the controls; however, we must avoid creating pressure spots (Pheasant, 1996). In addition, it is important that the hand is not impeded by any obstruction. For instance, if a user's hand has to fit through an opening to turn a control, the aperture has to be large enough to allow the user to grasp and turn the control. Similarly, if individuals are wearing protective gloves, the controls, tool handles, and apertures must all accommodate the extra bulkiness of the gloves. Finally, when considering the design of controls, it is critical that the surface is not too slippery or too rough, but has an appropriate texture for ease of gripping. Because the use of controls can also involve biomechanics, we will first discuss biomechanics and then look into issues related to the biomechanics of the hand.

9.11 BIOMECHANICS

Along with a good understanding of anthropometry to design an environment, we need to understand the required body movements within that space such that the task does not place excessive strain on the body. For example, we should not have our wrists in a continuously bent position while typing nor should we have to move our arms behind our backs in an awkward position to flip a switch or pick up materials. In the case of Jeff's mom, she should not be lifting or handling heavy dishes well above her head. These issues involve biomechanics.

Biomechanics explain the movements of the human body such as the directions and extent to which a joint or the back bends. If you consider the human body as a machine, biomechanics describe the appropriate uses and limits of this machine. If the limits are exceeded, then it is likely that there will be injury or damage to the machine. Industrial engineers apply their knowledge of biomechanics to simplify motion and reduce movement redundancies for a task, whereas the ergonomist applies biomechanics to reduce biomechanical stress (Chaffin, 1987). Hence, biomechanics explain in physical terms what a human body can and cannot do (Ayoub et al., 1987), and understanding biomechanics in addition to anthropometry can help us prevent further injury in the proper design of our environment.

9.11.1 Types of motions

Just as landmarks must be clearly specified in anthropometry, body movements must be clearly specified in biomechanics. Some basic movements include flexion, extension, abduction, and adduction. When we reduce the angle at the elbow or knee, for example, by bending our arm or leg, this is flexion. In contrast, extension occurs when we increase the angle such as at the elbow or knee by extending our arm or leg (see the top images in Figure 9.8). Now, if we were to hold our arm straight out in front and move it across your body, this would be adduction, while moving the arm out to the side would be abduction, as demonstrated by the bottom right image in Figure 9.8.

9.11.2 Hand biomechanics and hand tools

In many types of work, besides moving and bending various parts of our bodies such as our arms, legs, and back, we are often turning and twisting our wrists in different ways.

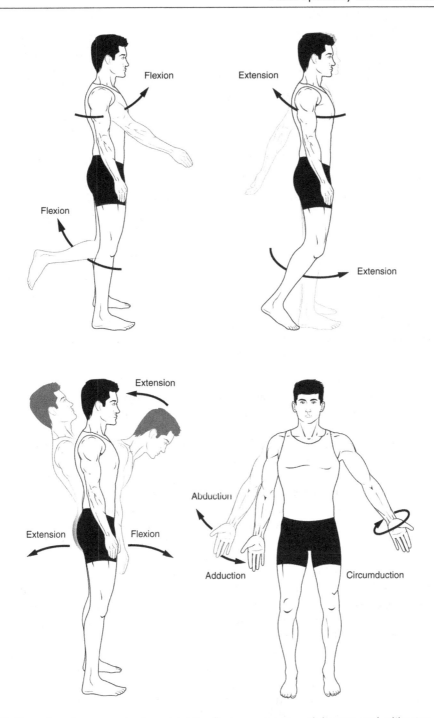

Figure 9.8 Examples of movements demonstrating flexion, extension, abduction, and adduction. The top two images demonstrate flexion and extension in the knee and shoulder. The bottom left image demonstrates flexion and extension at the hip. Abduction (away from the body center) and adduction (toward the body center) are demonstrated in the bottom right image. (Adapted Tonye Ogele CNX, CC BY-SA 3.0 <https://creativecommons.org/licenses/by-sa/3.0>, via Wikimedia Commons.)

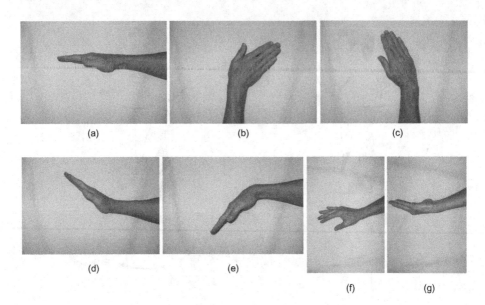

Figure 9.9 Examples of wrist movements. There are numerous motions possible from the neutral position with a straight wrist (a). These images demonstrate (b) ulnar deviation, (c) radial deviation, (d) dorsiflexion, (e) palmar flexion, (f) pronation, and (g) supination. (© Nancy J. Stone.)

Therefore, we will spend a little extra time with the biomechanics of the hand. The hand is an extremely flexible appendage that can move in many directions, as demonstrated in Figure 9.9. The **neutral position** of the hand is when it extends from the wrist without a bend in the wrist. From this neutral position, we can move our hand in the direction of the little finger, known as **ulnar deviation. Radial deviation** occurs when we move our hand toward the thumb. Besides moving the hand sideways, we can also move it up and down. With the hand palm side down, when we move the hand up and extend the wrist, we create **dorsiflexion.** When we move the hand down, we cause **palmar flexion.** We can also twist the wrist. With the palm facing down, turning the outer side with the pinky finger up is **pronation,** and twisting the side with the pinky down and around is **supination.**

Ideally, to reduce the chance of injury, we want to keep "straight" wrists whereby the hand and wrist are in the neutral position. Deviations from the neutral position increase the risk of injury. Wrist deviation, repetitive actions, or excessive force can cause inflamed tendons or **tendonitis.** If the actions continue without allowing the tendons to heal and if the swelling continues, there is a chance for greater injury. The swelled and inflamed tendons pass through the bone structure of the wrist, causing the tendons to rub against the bones and damage the tendons, which can lead to pain, reduced motion, and over time, permanent damage (Kroemer, 1987).

We are all susceptible to this type of injury. When using our computer keyboards, our wrists often have both dorsiflexion and ulnar deviation. In addition, keyboarding requires repetitive motions, which increases the chance for carpal tunnel syndrome. **Carpal tunnel syndrome** involves more than inflamed tendons, but compression damage to the median nerve that goes through the wrist's carpal tunnel (Eleftheriou et al., 2012). An individual who suffers from carpal tunnel syndrome might experience tingling or prickly feelings, numbness, an inability to grasp, and other losses of hand functions (Johnson, 1985). With the use of wrist rests and an ergonomically designed keyboard, it is possible to place the hands in their neutral position, reducing the chance of injury.

It is possible to observe wrist deviations in other tasks besides keyboarding. For example, if you were to use a screwdriver or certain types of pliers, both lead to ulnar deviation along with using force, and the screwdriver also includes twisting of the wrist in this deviated position. Therefore, a number of new designs, as well as electric tools, are intended to eliminate these deviations, placing the wrist in a more neutral position and reducing the chance for injury.

BOX 9.3:   WHEELCHAIRS—SEATING AND TRANSPORTATION FOR INDIVIDUALS WITH DISABILITIES

Wheelchair users either have a temporary injury and spend limited time in a wheelchair or have chronic issues that confine the individual to the wheelchair. Although seating and propulsion issues are a concern for all wheelchair users, they are more severe for the long-term user, which is our focus here.

One of the main problems for individuals confined to a wheelchair is pressure ulcers or sores, also known as bed sores. Although pressure ulcers can occur in many places, long-term wheelchair users are highly susceptible to pressure ulcers on the ischial tuberosities, at the back of the knee (popliteal area), on the protruding bumps of the spine known as bony prominences, on the shoulder, and on the back of the heels (Stockton et al., 2009).

To avoid pressure sores, wheelchair seating needs to meet the same seating standards as discussed in the text relative to seating height, seating depth, backrest incline, and so on, as long as it also accommodates the conditions of the user. For example, if the backrest leans back about 6°, this reduces disk pressure in the back, but we do not want a user to slide out of the chair, as some users do not have this control (Andersson & Ortengren, 1974). Armrests also help reduce disk pressure (Andersson & Ortengren, 1974) and assist the user in necessary movements that reduce pressure (Stockton et al., 2009). Regardless of the quality of the seat cushion, movement is needed to reduce pressure. Some suggest movements every 15–30 minutes (Stockton et al., 2009). Because some wheelchair users do not have the arm strength or control, they are often assisted in these movements by nurses to reduce pressure. To reduce the need for nursing assistance, a depressurization motion assistance device can assist the movement if a user has minimal strength to initiate the movement such as using the armrest to roll to the sides, forward, or backward (Chugo et al., 2011).

As wheelchair users differ in their needs and their distribution of pressure when seated, different types of pressure mapping are helping to better identify the pressure distribution (e.g., Eitzen, 2004). In addition, besides the material of the seat cushion, boney prominences can still become sore with long-term sitting. Air chambers in the seat can be used to relieve excessive pressure by identifying location of the excessive pressure and then using the air chambers to redistribute the pressure (Yang et al., 2010). Using vibrating grains is another method for creating individualized seat pan designs for better weight distribution (Liu et al., 2018).

The seating also can impact the user's mobility or propulsion, as the majority of wheelchair users are manual wheelchair users (MWUs). Although a lower seat and a seat placed in front of the axle of the main wheel create better stability, it also creates more biomechanical problems such as ulnar and radial wrist deviations and a greater chance of upper extremity (e.g., shoulder) musculoskeletal problems (Gorge & Louis, 2012). This could be a greater problem for children or adolescents whose bodies are growing and changing (Schnorenberg et al., 2014). **Critical Thinking Questions**: What are some possible seating design challenges for someone with various disabilities (e.g., missing limb, multiple sclerosis, or broken leg)? How does the seating cushion material impact not only seating, but the ability to move? What about the ability to clean the seat or the durability? Consider different seating positions and identify different biomechanical stressors that are likely. How might you accommodate the wheelchair user in a classroom or at a workstation?

9.11.3 Manual materials handling

We might not be concerned if we infrequently use a screwdriver or computer; however, a single event can cause an injury. In fact, some jobs put excessive stress on the body joints and the back that can lead to musculoskeletal disorders. Consequently, biomechanics are a large concern in industry where individuals must manually move materials, and the movements include lifting, bending, and often a combination of lifting and bending along with twisting. When an individual is physically moving materials, this is known as **manual materials handling** (MMH). MMH occurs in a variety of industrial tasks, is a hazard to workers, and can lead to injuries that are an enormous expense to companies (Ayoub et al., 1987). Fortunately, biomechanical data can determine the amount of physical stress placed on the body during certain activities such as lifting. These data can then be used to identify a form of handling materials that place less stress on the body (Ayoub et al., 1987).

Because MMH generally includes some form of lifting, one of the main concerns within the area of biomechanics is lower back pain (LBP). To avoid LBP, it is best to reduce the load or stress to the lower back (i.e., the lumbar region). Generally, the leg lifting technique where one bends the knees and keeps the back straight is presented as the best technique to reduce back strain. Yet, in a review of the literature focusing on tasks requiring lifting and lowering, some findings suggest that the back lifting technique, or bending at the waist, may be best under some circumstances (Cole & Grimshaw, 2003). In many cases, though, individuals often adapt to the situation and use a free form technique that may be perceived as beneficial, but is not mechanically sound (Cole & Grimshaw, 2003).

These contradictory findings about lifting techniques might reflect differences in lifting tasks including the absence or presence of handles, task symmetry or asymmetry, and the weight of the object being moved. A symmetrical task involves the movement of the object only in front of the individual (e.g., lifting and placing the object on a shelf). An asymmetrical task involves twisting of the trunk or the leaning of the torso to one side or the other. Not having handles on the object is the worst condition, regardless of the task symmetry or weight of the object (Drury et al., 1989). The best location of the handles was influenced by whether the task was symmetrical or asymmetrical.

When individuals performed a symmetrical task, placing the handles in the middle of the top side edges was best (Drury et al., 1989). For the asymmetrical task, individuals moved boxes between a conveyor belt and a pallet that was perpendicular to the conveyor belt. In this condition, the best handle placements varied depending on the box weight. Boxes with lower weight were lifted with less stress on the back if one handle was in the middle of the back edge of the box and the opposite side handle was in the middle of the bottom edge of the box. When the box weight was heavy, it was best if both handles were in the middle of the bottom edge of the box. These handle positions apparently helped the individuals keep the heavy box closer to their bodies. Further, these handle positions helped reduce ulnar deviation in the hands (Drury et al., 1989). Because asymmetrical tasks are considered to be more realistic to industry tasks (e.g., Cole & Grinshaw, 2003; Drury et al., 1989), it is important to understand these implications.

Even if you have not worked in a warehouse, you have probably witnessed an asymmetrical MMH task. Consider the cashier position at a grocery store, a type of MMH task, which is also a highly repetitive task. Most cashiers in the United States stand; however, many in Europe sit while performing their jobs. Generally, there is less stress when standing than sitting and when the individuals use a bi-optic scanner—the scanners are on both the vertical and horizontal planes—rather than the single vertical window scanner (Lehman et al., 2001), even though many individuals preferred the seated position. The lowered stress from standing was related to a reduction in the amount of arm abduction, a more natural level of neck flexion, and less disk pressure caused by sitting, especially given that the bi-optic scanner made it easier to read the bar codes, requiring less turning of the object. The worst condition occurred when the individuals were seated using the single vertical window scanner, as there was more stress on the neck and shoulder areas due to greater shoulder abduction (Lehman et al., 2001).

Another repetitive MMH profession that is affected by working height is bricklaying. Because of the high number of complaints of pain in the lower back and shoulders, Henk Van der Molen and colleagues evaluated the working environment of Dutch bricklayers and bricklayers' assistants (Van der Molen et al., 2004). Bricklayers tend to move bricks and mortar using an awkward one-handed position. Bricklayers' assistants, who are responsible for transporting the bricks and mortar, perform a great deal of lifting as well as pushing and pulling of wheelbarrows (Van der Molen et al., 2004). Because of these actions, the assistants experience a great deal of back flexion (>60°), as well as the handling of objects above and below shoulder height.

When the bricklayers used a scaffolding, this did not affect performance; however, it did significantly reduce the frequency and duration of back flexion (>60°). Further, self-reports indicated that the use of the scaffolding significantly reduced discomfort in the lower back (Van der Molen et al., 2004). Similarly, the bricklayers' assistants' performance was largely unaffected when using a crane for transporting the bricks and mortar, but back flexion frequency and duration decreased significantly. In addition, there was a significant reduction in the frequency of handling objects below shoulder level, but there was no change in the frequency of handling objects above shoulder level, possibly due to the small number of activities (Van der Molen et al., 2004). Recall in the opening vignette how Jeff placed dishes in the highest cupboards his mother could not reach. Handling heavy dishes high above her shoulder height could cause her to experience shoulder or other biomechanical stresses and strains.

From the examples above, it is possible to reduce biomechanical stress with fairly simple modifications. It is important, though, that one type of biomechanical stress does not replace another. Although the bricklayers' assistants had reduced back flexion when using the crane for transporting the bricks, they were now sitting longer while standing and walking less (Van der Molen et al., 2004). It is necessary to ensure that the new design does not lead to problems associated with improper seating.

9.12 WORKSPACE DESIGN

It should now be apparent that if we want to design a workspace, we must consider the anthropometric dimensions for sitting, standing, and work surface height. It is also important to consider the task and biomechanics (Kroemer, 1987; Kroemer & Grandjean, 1997). We will consider two important workspace design aspects that concern anthropometry: reach envelopes and line of sight. Additional environmental design issues are addressed in Chapter 10.

9.12.1 Reach envelope

Sitting at your desk or table, place your elbows on the edge of the surface and lay your forearms flat and straight out in front of you. Now, create a windshield wiper motion. The motion of your hands defines a region of space called your normal **reach envelope**, a space within easy reach for you without changing your seated or standing position. There is also an *extended reach envelope* such as when you lean forward and extend your arm to turn on your lamp or reach for another resource. In Chapter 10, we discuss how reach envelopes are used to determine where to place controls or items, placing the most important or more frequently used items closer to you. When studying, you will want to place the most important materials for studying within the reach envelope (e.g., computer, books, and paper) and those that are less important or used less frequently in the extended reach envelope (e.g., supplemental books, food, and beverage).

Now, consider the garage where people get a quick oil change. If the garage is designed well, the tools will be within the reach envelope of the employee. Of course, the envelope extends beyond just a flat working surface. The reach envelope is a three-dimensional space, but it is not symmetrical. Your reach envelope does not include items located behind you because our biomechanics do not easily allow for this type of movement. The three-dimensional reach envelope allows the garage to hang hoses overhead that can easily be pulled and used to fill the oil or other fluids, but there are few items, if any, behind the individual. Obviously, the worker could turn around or walk to another area, but that would redefine the workspace and reach envelope. For easy use, the 5th percentile reach envelope should probably be designed for the population of users.

Biomechanically, the tasks required in the extended reach envelope should be considered. Individuals can reach, grasp, and pull a cord such as in our oil change example, but should not be expected to handle heavy loads high above one's head, such as Jeff's mom. If the task requires the use of force or if someone must grasp an object with some force or turn a dial or knob, then the reach envelope should be shorter than for reaching and pushing a button. It is also important that the hands are not required to be held above one's shoulders for long periods of time (Kroemer, 1987). If this is necessary, some type of support is recommended to ensure a safer design.

BOX 9.4:   WORKSPACE DESIGN FOR FARMERS

When we consider workspace design, we might naturally think of the person working at a desk all day, using a computer. Many people, though, have other workspaces that can be addressed with anthropometric data to ensure a good fit between the worker and the environment. One of these environments is the tractor cab for farmers. H. Hsiao and his colleagues investigated the anthropometric dimensions of tractor cabs for farmers in West Virginia (Hsiao et al., 2005). Hsiao et al. collected traditional and 3D full-body scanner measurements on 100 individuals, 88 of them were men. These farmers in West Virginia tended to be bigger in sitting height and stature than the national mean for farmers, but they were near the national mean for the civilian population. Based on these data, they found that the current standard for tractor cab height was too short for the 99th percentile in this study. According to Hsiao et al., the current minimum clearance standard is 90 cm (~36 in.), but their calculated minimum clearance is 100.6 cm (~40 in.). This created a safety problem given that the farmers who had a greater sitting height often folded back the protective frame, most likely to give themselves more head room. If the tractor were to roll, these individuals would have no roll-over protection. **Critical Thinking Questions**: If the cab height is too short, what are some other "adjustments" the farmer is likely to make? How will this impact posture and biomechanical movements? How is safety impacted?

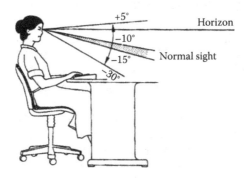

Figure 9.10 Line of sight. (From Taylor & Francis Group LLC Books.)

9.12.2 Line of sight

Our line of sight also should be considered because it influences degree of neck curvature, which can have long-term health implications. The **line of sight** refers to where we naturally look when we are in our natural standing or sitting position. The head should not be leaning too far forward or back, as this can cause injuries in our neck (Kroemer & Grandjean, 1997). If we are seated upright, our line of sight is the horizon if we are looking at objects at a far distance. Except for when we are daydreaming, we generally are not looking at an object located far away.

Our line of sight for objects that are relatively close (e.g., keyboards and documents) tends to be quite a bit below the horizon (see Figure 9.10). Therefore, when setting up our computer work stations, the computer screen should not be much higher than the keyboard (assuming the keyboard and computer screen stand are placed on the same surface); otherwise, our necks will be bent, causing pressure in the neck area (Kroemer et al., 2001). Given that our natural line of sight is close to the keyboard or just above it, the computer screen should be placed as close to and just behind the keyboard as possible (Kroemer et al., 2001, p. 419). Keep in mind, if you must recline when keeping your neck straight to see the computer screen, the screen is too high!

9.12.3 Computer workspace design

Because of the prevalence of computer use by children and adults alike, let's take a moment to consider the specifics of computer workspace design, pulling together what we have discussed. Beginning with the work surface height, recall that the home row of the keyboard should be at the same height as the individual's sitting elbow height. If the work surface height cannot be changed, then the seat may need to be raised or lowered for appropriate alignment.

Once the seating height is determined, we can focus on the seating itself. Traditionally, the proper posture for computer work was the perpendicular position, but more recently, people have supported the "laid-back" position (Pheasant, 1996). Recall that if people lean back a bit in their chairs, the backrest and lumbar support will reduce the load on the back. Furthermore, it is important that the individual can place both feet on the ground. If the chair has been raised to fit the work surface height, the individual might need a platform for support. Next, we can consider our wrist positions. We do not want our wrists bent, leaning on the table edge, and we want to avoid ulnar deviation. Finally, we want to check to see that our computer screen is not too high for our line of sight and that our keyboard and mouse are within our reach envelope.

9.13 APPLICATIONS TO SPECIAL POPULATIONS

By implementing these design changes to our computer workstations, we can reduce the stress on our back, neck, and shoulders by creating better biomechanical alignment. Although young people might not "feel" the effects, these changes can reduce the cumulative trauma that begins to appear later in life. For some people, there may need to be other or additional types of adjustments. As you may recall, much of the anthropometric data are used for designing workspaces for working adults between the ages of 20 and 65 years (Kroemer & Grandjean, 1997). Also, adjustability helps us address much of the variability in the working adults' body dimensions from the 5th to the 95th percentile. Nevertheless, these designs tend to focus on adults without disabilities. We need to determine ways to design environments to meet the needs of individuals with disabilities or who are not adults.

Some special populations include ethnic groups, age groups, individuals with disabilities, and pregnant women (Kroemer, 1997). Age groups are important because children are not only smaller than adults, but their dimensions or proportions are different from adults. Furthermore, with the increase of retirees returning to work and an aging workforce, in general, we must consider the total continuum of ages. For example, individuals more than 60 years of age often have a decrease in stature and reach, but an increase in waist measurements (Kroemer & Grandjean, 1997). These issues are also concerns for the design of non-work environments that are used by the elderly including assisted living environments. Areas of concern for the elderly include bathrooms. In particular, elderly adults often have trouble getting in and out of bathtubs, reaching the facet, or using the sink (Kroemer, 1997).

Because these various populations have different needs, we should consider universal design. **Universal Design** means designing for all potential users and was coined at North Carolina State University.

9.14 SUMMARY

In this chapter, we highlighted some of the critical areas of anthropometry such as basic measures, anthropometric tools, and how anthropometric data are collected and applied along with various aspects of biomechanics and workspace design. Anthropometric measurements are needed to understand human dimensions used in designing spaces, equipment, and tools humans use. Collecting anthropometric data requires the use of specified landmarks and the understanding of the orientation planes.

Using anthropometric data to design the environment or tools to fit the users better helps reduce fatigue and injury. Two important measures include clearance and reach, which are often measured using the limiting user. In addition, anthropometric data also helps create a better fit to ensure proper posture. Although we often collect static anthropometric measures, we also need dynamic measures to better understand the range of the design needs. Further, we include corrections for the type of clothing a person must wear and secular growth to achieve the best fit. Ideally, we design for adjustability and we can refer to various standards to help guide our design decisions. With much of the workforce working in sedentary jobs, seating is one of the main concerns. Proper seating can reduce loading on the spine and various joints.

Additional reductions in fatigue or loading can be achieved by considering biomechanical data, which helps reduce immediate or cumulative trauma such as carpal tunnel syndrome. Proper design can reduce injury, especially lower back problems, for individuals working in manual materials handling. The design of reach envelopes and line of sight also impact posture and biomechanical function. Next time you sit down at your desk to study or work on the computer, think about how the anthropometric data were or were not applied to this design. Using the basic information presented in this chapter, it should be possible to design or redesign your workspaces to be a better fit whether it is a kitchen, a study space, or an office space at work.

LIST OF KEY TERMS

Abduction

Adduction

Anthropometry

Ascending trial

Biomechanics

Breadth

Carpal tunnel syndrome

Circumference

Clearance

Cones

Coronal plane

Curvature

Depth

Descending trial

Distal

Dorsiflexion

Dynamic measures

Extension

Fitting trial

Flexion

Height

Ischial tuberosities

Landmarks

Limiting user

Line of sight

Load

Manual materials handling

Method of limits

Normal distribution

One-way constraint

Palmar flexion

Percentile point

Percentiles

Popliteal

Pronation

Proximal

Radial deviation

Rake

Reach

Reach envelope

Sagittal plane

Secular growth

Sliding caliper

Spreading caliper

Static measures

Stature

Supination

Tape measure

Tendonitis

Tilt

Transverse plane

Two-way constraint

Ulnar deviation

z-score

SUGGESTED READING

Chaffin, D. B. (1987). Biomechanical aspects of workplace design. In G. Salvendy (Ed.). *Handbook of Human Factors*, pp. 601–619, Chapter 5.4. Hoboken, NY: John Wiley & Sons.

Don Chaffin is a leader in the field of biomechanics. This chapter provides a good foundation on biomechanics.

Delleman, N. J., & Dul, J. (2006a). Evaluation of static working postures. In W. Karwowksi (Ed.). *Handbook of Standards and Guidelines in Ergonomics and Human Factors*, pp. 169–196. Mahwah, NJ: Lawrence Earlbaum.

If you want to gain a better understanding of how static measures are evaluated, this chapter provides an excellent review.

Delleman, N. J., & Dul, J. (2006b). Evaluation of working postures and movements in relation to machinery. In W. Karwowksi (Ed.). *Handbook of Standards and Guidelines in Ergonomics and Human Factors*, pp. 169–196, Mahwah, NJ: Lawrence Earlbaum.

If you are more interested in how to assess or evaluate dynamic movements, you should consider reading this chapter.

Delleman, N. J., Haslegrave, C. M., & Chaffin, D. B. (Eds.). (2004). *Working Postures and Movements: Tools for Evaluation and Engineering*. Boca Raton, FL: CRC Press.
  Written by leaders in the field, this book provides a thorough review of anthropometric and bio-mechanical issues encountered at work.
Kroemer, K. H. E. (1987). Engineering anthropometry. In G. Salvendy (Ed.). *Handbook of Human Factors*, pp. 154–168, Chapter 2.5. Hoboken, NY: John Wiley & Sons.
  All of Karl Kroemer's work provides valuable information on anthropometry. This chapter gives a good overview on terminology and the foundations of anthropometry.
Kroemer, K. H. E., Kroemer, H. B., & Kroemer-Elbert, K. E. (2001). *Ergonomics: How to Design for Ease and Efficiency* (2nd ed). Upper Saddle River, NJ: Prentice Hall.
  This text gives a good resource on anthropometry, especially for seating and workstation design.
Pheasant, S., & Haslegrave, C. M. (2006). *Bodyspace: Anthropometry, Ergonomics, and the Design of Work*. Boca Raton, FL: Taylor & Francis.
Stephen Pheasant was a leader in the field on anthropometry and the use of anthropometric data for designing workspaces. This book is still a great resource today even though Stephen Pheasant passed away in 1996.

CHAPTER EXERCISES

1. Using Table 9.1, calculate the 5th percentile for women's
   a. Stature
   b. Leg length
   c. Resting elbow height
2. If your distribution has a mean of 78 and a standard deviation of 8, calculate the 5th, 50th, and 95th percentile point.
3. Determine the ideal desk height and seating height for the 50th percentile male and the 5th percentile female when writing at a desk (using Table 9.1).
4. Calculate the ideal desk height for writing for the 50th percentile for all people (i.e., the appropriate surface working height including both men and women).
5. Observe individuals studying in the library. Identify and describe what needs to be changed for appropriate fit.
6. Observe individuals using the computers in the library. Identify and describe what needs to be changed for an appropriate fit.
7. Evaluate your seating when writing at a desk. Determine if
   a. Your seating is appropriate. Identify and describe what needs to be changed.
   b. Your posture is appropriate. Identify and describe what needs to be changed.
8. Evaluate your computer workstation and determine what is wrong with the current design. At a minimum, consider seating and surface height, reach envelopes, and line of sight. Identify and describe what you would change to make a better anthropometric fit.
9. Evaluate the seating in two different vehicles. Determine if the seating is appropriate for a large number of people (i.e., is it adjustable?). Consider issues of reach and clearance.
10. Observe at least one type of check-out counter (e.g., grocery store). Describe what is wrong with this design and what you might change
    a. For standing individuals
    b. For seated individuals (e.g., people in wheelchairs, seated workers)
11. Describe the differences in what you observed and what you would change after evaluating the sitting, reach, and clearance issues at a workstation for

a. Children
b. The elderly
c. Disabled individuals

REFERENCES

Albin, T., & Molenbroek, J. (2023). Introduction to the special issue, anthropometry in design. *Ergonomics in Design, 31*(3), 3–6.

Alemany, S., Vlero, J., & Ballester, A. (2023). Advanced processing of 4D body scanning technology for the ergonomic design of products and environments. *Ergonomics in Design, 31*(3), 30–37.

Andersson, B. J. G., & Ortengren, R. (1974). Lumbar disc pressure and myoelectric back muscle activity during sitting, III. Studies on a wheelchair. *Scandinavian Journal of Rehabilitation Medicine, 6*, 122–127.

Ayoub, M. M., Selan, J. L., & Jiang, B. C. (1987). Manual materials handling. In G. Salvendy (Ed.). *Handbook of Human Factors*, pp. 790–818, Chapter 7.2. Hoboken, NY: John Wiley & Sons.

Bradtmiller, B. (2023). 3D scanning and head-mounted products. *Ergonomics in Design, 31*(3), 26–29.

Chaffin, D. B. (1987). Biomechanical aspects of workplace design. In G. Salvendy (Ed.). *Handbook of Human Factors*, pp. 601–619, Chapter 5.4. Hoboken, NY: John Wiley & Sons.

Chugo, D., Fujita, K., Sakaida, Y., Yokota, S., Takase, K. (2011). Depressurization assistance according to a posture of a seated patient. In *4th International Conference on Human System Interaction, HSI*, art. no. 5937380, September 19–23, 2011. Las Vegas, NV. pp. 287–292.

Cole, M. H., & Grimshaw, P. N. (2003). Low back pain and lifting: a review of epidemiology and aetiology. *Work, 21*, 173–184.

Cotton, L. M., O'Connell, D. G., Palmer, P. P., & Rutland, M. D. (2002). Mismatch of school desks and chairs by ethnicity and grade level in middle school. *Work: Journal of Prevention, Assessment & Rehabilitation, 18*, 269–280.

Delleman, N. J., & Dul, J. (2006a). Evaluation of static working postures. In W. Karwowksi (Ed.). *Handbook of Standards and Guidelines in Ergonomics and Human Factors*, pp. 169–196. Mahwah, NJ: Lawrence Earlbaum.

Delleman, N. J., & Dul, J. (2006b). Evaluation of working postures and movements in relation to machinery. In W. Karwowksi (Ed.). *Handbook of Standards and Guidelines in Ergonomics and Human Factors*, pp. 169–196. Mahwah, NJ: Lawrence Earlbaum.

Delleman, N. J., Haslegrave, C. M., & Chaffin, D. B. (Eds.). (2004). *Working Postures and Movements: Tools for Evaluation and Engineering*. Boca Raton, FL: CRC Press.

Drury, C. G., Deeb, J. M., Hartman, B., Woolley, S., Drury, C. E., & Gallagher, S. (1989). Symmetric and asymmetric manual materials handling, part 2: biomechanics. *Ergonomics, 32*, 565–583.

Eitzen, I. (2004). Pressure mapping in seating: a frequency analysis approach. *Archives of Physical Medicine and Rehabilitation, 85*(7), 1136–1140.

Eleftheriou, A., Rachiotis, G., Varitimidis, S. E., Koutis, C., Mailzos, K. N., & Hadjichristodoulou, C. (2012). Cumulative keyboard strokes: a possible risk factor for carpal tunnel syndrome. *Occupational Medicine and Toxicology, 7*, article 16. doi:10.1186/1745-6673-7-16.

Geil, M. D. (2005). Consistency and accuracy of measurement of lower-limb amputee anthropometrics. *Journal of Rehabilitation Research and Development, 42*(2), 131–140. doi:10.1682/JRRD.2004.05.0054.

Gorce, P., & Louis, N. (2012). Wheelchair propulsion kinematics in beginners and expert users: influence of wheelchair settings. *Clinical Biomechanics, 27*, 7–15.

Grandjean, E. (1980). *Fitting the Task to the Man*. London: Taylor & Francis.

Herzberg, H. T. E. (1960). Dynamic anthropometry of working positions. *Human Factors, 2*, 147–155.

Hsiao, H., Whitestone, J., Bradtmiller, B., Whisler, R., Zwiener, J., Lafferty, C., Kau, T.-Y., & Gross, M. (2005). Anthropometric criteria for the design of tractor cabs and protection frames. *Ergonomics, 48*, 323–353.

Johnson, K. (1985). Analytical report on the causes and preventions of carpal tunnel syndrome. *Professional Safety*, 30, 48–51.

Knight, G., & Noyes, J. (1999). Children's behaviour and the design of school furniture. *Ergonomics*, 42, 747–760.

Kroemer, K. H. E. (1987). Engineering anthropometry. In G. Salvendy (Ed.). *Handbook of Human Factors*, pp. 154–168, Chapter 2.5. Hoboken, NY: John Wiley & Sons.

Kroemer, K. H. E. (1997). Anthropometry and biomechanics. In A. D. Fisk & W. A. Rogers (Eds.). *Handbook of Human Factors and the Older Adult*, pp. 87–124. San Diego, CA: Academic Press.

Kroemer, K. H. E., & Grandjean, E. (1997). *Fitting the Task to the Human: A Text of Occupational Ergonomics*. London: Taylor & Francis.

Kroemer, K. H. E., Kroemer, H. J., & Kroemer-Elbert, K. E. (1997). *Engineering Physiology: Bases of Human Factors/Ergonomics* (3rd ed). Upper Saddle River, NJ: Prentice Hall.

Kroemer, K. H. E., Kroemer, H. B., & Kroemer-Elbert, K. E. (2001). *Ergonomics: How to Design for Ease and Efficiency* (2nd ed). Upper Saddle River, NJ: Prentice Hall.

Lehman, K. R., Psihogios, J. P., & Meulenbroek, R. G. J. (2001). Effects of sitting versus standing and scanner type on cashiers. *Ergonomics*, 44, 719–738.

Liu, S., Qu, Y., Hou, S., Li, K., Li, X., Zhai, Y., & Ji, Y. (2018). Comfort evaluation of a subject-specific seating interface formed by vibrating grains. *Applied Ergonomics*, 71, 65–72.

Molenbroek, J. F. M. (Johan) (2018): DINED – anthropometric database. Version 1. 4TU.ResearchData. collection. https://doi.org/10.4121/uuid:199467d8-5c40-4a1f-a2f2-f2040db26270.

Nowak, E. (1997). Anthropometry for the needs of disabled people. In S. Kumar (Ed.). *Perspectives in Rehabilitation Ergonomics*, pp. 302–338. London: Taylor & Francis.

O'Brien, T. G. (1996). Anthropometry, workspace, and environmental test and evaluation. In T. G. O'Brien & S. G. Charlton (Eds.). *Handbook of Human Factors Testing and Evaluation*, pp. 223–264. Mahwah, NJ: Lawrence Erlbaum.

Oudenhuijzen, A., Essens, P., & Malone, T. B. (2002). In S. G. Charlton & T. G. O'Brien (Eds.). *Handbook of Human Factors Testing and Evaluation* (2nd ed), pp. 457–471. Mahway, NJ: Lawrence Erlbaum Associates.

Pheasant, S. T. (1990). Anthropometry and the design of workspaces. In J. R. Wilson & E. N. Corlett (Eds.). *Evaluation of Human Work: A Practical Ergonomics Methodology*. London: Taylor & Francis.

Pheasant, S. (1996). *Bodyspace: Anthropometry, Ergonomics and the Design of Work* (2nd ed). London: Taylor & Francis.

Pheasant, S., & Haslegrave, C. M. (2006). *Bodyspace: Anthropometry, Ergonomics, and the Design of Work* (3rd ed). Boca Raton, FL: Taylor & Francis.

Roebuck, J. A. Jr. (1995). *Anthropometric Methods: Designing to Fit the Human Body*. Santa Monica, CA: Human Factors and Ergonomics Society.

Schnorenberg, A., Slavens, B., Wang, M., Vogel, L., Smith, P., & Harris, G. (2014). Biomechanical model for evaluation of pediatric upper extremity joint dynamics during wheelchair mobility. *Journal of Biomechanics*, 47(1), 269–276. doi: 10.1016/j.jbiomech.2013.11.014.

Seidl, A., & Bubb, H. (2006). Standards in anthropometry. In W. Karwowksi (Ed.). *Handbook of Standards and Guidelines in Ergonomics and Human Factors*, pp. 169–196. Mahwah, NJ: Lawrence Earlbaum.

Sherehiy, B., Rodrick, D., & Karwowski, W. (2006). An overview of international standardization efforts in human factors and ergonomics. In W. Karwowksi (Ed.). *Handbook of Standards and Guidelines in Ergonomics and Human Factors*, pp. 3–46. Mahwah, NJ: Lawrence Earlbaum.

Stuart-Buttle, C. (2006). Overview of national and international standards and guidelines. In W. Karwowksi (Ed.). *Handbook of Standards and Guidelines in Ergonomics and Human Factors*, pp. 133–147. Mahwah, NJ: Lawrence Earlbaum.

Van der Molen, H. F., Grouwstra, R., Kuijer, P. P. F. M., Sluiter, J. K., & Frings-Dresen, M. H. W. (2004). Efficacy of adjusting working height and mechanizing of transport on physical work demands and local discomfort in construction work. *Ergonomics*, 10, 772–783.

Chapter 10

Environmental design

The issues Miguel confronted are related to environmental design. For instance, some individuals work outside in environments where the temperature might vary between bitterly cold in winter to hot and humid in the summer. Similarly, our office environments or other enclosed workspaces can range from extremely cool to comfortable to unbearably hot. It is also important to consider the lighting within the workspace given the tasks users have to perform, as there might not be enough lighting for maximum performance or, as in Miguel's case, glare can be annoying as well as debilitating. Similarly, noisy work environments may have a detrimental effect on performance. Finally, environmental issues can arise from the design and layout of rooms and buildings. Poor architectural designs can reduce individuals' wayfinding ability and affect the "feel" of the office space. To create effective environments, we need to understand the different aspects of environmental design.

10.1 CHAPTER OBJECTIVES

After reading this chapter, you should be able to:

- Describe the different aspects or factors of environmental conditions.
- Describe the potential positive or negative impact of the environmental temperature.

DOI: 10.1201/9781003515463-10

- Determine whether the environmental lighting conditions are appropriate and how to modify them for more efficiency.
- Describe the potential impact of noise and how this could impact the users.
- Use link analysis to determine a more efficient space layout.
- Understand how different environmental factors (e.g., windows or color) impact individuals.
- Understand the factors that affect wayfinding and how to redesign the environment to enhance wayfinding.

10.2 WHAT IS THE ENVIRONMENT?

Although Miguel's environment is an office, the environment could include any area or space in which a person works, lives, or plays. This could include small environments such as kitchens, bathrooms, or office spaces as well as larger environments such as whole buildings (e.g., schools or hospitals), parks, airports, or even cities. Furthermore, for some individuals, the relevant environments of concern may be the cab of a truck, the deck of a boat, a flight deck, or an oil platform.

The environment also often includes chairs or some type of seating, but this chapter goes beyond the discussion of anthropometry and biomechanics discussed in Chapter 9. The emphasis here is on understanding how temperature, lighting, or noise levels as well as the layout and design of the space impact performance. Although motion and vibration are often considered part of environmental design, those topics are covered in Chapter 4. Instead, this chapter will focus on the environment of smaller, individual spaces such as an office space or work environment, but will address issues pertinent to larger areas where appropriate.

10.3 ENVIRONMENTAL TEMPERATURE

Ideally, we want to determine the appropriate temperature for an environment. The problem is in defining "appropriate." If the temperature is extreme, either hot or cold, it is usually one of the first things we notice about a workspace. Miguel considered his environment to be cold, and the cold affected his comfort level. If Miguel's job required more physical activity, such as delivering packages, the temperature might have felt comfortable to him. Therefore, we need to consider our activity level, which affects whether we perceive an environment to be comfortable, too hot, or too cold. In addition, the level of humidity and amount of airflow impact our sensation of temperature and our perceptions of thermal comfort (Fanger, 1972). Professionals who study environmental temperature also consider the role of air velocity (e.g., wind chill) and humidity on our subjective experience of temperature.

10.3.1 Measures of temperature

When we read a thermometer, we are measuring air temperature. **Air temperature** excludes the impact of air velocity and humidity. To measure air temperature, the sensor is shielded and kept dry (Kroemer et al., 2001). Hence, air temperature is often referred to as the dry-bulb temperature.

As humidity affects our subjective sensation of heat and cold, we must assess the amount of water vapor in the air. Yet, the atmosphere can hold different amounts of moisture depending on the season. In particular, the atmosphere holds more moisture with higher temperatures and lower air pressure as in the summer. Therefore, we measure **relative humidity**, which is the amount of water vapor present relative to the amount of maximum water vapor possible at that time (Kroemer et al., 2001).

Besides relative humidity, **air velocity** or the amount of air flow or drafts can vary within our environments and have a significant impact on comfort. Air velocity lower than 30 ft/min (0.15 m/sec) is equivalent to "still air" or no perceptible air flow (Rohles & Konz, 1987) and has no impact on our sensation of temperature. Stronger air flows tend to cool us, as air flowing across our skin removes heat from our bodies creating a comfortable feeling in hot weather. Although this cooling is generally desirable in hot environments, it is not usually welcomed in cooler environments such as Miguel's office. As humidity levels affect the amount of cooling depending on the air flow (discussed in the next section), **effective temperature** reflects the combined effects of air temperature, relative humidity, and air velocity on us. Different combinations of air temperature, humidity, and air velocity can feel identical to us resulting in the same effective temperature.

Objects in our environment also influence our sensation of temperature. If we sit next to an outer wall during the winter, we might sense the coolness of the wall surface, or we might feel heat radiating from the surface of a building as we walk by it on a summer evening. What we sense coming off of a surface is referred to as the **mean radiant temperature** (Rohles & Konz, 1987, p. 697).

Finally, the measure called **wet bulb globe temperature** (WBGT) considers air temperature, radiant temperature, and humidity. This temperature is measured with a wet bulb thermometer. The bulb is kept wet, and as humidity levels decrease below 100%, cooling due to evaporation increases, lowering the temperature reading. Because WBGT accounts for some of the effects of air velocity as well, the WBGT is a better measure of hot environments (Parsons, 1991) and is especially helpful in determining if the hot environments will be too stressful on humans.

10.3.2 Heat exchange

Environments increase stress on our bodies when we cannot maintain our core body temperature around 98.6°F (37°C) through the process of homeostasis. If we get too hot, the body tries to cool itself by perspiring. When the perspiration moisture evaporates, it helps to cool the body. As cooling continues, our skin can become bumpy, sometimes referred to as "goose flesh" or "goose bumps," which reduces the amount of air flow and evaporation (Bensel & Santee, 1997, p. 927). If we get too cold, our body tries to warm itself by shivering (Kroemer et al., 2001). The process by which our bodies gain or lose heat is known as **heat exchange** and involves the process of the body either gaining heat from, or releasing heat to, the environment. Heat always transfers from the warmer to the cooler surface or object. If the environment is warmer than the body, or the environment is too humid for perspiration to evaporate, the body will gain and retain heat, becoming hotter. In contrast, if the environment is cooler than the body and the person is not wearing clothing to retain the heat, the heat will be released into the environment, cooling the body.

10.3.2.1 Process of heat exchange

Heat exchange takes place predominantly through the skin, although it also occurs through the respiratory system (Kroemer et al., 2001). The four main ways in which heat exchanges between the human body, specifically the skin, and the environment include radiation, convection, conduction, and evaporation. **Radiation** is the exchange of heat between two non-touching surfaces such as a person seated next to an outer wall or a windowpane. If the wall or windowpane is cold, the human's body heat will radiate toward the cold wall and the person will feel cooler or cold because the heat is being transferred from the body toward the colder wall or window. If the outer wall or windowpane is warmer than the person, then the heat from the windowpane will radiate toward the person, warming the person.

Radiation does not consider the air temperature between the two surfaces. In contrast, **convection** involves the transport of heat by movement of air or fluid molecules touching the skin. As with radiation, heat transfers from the warmer to the colder object. For example, if you were to step outside without a coat on a cold winter day, your body heat would be transferred to the colder air temperature. Similarly, if you decide to participate in the polar plunge (jumping into cold water, usually in the middle of winter), your body heat would be transferred to the cold lake water. In fact, convection in water is much quicker than in air. Convection also can occur in the opposite direction. As you step outside into the hot summer heat, the air temperature will transfer to your body, or if you step into a hot bath, the heat from the hot bath water will transfer to your body.

A more direct transfer of heat occurs through the process of **conduction** when an object touches your body. If the object is warmer than the body, then the heat will transfer to the body such as when you sit on a metal park bench in the summer. When you sit on the ice rink bleacher, though, the object's temperature is colder than the body's temperature and the body's heat is transferred to the object.

Finally, heat exchange can occur through **evaporation**, the removal of heat through moisture, often perspiration, drying on the skin. Even without perspiration, evaporation occurs with breathing, which is another form of heat loss (Bensel & Santee, 1997). The body can lose, but never gain, heat through evaporation. A summary of these types of heat exchange is presented in Table 10.1.

10.3.2.2 Clothing effect

As discussed above, heat exchange can occur through radiation, convection, conduction, and evaporation, and these processes are affected by air temperature, humidity, and air velocity. The clothing we wear also affects heat exchange because of its insulating effect. The amount of insulation clothing provides is measured using a **clo** unit (Gagge et al., 1941). An individual wearing one clo unit of clothing is likely to be comfortable at 70°F (21°C) in a room with no perceptible airflow (20 ft/min, 10 cm/sec) and relative humidity below 50% (Gagge et al., 1941). A change in one clo unit accounts for approximately 13°F (7°C) change in temperature (Gagge et al., 1941; Rohles & Konz, 1987). If the temperature dropped from 70°F (21°C) to 57°F (13.9°C) degrees, the individual would need 2 clo units of insulation to

Table 10.1 A summary of the heat exchange process

Type of heat exchange	Process	Example
Radiation	Exchange of heat between two objects not touching	Sitting near a radiator or a window in the winter
Convection	Exchange of heat between the body and air or fluids	Being surrounded by cold or hot air or water. Convection process is much quicker in water
Conduction	Exchange of heat between two touching objects	Sitting on a hot car seat in the summer or touching a cold metal seat in the winter
Evaporation	Cooling due to moisture drying on the skin	Cooling from perspiration drying or water drying after taking a shower or swim

be comfortable, but would be most comfortable in the nude if the temperature increased to 83°F (28.3°C). Yet, we wear clothes in most environments and also work in geographical locations that can get uncomfortably cold or hot.

In cold environments, we desire greater insulation so that our body heat does not exchange with the colder surrounding environment by means of radiation, convection, conduction, or evaporation. Approximately one quarter inch (0.62 cm) of insulation is equivalent to one clo unit of insulation (Bensel & Santee, 1997). Thicker clothing and less permeable clothing reduce heat loss through convection and reduced evaporation. The process of convection helps us reduce heat loss when we wear a sweater or jacket. When we first put on a sweater, the initially colder air is trapped within the sweater as well as between our sweater and body. Heat transferred by convection from our body warms this trapped air, which becomes comparable to our body temperature and insulates us. If the trapped air escapes too quickly, we lose this insulating effect. This trapped air effect is how divers' wet suits function. The wet suit traps a thin layer of water between the suit and the body, which is warmed by convection keeping the diver warm. The thickness of the wetsuits varies depending on the temperature of the water, but it is the thin layer of water between the wet suit and the diver's body that is critical for insulating the diver.

If our clothing does not allow for enough heat transfer from the body in warm or hot environments, this could lead to heat stress or illness. Similarly, not having the proper clothing in cold weather can lead to cold stress. Importantly, cold weather clothing should not be so thick and impermeable as to cause the person to sweat, which can lead to evaporative heat loss. Finally, wet clothing usually loses its insulation value (Parsons, 1991) allowing heat transfer toward (e.g., during a fire) or away from (e.g., during a snowstorm) the body via convection. A listing of clo values for various men's and women's clothing is presented in Table 10.2.

Table 10.2 Clo values for common men's and women's clothing items

Men's and women's clothing	Clo value
Women's underwear	0.03
Men's briefs	0.04
Sleeveless shirt	0.06
T-shirt	0.09
Women's top dress with long sleeves	0.15
Men's shirt with long sleeves	0.20
Shorts	0.06
Long pants—lightweight	0.20
Long pants—normal weight	0.25
Sweater—lightweight	0.25
Jacket—lightweight for summer	0.25
Jacket—normal weight	0.35
Winter coat	0.60
Socks	0.02
Nylons/pantyhose	0.02

Source: Adapted from http://www.engineeringtoolbox.com/clo-clothing-thermal-insulation-d_732.html.

10.3.2.3 Impact of humidity

In addition to clo units, humidity can impact heat transfer. As humidity increases, the water vapor content of the atmosphere increases. During the hot humid summer months, there might be too much moisture in the air to allow for evaporative cooling via perspiration. To increase evaporative cooling, we need to increase air velocity. When there is little or no airflow, the body's heat cannot be released into the environment, possibly leading to heat illness or heat exhaustion. This is why, during the summer months, the temperature often is reported with the **heat index**, a measure of how hot it feels given the humidity level. Looking at Figure 10.1, you see that the combined effects of an air temperature of 96°F (36°C) and a relative humidity of 45% are equivalent to an air temperature of 104°F (40°C) with minimal humidity. As the humidity increases, say to 75%, it feels like 132°F (about 56°C). Therefore, if you are in an environment that has an air temperature of 70°F (21°C) with low humidity, it is possible to create the same temperature feeling when the humidity increases if the air velocity also increases.

10.3.2.4 Impact of air velocity

Fans are often used during the hot, and especially humid, summer months to increase air velocity, as the increased air velocity helps with evaporation, increasing heat loss or cooling. In contrast, air velocity such as drafts in the cold winter months causes heat loss through convection or evaporation. Heat loss during the winter due to air velocity, otherwise known as the wind chill factor, can lead to severe consequences especially in extremely cold conditions.

NOAA's National Weather Service

Heat Index

Temperature (°F)

Relative Humidity (%)	80	82	84	86	88	90	92	94	96	98	100	102	104	106	108	110
40	80	81	83	85	88	91	94	97	101	105	109	114	119	124	130	136
45	80	82	84	87	89	93	96	100	104	109	114	119	124	130	137	
50	81	83	85	88	91	95	99	103	108	113	118	124	131	137		
55	81	84	86	89	93	97	101	106	112	117	124	130	137			
60	82	84	88	91	95	100	105	110	116	123	129	137				
65	82	85	89	93	98	103	108	114	121	128	136					
70	83	86	90	95	100	105	112	119	126	134						
75	84	88	92	97	103	109	116	124	132							
80	84	89	94	100	106	113	121	129								
85	85	90	96	102	110	117	126	135								
90	86	91	98	105	113	122	131									
95	86	93	100	108	117	127										
100	87	95	103	112	121	132										

Likelihood of Heat Disorders with Prolonged Exposure or Strenuous Activity

☐ Caution   ☐ Extreme Caution   ▨ Danger   ▪ Extreme Danger

Figure 10.1 NOAA's National Weather Service Heat Index. (From http://www.nws.noaa.gov/om/heat/heat_index.shtml.)

10.3.3 Excessive heat gain or heat loss

The body produces heat through metabolic activities, which is carried throughout the body by the blood (Kroemer et al., 2001). This blood flow makes us appear "flushed" when we get hot because the blood comes to the surface of the skin where heat is exchanged with the environment via convection or evaporation. In other situations, we might reduce (or should reduce) our physical exertion to minimize the amount of heat generated by the body to avoid overheating. In contrast, when we are cold, the body responds by reducing the flow of blood to the extremities to keep the core body temperature between 97°F (36.1°C) and 99°F (37.2°C). As we get colder, our bodies respond by shivering to generate heat, or we might intentionally increase our own physical exertion to increase heat generation.

It is critical that our core body temperature remains close to 98°F (36.7°C). Increases or decreases in core body temperature of 3.6°F (2°C) can affect body and task performance greatly, whereas a 10.8°F (6°C) change in core body temperature can be fatal (Kroemer et al., 2001). Although these values reflect rather large core body temperature fluctuations, even smaller fluctuations can cause problems. Therefore, we should be cognizant of our shelter, physical activity level, and clothing, and how these factors will affect our core body temperature.

10.3.3.1 Heat stress or heat illness

Exposure to extreme heat conditions can cause heat stress or heat illness. An individual might experience fatigue, cramps, heat rash (also known as prickly heat), or worse, heat exhaustion or heat stroke (Kroemer et al., 2001). Excessive sweating, increased heart rate, and an increased core body temperature are all signs of **heat stress**. It is critical that persons exposed to extreme heat drink plenty of water to reduce the likelihood of these problems, which are related to dehydration and an inability of the body to cool itself.

10.3.3.2 Cold stress

At the other extreme, exposure to severe cold might result in cold fingers or cold toes as the body reduces blood flow to about 1% of the normal blood flow when the person is at a comfortable temperature (Kroemer et al., 2001). Also, an individual is more likely to experience hypothermia in cold water because heat loss due to the convection process in water is about 20 times greater than the convection process of heat loss to air (Kroemer et al., 2001). This explains why it is critical to wear proper protective clothing in cold water. A drop in the core body temperature to 95°F (35°C) creates a dangerous hypothermic condition (Parsons, 1991).

Other side effects of cold stress include a loss of manual dexterity and the potential for frostbite. The loss of manual dexterity is potentially dangerous as the individual will be unable to perform lifesaving activities such as lighting a match (Kroemer et al., 2001). These reactions are often followed by apathy and then hypothermia (Kroemer et al., 2001). Although Miguel's office environment was not likely life threatening, it might have impacted the manual dexterity of his hands.

10.3.4 Acclimation and acclimatization

Individuals exposed to cold or hot conditions for prolonged periods show some acclimatization to the environmental conditions within 1–2 weeks (Kroemer et al., 2001; Rohles & Konz, 1987). Although many people talk about **acclimation**, this refers to our adjustment to one variable, generally temperature. **Acclimatization** is our adjustment to multiple environmental conditions such as temperature and relative humidity. (As different researchers use these two terms, the term used by the researchers is generally retained in the following discussion.) Acclimation is best when the acclimation process includes physical activity (Rohles & Konz, 1987). In the cold, acclimation tends to be more difficult, if at all possible.

Acclimatization to cold or hot environments can be challenging if exposure to these cold or hot temperatures is not continuous. Our exposure to these environments is often interrupted by our escape to heated or cooled environments where we live and work. When the acclimation process is interrupted, even for a few days, the acclimation process is less effective (Kroemer et al., 2001). As Miguel spends most of his day in an air-conditioned office, he would have trouble acclimatizing to the summer heat and humidity.

10.3.5 Impact of temperature on cognitive performance

As discussed above, extreme heat or cold can affect our body's functioning. In fact, a great deal is known about the body's reactions to these extreme conditions, but much less is known about how these conditions affect human cognition.

10.3.5.1 Heat and cognitive performance

The effect of heat on cognitive performance is still not fully understood. In some cases, there are performance decrements; in others, there are no effects or even increases in performance. Generally, it is necessary to consider the type of task being performed. For example, reaction time for some simple motor tasks is not usually affected by heat stress (Hancock & Vasmatzidis, 2003). More complex tasks such as vigilance (sustained attention), tracking tasks, and multi-tasking are more greatly affected by heat stress, with vigilance having the greatest decrements (Hancock & Vasmatzidis, 2003). Even when there were no differences in the number of correct or missed trials across all conditions, the number of incorrect responses (vigilance errors) was significantly higher in the 104°F (40°C) condition than either the 32°F (0°C) or 73.4°F (23°C) condition, which did not differ (Faerevik & Reinertsen, 2003). These cognitive decrements were associated with changes in core body temperature.

It was first thought that the effects of heat could be described by the inverse-U relationship of the Yerkes–Dodson law (Yerkes & Dodson, 1908). Heat was understood to function as a stressor, increasing arousal as heat increased in magnitude. Moderate levels of heat (i.e., arousal) would produce maximum performance, whereas lower or higher levels of heat produced lower levels of performance (see Figure 10.2).

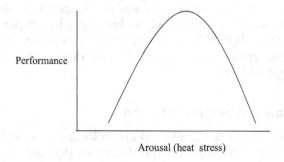

Figure 10.2 An example of the Yerkes–Dodson law inverse-U relationship.

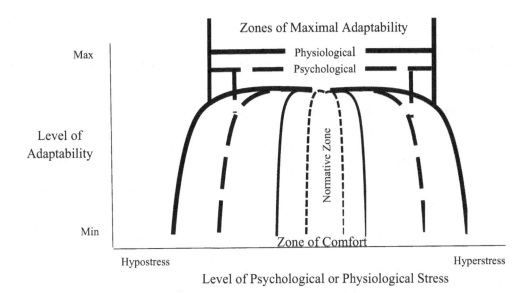

Figure 10.3 Hancock and warm's maximal adaptability model. (Adapted from Hancock & Warm, 1989.)

One argument for why more complex tasks are more greatly affected by heat stress is that heat might affect our ability to attend to various tasks (see Chapter 6 for a discussion on attention). In particular, Hancock and Warm (1989) proposed the Maximal Adaptability Model, which suggested that as stress (i.e., heat) decreases or increases, performance decreases as a result of a decrease in or the depletion of attentional resources. Looking at Figure 10.3, you will notice that the middle zones are referred to as the normative and comfort zones. In these zones, there is no or minimal heat stress, respectively, which requires no or little effort to adapt to the environment, eliminating the drain on the attentional resources. In turn, there are no effects on cognitive functioning.

As the heat stress increases toward hyperstress, this model proposes that attentional resources are taxed within the psychological maximal adaptability zone. Because the psychological maximal adaptability zone is suspected to occur at lower levels of heat stress than the physiological maximal adaptability zone, this maximal adaptability model suggests that reductions in performance will occur long before physiological problems occur. People will display various types of performance decrements long before the severe effects of heat stress appear.

Similarly, a decrease in heat stress toward hypostress could lead to decrements in cognitive performance. Yet, Hancock and Warm argued for a unipolar (one-sided) model whereby the zero level of heat stress could occur in the normative zone and would only increase as the stress level increased (the right side of Figure 10.3). This would help explain the absence of cognitive decrements as temperatures decrease from a comfortable level.

Data supporting this model come from research that varied the dry-bulb temperature and relative humidity (30% and 70%) to create six climates that represented three different WBGTs: 71.6°F (22°C), 82.4°F (28°C), and 93.2°F (34°C; Vasmatzidis et al., 2002). Individuals worked on a sequence of three paired tasks or one task composed of four activities under the different WBGTs. These tasks tapped different types of processing: encoding or sensory memory (visual and auditory input), working memory or central processing, and output or responses (manual and vocal output). The tasks included display monitoring (encoding), mathematical processing (central processing), memory searching (central processing), and tracking (output; see Chapter 6 for a discussion of these different types of attention and memory components). Given the task combinations such as display monitoring with mathematical processing, it was possible to test for input, working memory, or output decrements. Only when tasks are realistically complex can we appropriately test the maximal adaptability model (Hancock, 2022).

Heat stress affected working memory the least, affected visual and auditory input a bit more, and had the greatest debilitative effect on the manual tracking task (Vasmatzidis et al., 2002). The researchers referred to these varied effects as the "heat stress selectivity effect" (p. 235). They also suspected that, in accordance with the maximal adaptability model, working memory resources might be less susceptible to heat stress than input or output processes during time sharing tasks. When two working memory tasks (memory search and mathematical processing) were paired, only memory search experienced decrements in performance, which were greatest in the 93.2°F (34°C) WBGT relative to the 71.6°F (22°C) condition. In addition, compared to the climate of 71.6°F (22°C) created with 30% humidity, there were significantly more decrements at 93.2°F (34°C) WBGT when the climate was created with 70%, but not 30%, relative humidity. These data suggest that if individuals are working on multiple tasks that tap their working memory, the relative humidity level should be considered when addressing the effects of heat stress on performance.

10.3.5.2 Cold and Cognitive Performance

Similar to the impact of heat, we know relatively little about how cold impacts cognitive performance (Palinkas et al., 2005). Exposure to cold temperatures that do not produce a hypothermic condition might increase, decrease, or have no impact on performance. Yet, it is generally well accepted that once someone reaches a hypothermic state, one's mental performance declines rapidly.

To investigate the effect of cold temperatures on performance, Finnish individuals worked on various tasks during the summer and winter months (Palinkas et al., 2005). During the winter, external temperatures ranged from 10.4°F to 24.8°F (−12° to −4°C), while external summer temperatures ranged from 51.8°F to 57.2°F (+11°C to +14°C).

There was no evidence of seasonal effects when testing participants during the summer and winter (Palinkas et al., 2005). Regardless of season, cold temperatures increased performance on complex cognitive tasks involving short-term memory and logical reasoning. Palinkas and colleagues argued that these results supported the arousal theory. On the other hand, degraded performance was observed on simple tasks including sustained attention and visuomotor flexibility, which was argued to support the distraction hypothesis. Degradation of complex military cognitive simulations due to the cold further supports the distraction hypothesis (Martin et al., 2019). These results provide some support for different effects of cold on performance depending on the task.

Additional research evaluated how exposure to non-hypothermic conditions affected cognitive functioning (Mäkinen et al., 2006). For ten consecutive days, young men clad in minimal clothing performed cognitive tasks during the last 20 minutes of a 90-minute control condition (77°F ± 0.5°F; 25.0°C ± 0.3°C) and then repeated the cognitive tasks during the last 20 minutes of a 120-minute cold climate (50°F ± 0.5°F; 10.0°C ± 0.3°C) exposure. The relative humidity was held constant, and the air was still.

When assessing the performance results for all tasks combined, cold exposure tended to increase accuracy (percent correct), while increasing reaction times and decreasing efficiency (number correct divided by reaction time). When evaluating individual tasks such as those requiring sustained attention and working memory, however, there was an observed decrement in cognitive performance. Given the lack of consistent effects, Mäkinen et al. (2006)

argued that these results supported both the distraction hypothesis and arousal theory. In particular, the distraction hypothesis posits that negative effects result because individuals are distracted by the environmental conditions such as when decreased skin temperatures were associated with longer response times and decreased efficiency on a simple visuomotor reaction time task. When tasks are complex, the distraction could have a greater impact (Martin et al., 2019). Yet, arousal theory could justify both positive (i.e., reduced reaction times) and negative (decreased accuracy) effects (Mäkinen et al., 2006). Initially, reduced core temperatures were related to decreased reaction times, which increased efficiency. Other times, though, the decreased reaction times were related to decreases in accuracy, reducing efficiency. Could these results reflect the sample of only men (see Chapter 2 for issues related to sampling)?

10.3.5.3 General impact of heat and cold on cognitive performance

Although there appear to be some trends in the data, it is difficult to compare these studies because of differences in the tasks and WBGT used in the experiments. June Pilcher et al. (2002) used a statistical method called meta-analysis to analyze data from 23 studies to estimate the effects of heat and cold stress on cognitive performance. In a meta-analysis, the data from multiple research projects on the topic of interest are analyzed to determine an overall effect resulting in a statistical literature review. For this meta-analysis, the results from research reporting the impact of heat and cold stress on cognitive performance were analyzed. The temperatures were expressed as WBGT and sorted into five temperature groups: <50°F (10°C), 50°F–64.9°F (10°C–18.28°C), 70°F–79.9°F (21.11°C–26.61°C), 80°F–89.9°F (26.67°C–32.17°C), and ≥90°F (32.22°C) WBGT. For our discussion, consider the two lower temperature categories as "cold," the middle temperature range as "neutral," and the latter two as "hot" conditions.

Overall, similar decrements in performance were observed at the extreme WBGT temperatures relative to the neutral temperature range. Yet, the greatest decrements occurred under the hottest (≥90°F, 32.22°C) and coldest (<50°F, 10°C) conditions, while neutral temperatures had little effect on performance.

The performance decrements in the extreme temperatures varied depending on the type of task performed. Overall, the least affected were the simple reaction time tasks, and the most affected were tasks classified as "reasoning, learning, or memory tasks" (p. 690). Perceptual, attention, and mathematical tasks were moderately affected; however, different tasks were more greatly impacted depending on whether it was hot or cold. Attention, perceptual, and mathematical processing tasks were more greatly debilitated in the hot than cold conditions. In contrast, cold temperatures had a greater debilitating impact on reasoning, learning, and memory tasks, whereas there was a slight increase in performance on these tasks when performing in hot temperatures (Pilcher et al., 2002).

The duration of the task, experimental session, and pre-task exposure were also related to performance. Short task durations (<60 minutes) and short experimental sessions (<120 minutes) resulted in greater decrements in performance than long task durations and experimental sessions. These results suggest that people might be able to adapt to the temperatures when exposure times are longer. In contrast, when individuals were exposed to the temperature conditions for more than 60 minutes before the task, performance decrements were greater than when the pre-task exposure was shorter (<60 minutes). These data suggest that adaptation might only occur if the individuals are performing the tasks during the exposure time, but acclimation to cold appears to reduce the distracting effect (Martin et al., 2019). In conclusion, it is important to consider other variables or moderators when estimating temperature effects. In addition, it is recommended that heat and cold stress be considered as a form of cognitive load (Martin et al., 2019).

BOX 10.1:   USING HEAD GEAR TO REDUCE HEAT STRESS IN HOT CLIMATES

Clothing is one way to deal with either hot or cold environments. In particular, the use of head, hand, and foot coverings is essential in cold climates. Some type of headwear is extremely beneficial in hot weather. The headwear or head gear needs to reduce heat transfer to the body, as opposed to retain the heat. Prabir Mukhopadhyay (2009) evaluated lightweight head gear for youth who worked in the fields or walked long distances to school in India, as the summer temperatures could be over 104°F (40°C) and also humid. After males sat for 30 minutes in direct sunlight, ratings of discomfort and measured forehead skin temperature were significantly lower when either of two head gear prototypes made of polyurethane and aluminum foil were worn compared to no head gear. These prototypes had different designs but employed vertical or horizontal slits to allow for the natural air flow around the head. Although these prototypes were not rated highly on aesthetics (and only males participated, as the females refused to wear the head gear), they were lightweight and inexpensive. The material for these prototypes was selected intentionally to make the head gear lightweight and inexpensive to serve the needs of rural and poor individuals. **Critical Thinking Questions**: Given that the females would not wear the head gear, how might you address this problem, as the head gear seemed to reduce the chances for heat stress? To help protect individuals against extremely hot temperatures, what other variables of concern should be considered?

Because heat and cold have deleterious physical and possibly cognitive effects on performance, the International Organization for Standardization (ISO) has recommended guidelines for working in these various types of climates (similar to the guidelines for anthropometry in Chapter 9). The ISO has Technical Committees (TCs) that divide into Sub-Committees (SCs). The SCs also have working groups (WG). ISO's TC 159 is Ergonomics, and SC 5 is Ergonomics in the Physical Environment, which evaluates proposed standards related to environmental climate. Currently, there are 35 published ISO standards for TC 159/SC 5. Several of these are presented in Table 10.3 (www.iso.org).

10.4 ENVIRONMENTAL LIGHTING

Besides climate, another important environmental factor is the type and amount of lighting available. The type of lighting we use can impact our performance and safety. As discussed in Chapter 3, you might use additional lighting when studying in order to see the text on the pages. To increase lighting, you could use a lamp. The bulb releases light and the amount of light falling on various objects is **illumination**. When you illuminate the page of text, the page reflects some of the light. This reflected light is called **luminance**.

The rate at which light is emitted from a source is the **luminous flux**, which is measured in **lumens** (Cushman & Crist, 1987). **Lux** defines how many lumens are distributed per one square meter (i.e., 1 lux=1 lumen per square meter). (We use the International System [SI] units in this book, but you might see measures in the U.S. Customary System such as foot candle instead of lux.) **Luminous intensity** is the amount of luminous flux per area.

When selecting lighting sources, two important aspects to consider are efficiency and color rendering (Cushman & Crist, 1987). **Efficiency** is a measure of how many lumens are produced per watt of electricity. **Color rendering** is a measure used to express how well the

Table 10.3 Some ISO standards for dealing with hot or cold environments

ISO number	Title of ISO standard
ISO 11079:2007	Ergonomics of the thermal environment—determination and interpretation of cold stress when using required clothing insulation (IREQ) and local cooling effects
ISO 15743:2008	Ergonomics of the thermal environment—cold workplaces—Risk assessment and management
ISO 7933:2023	Ergonomics of the thermal environment—analytical determination and interpretation of heat stress using calculation of the predicted heat strain
ISO 9920:2007	Ergonomics of the thermal environment—estimation of thermal insulation and water vapor resistance of a clothing ensemble
ISO 8996:2021	Ergonomics of the thermal environment—determination of metabolic rate
ISO/TR 2801:2007	Clothing for protection against heat and flame—general recommendations for selection, care and use of protective clothing
ISO 7726:1998	Ergonomics of the thermal environment—instruments for measuring physical quantities
ISO 9886:2004	Ergonomics—evaluation of thermal strain by physiological measurements
ISO 9920:2007	Ergonomics of the thermal environment—estimation of thermal insulation and water vapor resistance of a clothing ensemble
ISO 10551:2019	Ergonomics of the thermal environment—subjective judgment scales for assessing physical environments
ISO 12894:2001	Ergonomics of the thermal environment—medical supervision of individuals exposed to extreme hot or cold environments
ISO 14505-3:2006	Ergonomics of the thermal environment—evaluation of thermal environments in vehicles—Part 3: Evaluation of thermal comfort using human subjects

Information from https://www.iso.org/search.html?q=&hPP=10&idx=all_en&p=0.

artificial light reproduces the true colors of objects compared to natural light. Some lights can alter the color appearance of objects, which can be an important consideration in the food, apparel, clothing, and medical fields. These issues are discussed further below.

10.4.1 Lamps and luminaires

Lamps (i.e., light bulbs) are sources that emit light energy. Efficiency and color rendering are affected by the type of lamp selected. Incandescent lamps ("typical" light bulbs) are frequently used and have good color rendering, but tend to be inefficient (Cushman & Crist, 1987). Because of this inefficiency, there is an increased emphasis on the use of compact fluorescent lamps. Fluorescent lamps, the most common type of a gaseous discharge lamp, have good efficiency and color rendering (Cushman & Crist, 1987).

When color rendering is not a primary concern, we might choose lamps that are more efficient. Because these lamps tend to have poorer color rendering capabilities, the light tends to "color" or change the color appearance of what we see. For example, streetlights often make objects look yellow and it might be difficult to determine a car's true color. Unless there is a hit and run, knowing the car's color is generally irrelevant to the task of driving or even walking. Yet, there is enough light to enhance the safety of both walking and driving.

A lamp is generally inserted into a **luminaire**, which is the light fixture and the electrical wiring associated with it. The type of luminaire influences the distribution of light and whether the light source is emitted upward or downward. Diffuse light sources produce relatively uniform illumination in all directions. Luminaires that shine the light downward are referred to as direct or semi-direct lighting, and if the light is directed upward, these luminaires are considered semi-indirect or indirect lighting (Cushman & Crist, 1987).

10.4.2 Glare and reflectance

When selecting a luminaire, it is important to consider the effects on glare. **Glare** occurs when light is shining into one's eyes producing annoyance, discomfort, disability, or performance decrements. When a light source, such as the sun, is in one's line of sight, this is **direct glare**. Miguel experienced direct glare from the windows when he looked outside.

Sitting opposite the windows during the PowerPoint presentation, Miguel experienced indirect glare and had difficulty reading the slides. We experience **indirect glare** when the light reflects off other surfaces such as our computer or TV screen, or even a window. Therefore, the amount of indirect glare will be dependent on the reflectance of various surfaces. **Reflectance** is measured as the difference between the illuminance and luminance levels. If the luminance level is nearly as high as the amount of light energy hitting the surface, then the reflectance is high.

Glare can be further classified according to how the light is reflected (Cushman & Crist, 1987). A surface that is smooth and polished reflects a direct line of light known as **specular glare**. When you use the glass face of your watch or phone to direct the sun into the eyes of a friend or to tantalize your cat, this is an example of specular glare. In contrast, **diffuse glare** spreads light in all directions and does not appear as a single point of light. It is also possible to experience **veiling reflections**, which are often reflected images. Older computer screens and large screen TVs often have veiling reflections, whereby you can see the room behind you reflected on the screen.

The effects of glare are classified as discomfort or disability (Cushman & Crist, 1987). As you might guess, **discomfort glare** is uncomfortable, but might not reduce visibility or performance. Unlike discomfort glare, **disability glare** decreases visibility, most likely decreasing performance. Disability glare that is so bright one cannot see after the glare has been removed, as in the case of the high beams of an oncoming car at night, is referred to as **blinding glare**.

It is important to reduce or prevent glare, as glare can reduce visibility. **Visibility** is our ability to discern what is presented to us visually. If something is visible, we can read the words on the page or see the critical parts of the machinery. As noted in Chapter 3, visibility is affected by contrast, luminance, exposure time, size of target, and the observer age. Although increased levels of illumination tend to enhance visibility and performance, there is a point of diminishing returns whereby more lighting will not enhance visibility and performance.

One of the critical factors of visibility is contrast. Contrast represents the difference in the amount of light reflected by say a black letter and the white paper on which it is printed. The greater the contrast in darkness between the letters and background paper or screen, the more visible the words should be. Yet, when the luminance levels of surrounding areas are low, creating contrast, performance often decreases. Manipulating the luminance levels in the work area around a computer screen determined that the surrounding luminance did not affect visual acuity, but it did affect performance when individuals had to shift their gaze between the screen and the darker surrounding area searching for objects (Sheedy et al., 2005). The decrements in performance were related to **transient adaptation**, whereby the eyes must continuously adjust between light and dark environments.

As long as the surrounding luminance was about the same level of the screen luminance or greater, problems due to transient adaptation were reduced or eliminated. Only with lower luminance values in the surrounding area was there a performance decrement. It was also noted that the luminance level could be lower for individuals younger than 40 years of age, as their eyes adjust more quickly, but individuals above 45 years of age needed greater luminance values for optimal performance (Sheedy et al., 2005).

10.4.3 Task lighting

Task lighting is the specific lighting required for a particular task or activity. Certain types of tasks require higher or lower levels of illumination. For example, reading requires greater illumination than socializing and could be enhanced with task lighting that directly illuminates the reading material. Further, some tasks may need shadowing to allow the observer to detect flaws in materials as in the case of inspection lighting. Consider the case when you dropped a contact or a pushpin on the floor. Looking straight down is often not helpful in locating the object. On the other hand, if you get down on the ground and look across the floor, you will see an assortment of shapes, which cast shadows and help you find various objects. This activity is not recommended on kitchen floors!

When the task involves the use of computers, remember that computer monitors emit light. Hence, the amount of ambient lighting can be reduced. Table 10.4 presents a list of luminance levels for various common tasks.

10.4.4 Light distribution

Another aspect of lighting that should be discussed is the distribution of light or illumination. Imagine a dark street illuminated by streetlamps spaced every half block or so. As you walk down the street at night attempting to see the curb, you will have the greatest illumination on the sidewalk as you approach the streetlamp, but this will fade as you walk past the streetlamp. At some point, you will leave the light from this first lamp and then enter the light from the next lamp. The light distribution, the scattering of illuminance over an environment, can be as important as how much light is available. Distributions of light often appear as rings of successively reduced illumination levels, as demonstrated in Figure 10.4.

Uneven light distributions, as with the streetlamp example above, create visual difficulties for people traversing the space. As people move into and then away from lighting, they experience transient adaptation. Due to the time it takes to dark adapt, an individual is likely to miss something while in the darker environment. This is a greater problem for elderly individuals who generally need more lighting to see and are slower to adapt. Poor light distributions often make elderly individuals more prone to falling at the bottom of stairs because the lighting fades and they cannot see the last step (see Box 10.2).

Table 10.4 Recommended lux levels for various common tasks

Task	Age groups (in years)		
	Under 25	25–65 (inclusive)	Above 65
Studying	150 lx (min)	300 lx (min)	600 lx (min)
Reading analog (i.e., paper)	250 lx (ave)	500 lx (ave)	1000 lx (ave)
Reading digital	150 lx (ave)	300 lx (ave)	600 lx (ave)
Taking tests on paper	200 lx (ave)	400 lx (ave)	800 lx (ave)
Taking tests on laptop	75 lx (ave)	150 lx (ave)	300 lx (ave)
Dining (informal)	50 lx (ave)	100 lx (ave)	200 lx (ave)
Cooking on kitchen cooktop	150 lx (ave)	300 lx (ave)	600 lx (ave)
Preparing food	250 lx (ave)	500 lx (ave)	1000 lx (ave)
Washing, ironing, drying laundry	100 lx (ave)	200 lx (ave)	400 lx (ave)

Source: Adapted from DiLaura et al. (2011).
Minimum (min) or average (ave) lux levels reported for various common tasks when more than 50% of the users represent one of the three different age groups.

Figure 10.4 An example of light distributions. (© Nancy J. Stone.)

BOX 10.2:   PREVENTING SLIPS, TRIPS, AND FALLS IN POOR LIGHTING

Depth perception allows us to perceive, understand, and navigate our environment (see Chapter 3 for a discussion on depth perception). If there are changes in the light distribution whereby our eyes experience transient adaptation or if there is not sufficient lighting, our depth perception can be debilitated. In addition, as we age, our ability to dark adapt slows and we require higher illumination levels (Kroemer, 2006). Falling off steps or tripping on a riser is often related to poor illumination. Inappropriate illumination reduces our visual cues, impacting our depth perception that helps us identify the change in the terrain such as a step or riser (Clark et al., 1996). Slips, trips, and falls can occur anywhere such as in stairwells or along walkways, inside or outside. In one reported case of a fall, the individual who tripped had been sitting outside at a mall restaurant during the late afternoon and then walked inside after it started to become dark (Clark et al., 1996). The individual thought the walkway was a ramp, but it was actually a set of small steps spread across the patio area (i.e., short risers separated by a long flat space). Although other environmental design factors (e.g., flooring design, plant or furniture location) also impact slips, trips, and falls because of our affected perception of the environment, visual limitations and poor lighting are often involved. When we implement lighting to enhance illumination with the intent of preventing slips, trips, or falls, we must consider how the reflected light might affect others in the environment relative to glare or other contrast effects. **Critical Thinking Questions**: How would you design a path at an outdoor restaurant that is well lighted, but does not reduce the tranquility of the outdoor setting? How could you use landscaping to increase awareness of steps?

10.5 ENVIRONMENTAL NOISE

Noise, another environmental factor, was defined in Chapter 4 as any sound unrelated to the task. In addition, Chapter 4 reviewed different types of noise (e.g., impact or impulse), which can lead to hearing loss (presbycusis and sociocusis), and temporary and permanent threshold shifts. As threshold shifts and hearing loss stress the importance of hearing protection, the following discussion will focus on the impact of noise and how to cope with it, as opposed to the definition of noise.

10.5.1 Hearing protection

We are likely to operate different types of tools or equipment that are noisy such as lawn-mowers, saws, or drills, or to expose ourselves to high levels of music that can lead to a *noise-induced permanent threshold shift* (NIPTS). We can safeguard our hearing using hearing protection. Two types of hearing protection include earplugs and earmuffs (Jones & Broadbent, 1987). *Earplugs* are inserted into the ear canal, whereas *earmuffs* cover the outer ear (Jones & Broadbent, 1987). The best protection occurs when both earplugs and earmuffs are used.

The Occupational Safety and Health Administration (OSHA) defines exposure limits for various sound levels to protect workers from too loud or too long of an exposure. OSHA Standard 1910.25 is based on data from NIOSH, but are not as stringent as the NIOSH recommendations presented in Table 10.5. Keep in mind that these regulations are only applied on the job. There is no one overseeing what types of exposures we experience during our leisure time.

10.5.2 Sounds, noise, and performance

Even though we can protect our hearing and reduce our exposure time to loud sounds, being in noisy environments can still affect us. As you know from Chapter 4, sounds from auditory displays are often used as warning signals. These signals, of course, must be iden-tifiable, but hopefully not deafening. If these warning signals are presented in a similar tone or loudness as other sounds within the environment, it may mask or reduce one's ability to detect the other sounds. This masking (see Chapter 4) can greatly impact the detection of speech (Jones & Broadbent, 1987). As any auditory display or sound, not just warnings, can mask another sound, it is critical to account for potential masking when designing auditory systems to avoid potentially debilitating performance.

Besides masking, it is possible that environmental noise can increase one's perception of workload (Becker et al., 1995) and be distracting, causing interference with various tasks. High school students who listened to irrelevant but meaningful speech demonstrated poorer long-term memory recall of a written passage than when the environment was quiet (Knez & Hygge, 2002). Similarly, when learning and recalling a written passage, individuals' per-formance was significantly worse when there was office noise with understandable speech compared to office noise without speech or speech in a foreign language during the learning

Table 10.5 NIOSH and OSHA recommended exposure time for various noise levels at work

Exposure time (in hours)	Sound level in dBA	
	NIOSH recommendations	OSHA recommendations
8.00	85	90
4.00	88	95
2.00	91	100
1.00	94	105
0.50	97	110
0.25	100	115
<0.02	111	–
Never	>140	–

Source: Adapted from US Department of Health and Human Services, 1998 (https://www.nonoise.org/hearing/criteria/criteria.htm) and http://www.osha.gov/pls/oshaweb/owadisp.show_document?p_table=standards&p_id.

phase (Banbury & Berry, 1998). When these noise conditions occurred during the learning and recall phases, performance in all conditions dropped relative to the quiet condition. The fact that performance dropped over time with noise without speech present does not support the argument that the phonological store (see Chapter 6) was affected, but that the disruption arose due to the constantly changing sounds. The variable noise better explained the performance decrements.

Similarly, when individuals worked on a mental arithmetic task in the presence of foreign language speech (Greek), random speech, or number speech, performance decreased the same extent in all three conditions relative to the quiet condition (Banbury & Berry, 2005). These findings further support the notion that the sounds or noise do not have to contain understandable speech to be distracting. It appears to be the variation in sound that is distracting. When there are many sounds, it is harder to distinguish and notice the variation, which is less distracting. This explains why office workers reported that completely unnecessary noises such as a phone endlessly ringing was perceived to be extremely disruptive as were conversations (Banbury & Berry, 2005). Sounds that were perceived as necessities such as printer, computer, and typing or keyboarding sounds were rated as less disturbing.

Variable noise also might explain why jet engine noise from a jet flying overhead led to a decrease in performance on a vigilance task (Becker et al., 1995). Also, individuals exposed to a low frequency noise that could vary between 20 and 250 Hz demonstrated a slower reaction time on a vigilance task compared to individuals exposed to a relatively stable frequency noise (Pawlaczyk-Luszczyńska et al., 2005).

Given the effects of sounds on performance, and potentially decreased persistence on a task when exposed to noise (Wohlwill et al., 1976), what might be the effects of music on learning? When listening to calming music, 10- and 11-year-old students performed better on arithmetic and reading comprehension (word recall) tasks (Hallam et al., 2002) compared to children not listening to any music. Listening to aggressive music significantly impaired performance on the reading comprehension task relative to calming music, and these students also reported lower levels of altruistic behaviors. These findings suggest that music of different types might be more beneficial or detrimental to learning.

BOX 10.3: ARE OUR SCHOOLS TOO LOUD FOR OUR CHILDREN TO LEARN?

Because young children under 15 years of age and children with various hearing or learning disorders may have a more difficult time deciphering spoken language, it is important that what is spoken is distinguishable from background noise. Ching Yee Choi and Bradley McPherson evaluated the noise levels within 47 primary schools in Hong Kong to determine if the spoken word was loud enough to be heard over noise generated within or outside the classroom (Choi & McPherson, 2005). The noise level within the majority of classrooms was between 55 and 65 dB(A), and all the classrooms had noise levels above the recommended limit of 50 dB(A). Further, when teachers did not use a microphone to amplify their voices, the speech-to-noise ratio was generally below the recommended 10 dB(A) for children with normal hearing and never met the recommended 15 dB(A) for all children. When amplification was used, it was possible to meet the 15 dB(A) recommendation for all children. Unfortunately, amplification was uncommon (Choi & McPherson, 2005). Teachers cannot be expected to project loudly enough during a full day of teaching when the environmental noise is high. This could cause hoarse voices and other problems. **Critical Thinking Questions**: What are the likely educational impacts on the children's learning when the speech-to-noise ratio is lower than 15 dB(A)? How could the environment be redesigned to offset the low speech-to-noise ratio or to reduce this impact?

10.6 ARRANGEMENT AND DESIGN OF THE (INTERIOR) ENVIRONMENT

In addition to temperature, lighting, and sound effects, the layout of a physical space is another important component of environmental design. Just as you want to arrange your displays and controls for greater efficiency, as discussed in Chapters 3 and 8, the arrangement of your environment also influences efficiency, effectiveness, and safety.

In Chapter 9, we also discussed design issues related to anthropometric data in order to accommodate the physical range of expected users. When designing environments, we need to consider anthropometric data such as reach envelopes and clearances, but also include information on how specific arrangements of the environment impact individuals' interactions with other people and equipment or machines within the space. The interaction of machines also can be a concern, but our focus is on interactions people have with people and equipment.

10.6.1 Link analysis

To determine appropriate layouts and arrangements of spaces, one must first determine how individuals interact with the space. **Link analysis** is a common method for determining the movements of people within a workspace. Two key components of link analysis are frequency and importance of these movements (Chapanis, 1959). When a person must interact with certain equipment more frequently, that equipment should be positioned closer to the user. Similarly, if one of the controls, such as an emergency shut off switch, is critical, it also should be located closer to the user, even if the frequency of use is low. As discussed in Chapter 8, though, these emergency controls need to be placed strategically to prevent accidental activation. Controls of high frequency and importance are ideally located within one's reach or extended reach envelope (see Chapter 9).

To determine the *frequency* of use, we generally start with a simulation. While individuals perform the tasks, observers record the frequency with which an individual interacts with the environment. Interactions include people's movements from one station to another or the movement of an operator's hand or foot from one control to another. In the case of an office setting, individuals generally work on computers, use phones, access resources from bookshelves, and deposit materials into filing cabinets. These interactions, or connections, are *links* that can be recorded on a diagram by drawing these movements each time they occur while individuals perform the task. It is important to record the directional flow of these interactions, which might impact the sequential layout of various stations, controls, or equipment. Once the totals are determined, the links can be rank ordered.

The *importance* of a particular interaction can be determined by subject matter experts. *Subject matter experts* (SMEs) are individuals who are experts in the field and can assess how critical each process is. Another source of data is the task analysis (see Chapter 5). A task analysis captures information about how the equipment works and identifies critical processes related to the use of equipment such as the relationships, or links, between various controls or displays on the equipment.

Once the frequency and importance ranks are determined, these scores can be summed to determine an overall rank (Chapanis, 1959). It is helpful to create a flow diagram using these data indicating the various linkages with their respective rankings. Flow diagrams often indicate the links with higher rankings (e.g., greater frequency) as thicker lines, as there would be more lines generated during the activity (see Figure 10.5a). Flow diagrams help us identify potential problems with a current layout, as it is possible to visually determine if certain controls, components, or workstations that have high frequency and high

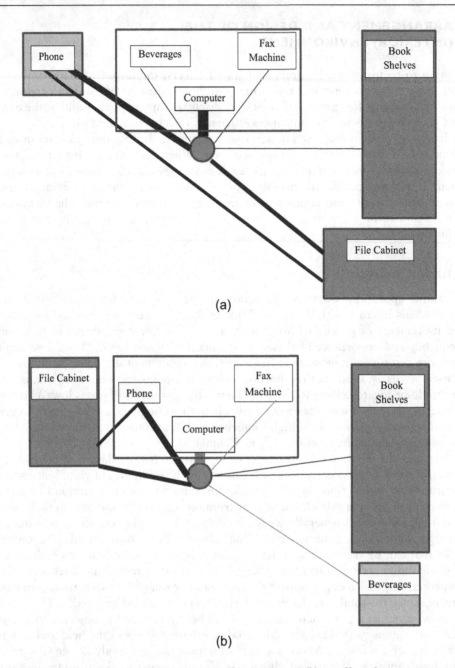

Figure 10.5 Link analysis diagram of (a) an original office layout and (b) an office after rearrangement. (© Nancy J. Stone.). The circle represents the individual working in the office.

importance links are too far apart, reducing efficiency. In addition, if certain paths cross, the order of the layout might be wrong for a particular sequence of actions. If the crossing paths involve the movement of different people, this could create a physical clearance issue, whereby two people might be delayed if they need to pass through this point at the same time. Ideally, we want to place those components with higher ranks closer together and those with lower ranks (lower frequency, lower importance) farther apart.

As noted above, link analysis also can be used to determine movement between controls, but it is also important to consider the movement to the controls (Lin & Wu, 2010). Even if the layout of the controls on a car dashboard is efficient, this efficiency will be decreased if the time it takes to access these controls is increased because the controls are in a bad location. When a sequence of movements is initiated, the control that is activated first should be close and easily accessible. Other factors to consider when designing a workspace include determining control-display compatibility (see Chapter 8), grouping together controls with a common function, placing controls in proper sequence, and ensuring design consistency when there are multiple control functions that are similar.

Looking at Figure 10.5a, we see the link between the worker and the computer has the highest frequency and/or importance (i.e., the thickest line). Therefore, the computer is positioned well. The link with the next highest frequency and/or importance is between the worker and phone followed by the link between the worker and file cabinet and, finally, between the file cabinet and phone. The remaining links are weak or minimal.

Once the link analysis of the current space is complete, we can use scaled cutouts or a computerized program to create multiple designs to determine what layout is best given these data and to ensure everything will fit into the space. It is generally not possible to design the perfect layout (i.e., optimize) with no conflicts or with everything in the perfect location so we work to satisfice as best we can. When we **satisfice**, we attempt to determine the "best" layout so each component in the design is placed in the best position possible without sacrificing the placement of any one component (see Figure 10.5b).

Link analysis can be used to determine the layout of a single office, as in the example above, or it can be expanded to include the layout of multiple offices or rooms as well as items within a larger space. In other words, we can determine the appropriate layout of individual offices, whole buildings, or even university campuses. As we include more components in our space, though, the link analysis process can become quite cumbersome.

10.6.2 Open-plan versus cellular offices

Another design factor includes whether each workspace is within an open-plan or cellular office. At one extreme are the **open-plan** designs, whereby there are no dividers separating different work areas other than individual desks or workstations. In many open-plan environments, individuals use partitions to create a more identifiable workspace. A **cellular office** is at the other extreme and consists of a completely enclosed workspace with a door that closes.

Advantages to an open-plan workspace include an increase in visual and auditory communication creating a more flexible working environment, as well as a more interactive or social environment (Hedge, 1982). Although potentially advantageous for certain jobs, the flexibility and increased flow of information within an open-plan office can reduce visual and auditory privacy. Individuals working on more complex jobs or tasks tend to be more dissatisfied with open-plan offices because the environments are perceived to be more distracting (Hedge, 1982). This explains the use of partitions.

Partitions can increase one's sense of privacy without eliminating the ability to visually and orally communicate easily. Yet, the height of partitions tends to impact one's satisfaction with the workspace (Yildirim et al., 2007). Individuals were more satisfied with their workspace when the partition was 4.6 ft (140 cm) high, whereby a person could not look over the partition when seated than with a shorter partition (3.9 ft, 120 cm). A sense of privacy is strongly and negatively related to one's ability to have confidential conversations (Sundstrom et al., 1982). Due to the distractions within an open-plan office, having quiet spaces that people can use and easily access when needed can reduce negative perceptions about the open-plan office design (Haapakangas et al., 2018).

In contrast, the cellular office allows for high levels of visual and auditory privacy, but greatly reduces the ability to communicate easily. The cellular office is beneficial when individuals need limited distractions and the opportunity to concentrate on a task that has a high cognitive demand, or there is a need for confidential conversations. However, cellular offices also create a boundary or identifiable territory, whereby these offices might be perceived as less welcoming than an open-plan or cubicle office. Because there is more of a boundary with cellular offices than open-plan spaces, individuals are less likely to just walk into the space.

10.6.3 Windows and views

Whether in a cellular office or open-plan environment, we have the technology to create sufficient and proper lighting for people to do their jobs; however, individuals prefer having access to windows (Yildirim et al., 2007). Preferences for windows could be confounded (see Chapter 2) with the likelihood that windowed workspaces are closer to the outer edge of the building and possibly in a corner, farther away from high traffic areas, equipment rooms, stairwells, and the air conditioning or heating system.

In windowless offices, the median number of decorations or pictures on the walls tended to be greater than in windowed offices. In addition, these items were generally of a more landscape or "natural" décor, whereas the number of cityscapes was about the same. This created a ratio of landscape to cityscape décor that was greater in windowless than windowed offices (Heerwagen & Orians, 1986). Individuals like to have windows to view the outside (Butler & Biner, 1989), especially a view of nature (Kaplan et al., 1972).

Stephen Kaplan (1983) proposed that if there is person-environment compatibility, the environment supports what the individual needs and wants to do, while also allowing for reflection. Similarly, if the individual needs and wants are supported by the environment, there is person–environment compatibility. If the person–environment relationship is not compatible, more energy or cognitive demand is needed to perform the necessary tasks and restoration is needed.

Kaplan argued that nature or images of nature or wilderness were restorative. Students with a mostly natural view from their dorm rooms, compared to students whose views were more of a built environment, tended to have better directed attention (Tennessen & Cimprich, 1995; see Chapter 6 for a discussion on attention). Similarly, selective attention was better after viewing an image of a green or monochromatic tree, and anger decreased after viewing the color green (tree form or formless; Zhang et al., 2023). Although attention was not affected, virtual reality simulated windows with a view of nature resulted in a greater number of creative responses (but not responses that were more creative), more positive affect, and less negative affect than either no windows or shuttered windows (Sharam et al., 2023). As the Kaplans suggested, nature could be restorative depending on the task and the view of the natural scene.

10.6.4 Environmental color

Besides a desire for windows, individuals often have preferences for different office color schemes. The impact of environmental color is less well understood than other environmental factors due to other interacting variables such as one's personality or job demands. When considering the spectrum of color (see Chapter 3), it is suggested that the blue and violet colors (the high frequency end of the spectrum) are calming and the reds (the low frequency end of the spectrum) are stimulating colors; however, these colors likely interact with perceived task demand. For instance, when performing a boring task, it might seem reasonable that the stimulation from the color red might help the situation. In contrast, when performing a high demand task, a calming color such as blue might help offset the effects of the high demand task.

Students performing a low-demand simulated telemarketing task did perform better in a red compared to a blue environment (Stone, 2003). In addition, performance in a red environment was lower when working on a reading task perceived to have a higher task demand (Stone, 2001). Yet, students performing an extremely monotonous task became frustrated and made more errors in a red compared to a blue cubicle (Stone & English, 1998). The combination of task frustration and color stimulation might have led to this negative outcome.

In a cross-cultural evaluation of workspaces, individuals rated their own mood, the perceived level of lighting (e.g., too dark, just right), and decoration (e.g., no color, neutral, colorful; Küller et al., 2006). Moods tended to be more positive when individuals rated the work environment as having color than either being a neutral or no color situation. Although environmental color appears to impact mood and possibly performance, and we can make some general statements about these effects, more research is needed to explain some apparently contradictory results. The color-in-context theory (Elliot & Maier, 2012) might contribute to our understanding; however, research has not consistently supported this theory (e.g., Lipson-Smith et al., 2021).

10.6.5 Wayfinding

Up to this point, we have focused primarily on one workspace. As we consider multiple spaces, we need to ponder the relative layout and how windows and color will impact the usefulness of this whole space whether a suite of offices, a whole building, or a city. In particular, navigating among spaces, or **wayfinding**, can be greatly impacted by the environmental design including layout of space, signage, and color. Wayfinding success is related to our *cognitive maps*, or our mental image of the layout of a space. If you were to close your eyes and visualize a particular place such as your home, school, or workplace, and visually travel through the house or across campus, the images you "see" reflect your cognitive map. As a side note, aphantasia is the condition of not being able to visualize. If you were to draw the visualized map, a *sketch map*, this reflects your perception of how things are placed in relationship to each other. Your cognitive map is your understanding of how things are related in space and is impacted by your experience with the area.

If you are new to campus, your cognitive map will be weaker, have more blank spots, and have more distorted distances than the cognitive map of someone who has been on campus a couple of years and really knows how to navigate the campus. As you might guess, the quality of one's cognitive map impacts decision-making (see Chapter 7) when deciding which route to take to get to a particular location or if there is a need for an emergency egress route.

Recall that Miguel had trouble finding his way in a new building because the room numbers were not in a logical order and the design did not help differentiate areas. The use of landmarks or environmental differentiation such as different architectural styles or color as well as signage (see Chapter 3) can increase wayfinding success and proficiency; however, using just one of these environmental design aspects might not be sufficient. As the complexity of the building design increases, people make more wrong turns, backtrack more often, and travel at a slower pace. To increase wayfinding success in complex environments, signage is most helpful; however, using multiple cues is best (O'Neill, 1991). Although textual

Figure 10.6 An interior of a casino. Casinos are a maze of colors, lights, and décor, with no clocks to keep people focused on gambling. (TastyPoutine, CC BY-SA 3.0 <https://creativecommons.org/licenses/by-sa/3.0>, via Wikimedia Commons.)

signage is a bit distracting when the environment is simple, it is more helpful than a graphic sign when the design is more complex.

Environmental design can be and is used in multiple settings to control or influence movement. For example, environmental design has been used to deter criminal or unwanted behaviors (Moffatt, 1983). In particular, environmental designs can create spaces that are easier to defend, increase ability for surveillance, deter criminal activity with obstacles, or create more ownership by enhancing territoriality. Making stairs more accessible or making individuals aware of the stairs with signage can increase healthier behavior (Eves et al., 2008). Casinos use knowledge about how the environment can impact behavior (to make more money, of course) by varying the layout, lighting, color, and décor (Finlay et al., 2010). One casino design is shown in Figure 10.6.

Although we have focused primarily on interior designs and single buildings, environmental design can be applied to larger environments such as university campuses, shopping malls, and cities.

BOX 10.4: UNDERSTANDING WAYFINDING IN VARIOUS SETTINGS FOR VARIOUS USERS

Wayfinding, our ability to navigate the environment, can be diminished when the environment is unfamiliar or poorly designed. Common wayfinding problems include getting lost, backtracking, taking more time, and not finding the desired destination. Add in any cognitive or physical limitations such as dementia or visual impairments and these problems can be more severe. Regardless of the setting (e.g., residence or outdoors), signage and landmarks are critical for effective wayfinding in unfamiliar environments.

A hospital environment needs to accommodate medical personnel, patients, and visitors. Research participants performed wayfinding tasks wearing one of five goggles that simulated some of the most common visual impairments (e.g., glaucoma), as well without the goggles (i.e., normal vision; Rousek & Hallbeck, 2011). As expected, there were more wayfinding issues when the participants wore the visual impairment goggles. Concerns included poor illumination of the signs, which is also a problem with normal vision, but to a lesser extent. The sign design relative to letter size, contrast, and location was problematic (see Chapter 3 for information on signage). Open spaces also caused more problems when wearing the goggles, as there were no cues, and flooring was often misperceived as steps, or as slippery or wet.

Similarly, wayfinding can be a challenge for Alzheimer's patients living in a healthcare facility, especially at advanced stages of the disease (Passini et al., 2000). Although individuals at the earlier stages of the disease might have no trouble developing cognitive maps, this ability gets progressively worse, and Alzheimer's patients have difficulty identifying spaces when it all looks too much the same or monotone. There is a need for identifiable landmarks and signage that serve as clues, but some displays (signage) can be distracting if not relevant.

When outdoors in unfamiliar neighborhoods, common landmarks tended to have contrast with the environment such as buildings (not houses) in a neighborhood (Ishikawa & Nakamura, 2012). In addition, landmarks tended to be at intersections helping people know whether to go straight or turn. Landmarks were also helpful along a straight path, probably to give people some confidence they were on the right track. Signage and landmarks were also important to college students navigating through a natural park environment (Soh & Smith-Jackson, 2004). Signage at junctions was critical and participants wanted more conspicuous signs, but this design would contrast with the desired natural look. Although signs were best when individuals came to a junction, landmarks also helped to orient the hikers. **Critical Thinking Questions**: How could you use landscaping, flooring design, or signage to better demarcate paths without causing unwanted visual illusions? What possible problems might arise? What makes a good landmark? What are some of the design differences to consider when designing different environments such as a college campus, hospital, or assisted living environment?

10.7 APPLICATIONS TO SPECIAL POPULATIONS

When designing environments, we should consider the potential range of users, which might include individuals with hearing, sight, maneuverability, or cognitive impairments. If individuals have some degree of hearing loss, the environmental design should help individuals hear essential information (or receive it in another format). In educational settings, we need to ensure that all learners can access the transmitted information. It is also important to ensure noise levels are low enough so that they do not impact the learning process.

Lighting level also could be an important environmental factor if someone has visual impairments. In particular, elderly individuals tend to have greater problems with varying light distributions. If a floor is uneven or there is a step that is hard to see, more lighting is beneficial to avoid slips, trips, and falls. Similarly, anyone with visual impairments might need more task lighting to complete a particular task. When using task lighting, one needs to consider and address possible problems arising due to glare or uneven light distributions.

To accommodate individuals who use assistance such as a walker or wheelchair when ambulatory, the layout of the space needs to include additional space allowance for this equipment and any personnel who might assist. One of the greatest challenges for the elderly is maneuvering safely within a bathroom. Similarly, we must consider all users such as visitors and healthcare providers when designing for individuals who are in facilities such as nursing homes or extended living environments. Many of these issues are addressed with universal design discussed in Chapter 9.

Another environmental design issue concerns individuals experiencing some level of dementia. Wayfinding for individuals with dementia might require more landmarks than signage (e.g., O'Neill, 1991). Multiple coding methods such as landmarks, signage, and color could help with wayfinding. One of our great challenges is designing for an aging population so these individuals can still have freedom and dignity (e.g., Charness & Holley, 2001)

10.8 SUMMARY

Environmental factors include temperature, lighting, noise, and arrangement of the space. Our experience of the environmental temperature is affected by not only air temperature, but also humidity, air flow, and clothing. The relative humidity, level of air flow, and amount of clothing impact heat transfer due to conduction, convection, evaporation, or radiation. As we move from a neutral temperature to either extreme, we begin to see physiological effects such as heat and cold stress, which can become life threatening if not addressed, especially if individuals are not acclimatized. In addition, there tend to be cognitive decrements as we deviate from neutral temperatures.

The amount of lighting required for an environment depends on the task. In some cases, additional task lighting is needed when reading or high visual acuity is needed. Concerns that can arise from lighting include glare and transient adaptation. Glare can be classified as discomfort or disability; direct or indirect; or specular, defuse, or veiling reflections. Performance decrements can arise due to glare, as visibility is reduced. Similarly, varying light distributions can increase transient adaptation, which can reduce performance if individuals have not dark adapted in environments where light levels are lower.

Because of the potential for temporary or permanent threshold shifts with exposure to noise, individuals are encouraged to wear ear plugs or earmuffs, a form of hearing protection when sound levels increase, as specified by OSHA. Cognitively, ambient sounds become noise when they are unrelated to the task. All sounds, not just noise, can potentially mask necessary auditory information, debilitating performance. In addition, noise tends to have a greater negative impact on performance when it is easily identifiable or distinguishable (e.g., a solitary phone ringing). With more sounds, a single sound is less identifiable and performance decrements appear to be fewer.

Finally, there is the physical arrangement and design of the environment. Link analysis helps us identify best layouts by determining which components are frequently used or have high importance, even if used infrequently. As we cannot place all of these critical components next to the user, we attempt to satisfice this layout.

To enhance the environment, people often desire windows because they offer a view, and views might be restorative or distracting depending on one's task. Similarly, individuals have different color preferences for their spaces, but our understanding of the impact of color on performance is limited. Generally, blue colors at the end of the spectrum with shorter wavelengths tend to be calming and the red colors (longer wavelengths) tend to be stimulating; however, these colors appear to interact with additional variables such as the task and other environmental factors (e.g., windows or pictures).

Lastly, how well we design the environment can impact our wayfinding success. The use of signage helps individuals navigate their environment and color can help code spaces, especially for individuals who might be experiencing varying levels of dementia, to increase the identifiability of a space.

LIST OF KEY TERMS

Acclimation
Acclimatization
Air temperature
Air velocity (flow)
Blinding glare
Cellular office
Clo
Color rendering
Conduction
Convection
Diffuse glare
Direct glare
Disability glare
Discomfort glare
Effective temperature
Efficiency
Evaporation
Glare
Heat exchange
Heat index
Heat stress
Illumination
Indirect glare
Lamps

Light distribution
Link analysis
Lumen (lm)
Luminaire
Luminance
Luminous flux
Luminous intensity
Lux (lx)
Mean radiant temperature
Open-plan
Permanent threshold shift
Radiation
Reflectance
Relative humidity
Satisfice
Specular glare
Task lighting
Transient adaptation
Veiling reflection
Visibility
Wayfinding
Wet bulb globe temperature
Wind chill

SUGGESTED READINGS

Fanger, P. O. (1972). *Thermal Comfort: Analysis and Applications in Environmental Engineering.* New York: McGraw-Hill
   This is a classic work on thermal comfort and recommended reading for someone interested in the foundation of thermal comfort.
Kroemer, K. H. E., Kroemer, H. B., & Kroemer-Elbert, K. E. (2001). *Ergonomics: How to Design for Ease and Efficiency* (2nd ed). Upper Saddle River, NJ: Prentice Hall.
   This book contains the essentials of the mechanics of heat transfer and the effects of heat and cold on the human body.
Pheasant, S., & Haslegrave, C. M. (2006). *Bodyspace: Anthropometry, Ergonomics and the Design of Work.* Boca Raton, FL: Taylor & Francis
   Before his untimely death, Stephen Pheasant was a leader in the field of anthropometry. This work with Christine Haslegrave is an updated version of his early books and provides a wealth of anthropometric data.

CHAPTER EXERCISES

1. You oversee the landscape personnel for the university. They have been working in temperatures just above 80°F (26.7°C) with a relative humidity of 45% for the past several weeks. This coming week the temperature is to move into the upper 80's (31.7°C or higher) with a 70% relative humidity.
   a. Explain what types of physiological effects are likely to occur.

   b. What will you recommend that the workers do to avoid negative effects from the temperature?

   c. How might you modify their work schedule, if at all?

2. Road crews are working 24-hour shifts to clear the roads during a strong snowstorm that is likely to last another day or two. The temperature has dropped to below 30°F (−1.1°C) during the day and the wind has increased causing large drifts to clear.

   a. Explain what types of physiological effects are likely to occur.

   b. What recommendations will you give your crews to avoid any negative effects due to the cold climate?

   c. How might you modify the work process or schedule to reduce the number of cold-related effects?

3. For the conditions identified in exercises 1 and 2 above, explain what types of cognitive changes the workers might experience.

4. Explain what the appropriate illumination and set up of the luminaires should be for the following settings:

   a. Studying in a library, in a dorm room, or outside in a park.

   b. Looking for defects in glass bottles.

   c. Preparing dinner (and the users/preparers are of all ages).

5. Identify various sounds in the following settings that could mask or cause interference with another sound and explain how you could reduce the masking or interference.

   a. A cockpit or flight deck

   b. A car

   c. An office

   d. A classroom

6. For the sounds identified in Exercise 5, explain what the performance effects are likely to be and explain what the likely underlying cause is for this change in performance.

7. If individuals working in loud environments (e.g., engine rooms or airport tarmac) wear ear plugs and/or earmuffs, explain how these individuals can best communicate with each other.

8. Select a space you use frequently (e.g., dorm room, kitchen, or office space) and conduct your own link analysis.

   a. Identify problem areas with the layout.

   b. Without concern for cost, redesign (but be able to justify) how you would change the layout to make it more efficient.

   c. Explain what you are satisficing.

9. In designing workspaces, you have the opportunity to select either cellular or open-plan offices.

   a. Explain under what conditions you would select a cellular or open-plan office.

   b. If you select an open-plan office, explain whether you would or would not use partitions and what effect this decision might have on performance.

   c. Explain how windows and the office color scheme might impact the workers in this space.

10. Explain how you would enhance wayfinding in the following settings and justify your designs.

   a. A healthcare facility for the elderly

   b. An elementary school

   c. A shopping mall

   d. A single-family dwelling with four bedrooms

   e. A large high-rise apartment or office building

REFERENCES

Banbury, S., & Berry, D. C. (1998). Disruption of office-related tasks by speech and office noise. *British Journal of Psychology*, 89, 499–517.

Banbury, S. P., & Berry, D. C. (2005). Office noise and employee concentration: identifying causes of disruption and potential improvements. *Ergonomics*, 48, 25–37.

Becker, A. B., Warm, J. S., Dember, W. N., & Hancock, P. A. (1995). Effects of jet engine noise and performance feedback on perceived workload in a monitoring task. *International Journal of Aviation Psychology*, 5, 49–62.

Bensel, C. K., & Santee, W. R. (1997). Climate and clothing. In G. Salvendy (Ed.). *Handbook of Human Factors and Ergonomics* (2nd ed.), pp. 909–934. New York: John Wiley & Sons.

Butler, D. L., & Biner P. M. (1989). Effects of setting on window preferences and factors associated with those preferences. *Environment and Behavior*, 21, 17–31.

Chapanis, A. (1959). *Research Techniques in Human Engineering*. Baltimore, MD: John Hopkins Press.

Charness, N., & Holley, P. (2001). Human factors and environmental support in Alzheimer's disease. *Aging & Mental Health*, 5(supplement 1), S65–S73.

Choi, C. Y., & McPherson, B. (2005). Noise levels in Hong Kong primary schools: implications for classroom listening. *International Journal of Disability, Development, and Education*, 52, 345–360.

Clark, M., Jackson, P. L., & Cohen, H. H. (1996). What you don't see can hurt you: understanding the role of depth perception in slip, trip, and fall accidents. *Ergonomics in Design*, 4(3), 16–21.

Cushman, W. H., & Crist, B. (1987). Illumination. In G. Salvendy (Ed.). *Handbook of Human Factors*, pp. 670–695. New York: John Wiley & Sons.

DiLaura, D. L., Houser, K. W., Mistrick, R. G., & Steffy, G. R. (Eds.). (2011). *The Lighting Handbook: Reference and Application* (10th ed.). New York: Illuminating Engineering Society of North America.

Elliot, A. J., & Maier, M. A. (2012). Color-in-context theory. In P. Devine, A. Plant, P. Devine, & A. Plant (Eds.). *Advances in Experimental Social Psychology*, Vol. 45, pp. 61–125. San Diego, CA: Academic Press. doi:10.1016/B978-0-12-394286-9.00002-0

Eves, F. F., Olander, E. K., Nicoll, G., Puig-Ribera, A., & Griffin, C. (2008). Increasing stair climbing in a train station: the effects of contextual variables and visibility. *Journal of Environmental Psychology*, 29, 300–303.

Faerevik, H., & Reinertsen, R. E. (2003). Effects of wearing aircrew protective clothing on physiological and cognitive responses under various ambient conditions. *Ergonomics*, 46, 780–799.

Fanger, P. O. (1972). *Thermal Comfort: Analysis and Applications in Environmental Engineering*. New York: McGraw-Hill

Finlay, K., Marmurek, H. H. C., Kanetkar, V., & Londerville, J. (2010). Casino décor effects on gambling emotions and intentions. *Environment and Behavior*, 42, 524–545.

Gagge, A. P., Burton, A. C., & Bazett, H. C. (1941). A practical system of units for the description of the heat exchange of man with his environment. *Science*, 94, 428–430.

Haapakangas, A., Hongisto, V., Varjo, J., & Lahtinen, M. (2018). Benefits of quiet workspaces in open-plan offices – evidence from two office relocations. *Journal of Psychology*, 56, 63–75.

Hallam, S., Price, J., & Katsarou, G. (2002). The effects of background music on primary school pupils' task performance. *Educational Studies*, 28, 111–122.

Hancock, P. A. (2022). In defense of the maximal adaptability model. *Physiology & Behavior*, 252, 113844.

Hancock, P. A., & Vasmatzidis, I. (2003). Effects of heat stress on cognitive performance: the current state of knowledge. *International Journal of Hypothermia*, 19, 355–372.

Hancock, P. A., & Warm, J. S. (1989). A dynamic model of stress and sustained attention. *Human Factors*, 31, 519–537.

Hedge, A. (1982). The open-plan office: a systematic investigation of employee reactions to their work environment. *Environment and Behavior*, 14, 519–542.

Heerwagen, J. H., & Orians, G. H. (1986). Adaptations to windowlessness: a study of the use of visual décor in windowed and windowless offices. *Environment and Behavior*, 18, 623–639.

Ishikawa, T., & Nakamura, U. (2012). Landmark selection in the environment: relationships with object characteristics and sense of direction. *Spatial Cognition & Computation*, 12, 1–22.

Jones, D. M., & Broadbent, D. E. (1987). Noise. In G. Salvendy (Ed.). *Handbook of Human Factors*, pp. 623–649. New York: John Wiley & Sons.

Kaplan, S. (1983). A model of person-environment compatibility. *Environment and Behavior*, 15, 311–332.

Kaplan, S., Kaplan, R., & Wendt, J. S. (1972). Rated preference and complexity for natural and urban visual material. *Perception and Psychophysics*, 12, 354–356.

Knez, I., & Hygge, S. (2002). Irrelevant speech and indoor lighting: effects on cognitive performance and self-reported affect. *Applied Cognitive Psychology*, 16, 709–718.

Kroemer, K. H. E. (2006). Designing for older people. *Ergonomics in Design*, 14(4), 25–31.

Kroemer, K. H. E., Kroemer, H. B., & Kroemer-Elbert, K. E. (2001). *Ergonomics: How to Design for Ease and Efficiency* (2nd ed). Upper Saddle River, NJ: Prentice Hall.

Küller, R., Ballal, S., Laike, T., Mikellides, B., & Tonello, G. (2006). The impact of light and colour on psychological mood: a cross-cultural study of indoor work environments. *Ergonomics*, 49, 1496–1507.

Lin, C.-J., & Wu, C. (2010). Improved link analysis method for user interface design – modified link table and optimisation-based algorithm. *Behavior & Information Technology*, 29, 199–216.

Lipson-Smith, R., Bernhardt, J., Zamuner, E., Churiloy, L., Busietta, N., & Moratti, D. (2021). Exploring colour in context using virtual reality: does a room change how you feel?. *Virtual Reality*, 25, 631–645.

Mäkinen, T. M., Palinkas, L. A., Reeves, D. L., Pääkkönen, T., Rintamäki, H., Leppäluoto, J., & Hassi, J. (2006). Effect of repeated exposures to cold on cognitive performance in humans. *Physiology and Behavior*, 87, 166–176.

Martin, K., McLeod, E., Periard, J., Rattray, B., Keegan, R., & Pyne, D. B. (2019). The impact of environmental stress on cognitive performance: a systematic review. *Human Factors*, 61(8), 1205–1246.

Moffatt, R. E. (1983). Crime prevention through environmental design – a management perspective. *Canadian Journal of Criminology*, 25, 19–31.

Mukhopadhyay, P. (2009). Ergonomic design of head gear for use by rural youths in summer. *Work*, 34, 431–438. doi 10.3233/WOR-2009-0943

O'Neill, M. J. (1991). Effects of signage and floor plan configuration on wayfinding accuracy. *Environment and Behavior*, 23, 552–574.

Palinkas, L. A., Mäkinen, T. M., Pääkkönen, T., Rintamäki, H., Leppäluoto, J., & Hassi, J. (2005). Influence of seasonally adjusted exposure to cold and darkness on cognitive performance in circumpolar residents. *Scandinavian Journal of Psychology*, 46, 239–246.

Parsons, K. (1991). Human response to thermal environments: Principles and methods. In J. R. Wilson & E. N. Corlett (Eds.). *Evaluation of Human Work: A Practical Ergonomics Methodology*, pp. 387–405. Philadelphia, PA: Taylor & Francis.

Passini, R., Pigot, H., Rainville, C., & Tétreault, M-H. (2000). Wayfinding in a nursing home for advanced dementia of the Alzheimer's type. *Environment and Behavior*, 32, 684–710.

Pawlaczyk-Luszczyńska, M., Dudarewicz, A., Waszkowska, M., Szymczak, W., & Śliwińska-Kowalska, M. (2005). The impact of low frequency noise on human mental performance. *International Journal of Occupational Medicine and Environmental Health*, 18, 185–198.

Pilcher, J. J., Nadler, E., & Busch, C. (2002). Effects of hot and cold temperature exposure on performance: a meta-analytic review. *Ergonomics*, 45, 682–698.

Rohles, R. H., & Konz, S. A. (1987). Climate. In G. Salvendy (Ed.). *Handbook of Human Factors*, pp. 696–707. New York: John Wiley & Sons.

Rousek, J. B., & Hallbeck, M. S. (2011). The use of simulated visual impairment to identify hospital design elements that contribute to wayfinding difficulties. *International Journal of Industrial Ergonomics*, 41, 447–458.

Sharam, L. A., Mayer, K. M., & Baumann, O. (2023). Design by nature: the influence of windows on cognitive performance and affect. *Journal of Environmental Psychology*, 85, 101923.

Sheedy, J. E., Smith, R., & Hayes, J. (2005). Visual effects of the luminance surrounding a computer display. *Ergonomics*, 48, 1114–1128.

Soh, B. K., & Smith-Jackson, T. L. (2004). Influence of map design, individual differences, and environmental cues on wayfinding performance. *Spatial Cognition and Computation*, 4, 137–165.

Stone, N. J. (2001). Designing effective study environments. *Journal of Environmental Psychology, 21,* 179–190. doi:10.1006/jevp.2000.0193

Stone, N. J. (2003). Environmental view and color for a simulated telemarketing task. *Journal of Environmental Psychology, 23,* 63–78. doi:10.1016/S0272-4944(02)00107-X

Stone, N. J., & English*, A. J. (1998). Task type, poster presence, and workspace color on mood, satisfaction, and performance. *Journal of Environmental Psychology, 18,* 175–185. doi:10.1006/jevp.1998.0084

Sundstrom, E., Herbert, R. K., & Brown, D. W. (1982). Privacy and communication in an open-plan office: a case study. *Environment and Behavior, 14,* 379–392.

Tennessen, C. M., & Cimprich, B. (1995). View to nature: effects on attention. *Journal of Environmental Psychology, 15,* 77–85.

US Department of Health and Human Services. (1998). Criteria for a Recommended Standard: Occupational Noise Exposure Revised Criteria. Cincinnati, OH, 1–122.

Vasmatzidis, I., Schlegel, R. E., & Hancock, P. A. (2002). An investigation of heat stress on time-sharing performance. *Ergonomics, 45,* 218–239.

Wohlwill, J. F., Nasar, J. L., DeJoy, D. M., & Foruzani, H. H. (1976). Behavioral effects of a noisy environment: task involvement versus passive exposure. *Journal of Applied Psychology, 61,* 67–74.

Yerkes, R. M., & Dodson, J. D. (1908). The relation of strength of stimulus to rapidity of habit-formation. *Journal of Comparative Neurological Psychology, 18,* 459–482.

Yildirim, K., Akalin-Baskaya, A., & Celebi, M. (2007). The effects of window proximity, partition height, and gender on perceptions of open-plan offices. *Journal of Environmental Psychology, 27,* 154–165.

Zhang, D., Jin, X., Wang, L., & Jin, Y. (2023). Form and color visual perception in green exercise: positive effects on attention, mood, and self-esteem. *Journal of Environmental Psychology, 88,* 102028.

Chapter 11

Human error

Causes and prevention

Chapter Vignette

Sandra is a chemist who works in a lab for a pharmaceutical company that is aiming to develop a drug to assist people with various levels of traumatic brain injury. The work is fascinating and relevant to her life, as Sandra cares for her aging father. Although her father does not have a traumatic brain injury, Sandra is familiar with the dementia he is experiencing related to the early stages of Alzheimer's disease. She is often called at work to comfort her father when he becomes agitated. His bad days can begin when he has forgotten to take his medication or has taken the wrong medications.

Sandra created a systematic plan for his medications, but if he cannot find the labeled pill boxes, gets distracted, or does not understand the boxes, he does not take the medication or the correct medication. On other occasions, the nursing staff changes his routine for the day, and he becomes a bit disoriented and performs his "usual" routine, which generally means he goes to the activity hall at the wrong time of the day or goes for a walk when he is supposed to be at an appointment. The nursing staff's rationale was that changing the routine was intended to keep people from getting bored by adding excitement to the day because of "new" things. Sandra has tried on several occasions to explain to the staff that for someone like her father, these changes actually contribute to and cause disorientation, confusion, and his "errors" during the day.

It is during times like this that she reminisces about the good ole days. Although she loves her job in Chicago, the stress of her job and of managing the care of her father is taking its toll. She misses her home state of Iowa—the open fields and quiet days on the farm. Yet, farming is not stress free. She learned about a former neighbor's son who was hurt using the combine the other day. The young man's clothes were caught by one of the augers that pulled him into the machine severely injuring one of his arms. The surgeons saved the limb, but the recovery will be slow and they are not sure if he will regain the full use of his arm.

When Sandra's father forgets to take his medication or goes for a walk when he should be at an appointment, or when accidents with a combine occur, "human error" is traditionally used to describe the incident. It is true that these incidents reflect undesirable actions performed by humans; however, in recent years, there has been a shift from viewing human error as the cause of the incident to viewing human error as a consequence of the situation. Keeping in mind that human error can either be a cause or a consequence of the situation, all of the events and demands Sandra experiences are related to human error. Although Sandra is a competent chemist, her task demands and stress of managing her father's care could contribute to the likelihood that she will make errors. The more demand or stress a person experiences, the greater the likelihood for error.

It is possible to reduce the likelihood of error if a person has more experience, but in some situations, more experience or expertise cannot help. Any errors Sandra might experience could

DOI: 10.1201/9781003515463-11

be due to the excessive number of deadlines placed on her, which reflects the expectations within the organizational culture or structure that create an environment for error. On the other hand, Sandra's father is losing mental abilities, whereby he does not remember the correct action to perform or the correct medication to take. It is important to determine ways to reduce error because error can often lead to accidents, which are negative consequences to equipment or people. Taking the incorrect medication could cause additional medical problems, and farming machinery errors can lead to severe bodily injury. Using human factors methods, we can understand what human error is, develop ways to assess human error, determine what factors influence human error, and reduce error, thereby increasing human reliability.

11.1 CHAPTER OBJECTIVES

After reading this chapter, you should be able to:

- Describe and define what human error is.
- Identify and describe the different classifications of human error.
- Define human reliability.
- Define what a near miss and an accident are.
- Describe the various factors that potentially impact error.
- Describe the difference between "stress" and "stressor."
- Describe the relationship between stress and human error.
- Identify problems with warnings and develop effective warnings.
- Identify human factors methods to reduce errors.
- Identify and understand the purposes of the various agencies that influence safety management.

11.2 WHAT IS HUMAN ERROR?

When a plane crashes, a driver follows the wrong highway exit, or Sandra's father forgets to take his medicine, we are likely to consider this human error. We consider these incidents human error because someone or a group of people, intentionally or unintentionally, acted in an inappropriate fashion or did not act as needed, in order to obtain the system's goals such as safely landing the plane, using the correct highway exit, or taking the appropriate medication. Therefore, **human error** is an action outside the bounds of expected or acceptable performance for the particular situation or system (Miller & Swain, 1987). Errors are usually unintentional such as intending to pick up the pen on our desk, but we accidentally pick up the highlighter next to it. Other times we make a conscious effort to perform a particular action, which is not appropriate for that situation such as intentionally selecting a highway exit, but realizing it is the wrong one after we exit. Similarly, individuals who take multiple medications might become confused and take the wrong one (see Figure 11.1). Finally, errors also arise when we fail to act appropriately such as forgetting to take our medicine.

Although not all errors put us in danger, to enhance safety, we must reduce human error. Returning to the concept of a system presented in Chapter 1, we need to appreciate the fact that some systems are more challenging for humans. When there is a system error, we might call it "operator error," but we need to determine all factors contributing to the error. Bad design is often the culprit of human error. Given the way humans make decisions (see Chapter 7), our memories work (see Chapter 6), and our motor control functions (see

Figure 11.1 The confusing world of medicine. Confusion about which medicine to take could be due to the individual's error in understanding what medicine to take or the error in the design of the labeling that confuses the person as to which medicine is correct. (© Nancy J. Stone.)

Chapter 8), humans are likely to make errors when the human-built system is not compatible with the users. Environmental designs (see Chapter 10) also can cause individuals to err.

When the operator or user believes there are no malfunctions and the system appears to be functioning correctly, the operator will make decisions to perform the behaviors that are expected in a functioning system. When the operator perceives a problem in the system, the operator will make decisions, problem solve, and act in ways to fix the perceived problem (Rouse & Rouse, 1983). Although we might conclude that an incorrect perception or interpretation of the system is human error, Donald Norman (1988) would argue that the error is possibly due to the design of the system. According to Norman, if the system is designed such that it is difficult to understand or it is not intuitive, the system requires what Norman would call "in the head" knowledge or something that must be learned. In this case, the design of the system might be the cause of the human error. Norman argues for designs that are more intuitive with "in the world" knowledge in order to reduce errors.

If errors do occur, they generally are not mechanical errors. Although there is a great deal of research on human error, engineers have tested machine error to a greater extent and have created machines that have minimal margins of error. Therefore, when accidents happen, our first reaction is often to look to the human as the cause of the error, even if the design of the system or the human–machine or human–system interface is the problem. The use of the term "human error" as the label exacerbates this problem. Erik Hollnagel (1993) uses "erroneous actions" to describe these behaviors, as there is no blame assigned, only an identification of the error.

Applying the systems perspective (see Chapter 1), the perception of error has changed (or is changing) from identifying the human as the cause of error (i.e., "operator error," e.g., Reason, 1990) to identifying problems within the total system design as the cause (e.g., Dekker, 2002). To reduce human error (i.e., erroneous actions) with quickly changing and advancing technology, we need to ensure that the machinery or equipment is compatible with the humans that operate it. Many errors do not occur because humans are able to adapt to cope with the complexities of the systems (Woods et al., 2010). Beyond training the human user, human factors specialists strive to create better designs that reduce the need to adapt, which involves modifying the equipment for easier human use. In turn, we can enhance performance and reduce error. To better understand error and its causes, we need to understand the classifications, categories, or types of errors that occur.

11.3 HUMAN ERROR CLASSIFICATIONS

11.3.1 Errors of omission and commission

Human error can be classified or categorized in a variety of ways. As mentioned earlier, errors may be intentional or unintentional behaviors, but this is an extremely broad category. From a systems perspective, error can occur at the input, process, or output phase. Originally, the focus was on the behavioral aspects of error or, more specifically, the human output. Human output error could be an incorrect or absent action leading to incorrect or absent input to another system. Alan Swain introduced a scheme in 1963 for classifying these human output errors that included errors of omission and errors of commission (Miller & Swain, 1987). **Errors of omission** occur when an action or behavior that should be performed is not executed (i.e., omitted, forgotten, or skipped). In contrast, when an action or behavior has occurred, but it is the wrong action or an action incorrectly executed (i.e., at the wrong time, in the wrong sequence, or lacks the necessary level of quality), this is an **error of commission** (see Table 11.1).

11.3.2 Slips, lapses, and mistakes

Although evaluating behavioral output error is important, this classification does not help us understand and address the underlying determinants of the error. The cause might actually occur due to poor design features, which influences how the input is processed. Hence, the evaluation of error now focuses more on the cognitive processes involved (Rasmussen, 1985). Norman (1981) focuses on the process behind the output and sometimes the input by making the distinction between slips (or lapses) and mistakes. A **slip** (or **lapse**) is an error in performing an intended action, whereas a **mistake** is an error in the original mental model or goal developed, which leads to the incorrect outcome for the system (see Table 11.1).

Table 11.1 Error classifications

Researcher	Year	Classification system of errors
Alan Swain	1963	Omission or commission Error of omission: a behavior not completed. Error of commission: a behavior that is incorrect, out of sequence, or not timely.
Donald Norman	1981	Slip or mistake: Slip: an incorrect action. Mistake: the correct action created by an incorrectly designed environment.
Jens Rasmussen	1974, 1983	Skill-, rule-, or knowledge-based errors: Skill-based: an overlearned, automatic behavior that did not initiate with a signal. Rule-based: learned processes for the current situation (i.e., the sign) are not effective. Knowledge-based: unable to develop creative solutions to unique problems (i.e., the symbols).

Slips are more unconscious in the sense that as a process becomes automated we often fall back into this automatic pattern when we are attempting to deviate from the typical pattern. Let's say on Wednesday mornings we go to class and then to work. On this coming Wednesday, though, we make plans to meet some friends after class and to go to work later. It is possible that after class we head off to work because this is our habitual or normal course of action, which would be considered a slip today because we did not meet our friends. Similarly, when the nursing staff changed Sandra's father's routine, his unintended action of missing his appointment would be classified as a slip. To avoid slips, we must make a concerted effort to do something unusual from our normal routine.

Mistakes are more deliberate and the result of conscious thinking as the original goal or direction was intentionally set, but inappropriate given the circumstances. An example of a mistake could be determining the best route for a day of shopping and running errands. Near the end of the day of shopping, we might realize that another route would have better served us in finding a particular shop that we still need to visit. Although we completed the originally planned route correctly, the plan did not serve us well in the end, as we had to backtrack to the one remaining store. This type of premeditated planning can be analogous to management making various procedural decisions that appear much later as "operator error." Processing errors such as misreading instructions or misinterpreting a road sign are human processing errors. Yet, bad design also could cause the input to be misunderstood, which would make this a design error. In summary, slips and lapses are execution errors, while mistakes are planning errors (Reason, 1990).

Although the distinction between slips and mistakes is helpful, there is still a need for better or more thorough distinctions between types of errors. Regardless of whether an action is the result of a slip or a mistake, the outcome (e.g., error) is often the same. For example, we might take the wrong highway exit because it is the normal exit we take (i.e., a slip) or because the sign for the exit is placed just after (as is often the case), not just before, the exit (i.e., a mistake). Because someone decided that the best placement for the sign was after the exit, identification of this mistake occurs sometime after people drive past the exit they needed. As mistakes are often more difficult to identify than slips, they can be more dangerous because they are "hidden" (Reason, 1990).

11.3.3 Skill-, rule-, and knowledge-based errors

For a better understanding of the cognitive causes of errors and for designing safer and more error-free systems, further distinctions tap into Jens Rasmussen's (1974, 1983) work on skill-, rule-, and knowledge-based errors. The type of cognitive processing required varies for these different levels of actions. Errors arise when the behaviors or actions do not serve the situation well. **Skill-based behaviors** are more along the lines of stimulus-response behaviors that require little conscious or attentional effort because the behaviors are over-learned and automatic. The stimuli in the environment merely act as guides, reminders, or what Rasmussen (1983) calls a *signal* as to what action is next.

Moving into **rule-based behaviors**, performance is based on learned rules and procedures, whereby the input from the environment serves as a *sign* (Rasmussen, 1983) as to what proper rule should be used at this time. The user is working with a set of pre-defined, learned options for actions. Finally, **knowledge-based behavior** is used when the individual encounters a novel situation. This behavior is significantly different from skill- and rule-based behavior, as it is much more conceptual. The input from the environment now serves as *symbols* (Rasmussen, 1983) that require better mental models, as no known skill or rule can address the current situation and the user must create a novel response. One way to distinguish between the skill-, rule-, and knowledge-based activities is that skill-based problems (slips) occur BEFORE a problem is identified, and rule- and knowledge-based activity occurs AFTER a problem has been identified. Therefore, skill-based errors tend to be slips and lapses, whereas rule- and knowledge-based errors would often be considered mistakes.

11.3.4 Generic error-modeling system

James Reason (1990) applied Norman's and Rasmussen's concepts to his generic error-modeling system (GEMS) for understanding human error (discussed in detail in the next section). Functioning at the skill-based level is primarily a monitoring activity, which requires what Reason calls *attentional checks* to ensure everything is functioning as expected. If everything appears correct from the use of attentional checks, the individual will keep working as usual using these skill-based actions. When there is lack of attention (a sign is missed) or overattention (checking at the wrong time), then there is the possibility of monitoring failures.

According to GEMS, only when there is a perceived problem will the individual move to the next level of rule-based functioning to seek preprogrammed solutions. Individuals search their memories for known methods they have effectively used for solving this type of problem in the past. When the problem is unique, however, this requires the individual to move into knowledge-based functioning that requires the creation of innovative and novel solutions. According to Reason, people prefer to work at the rule-based level because it requires less cognitive demand. Therefore, there are often problem-solving failures because knowledge-based functioning increases mental load.

Our discussion of human error classifications began with the focus on behaviors or outputs, as this was the original focus when studying human error. We just completed our discussion on the types of errors that might occur during the processing phase of a system. There are also errors that occur at the input phase of a system. These input errors generally arise because the individual cannot properly process the input. This could be due to a mistake in design whereby the visual input is illegible (see Chapter 3), the auditory input is masked (see Chapter 4), or tactile information is blocked because of the necessary equipment or clothing the user must wear (e.g., gloves; see Chapter 3). Input errors are usually not considered human errors, but are normally considered a bad design issue or a mistake in Norman's terms, which cause humans to act inappropriately.

BOX 11.1: UNDERSTANDING AND REDUCING HUMAN ERROR: COMMUNICATION IS KEY

In everyday language, we are not specific in how we define an error or mistake. In fact, we often use the same terms to mean several things such as (1) we did not do what we wanted or intended to do, (2) we did not do what we should have done, or (3) we did not do what we wished we had not done. For example, we often call things we do incorrectly errors, mistakes, or accidents. In the realm of human error, the distinctions are extremely specific. The statements "I made a mistake" and "That was an accident" can mean similar things to a layperson, but might not apply to the situation for a human error specialist. The specificity of the language is necessary to assist with the investigation of the error and in determining potential causes. Therefore, clear and unambiguous communication is necessary.

Forensic experts often find that the type of communication used can cause accidents because the proper information is not conveyed to those who need the information (Beck & Cohen, 1994). Beck and Cohen reviewed the communication breakdown that resulted in one man having his legs severed as he sat on the train tracks in protest of the activity at the Concord Naval Weapons Station in September 1987. One of the issues related to this tragedy was that there was only one-way communication and no feedback loop. Therefore, the train crew was unaware that the protesters were not going to move, as they always have. In addition, without the feedback loop, the protesters did not realize that the train crew had not been informed of this change.

Warnings are a form of communication, but not always understood. A beach sign intended to warn of falling rocks from the cliff was misinterpreted more by visitors than by locals, suggesting a need for pre-existing knowledge to interpret the sign (Aucote et al., 2012).

These examples of the importance of communication reinforce the notion that research is critical in determining ways to enhance communication. Although two-way communication and a feedback loop can enhance communication, this is not possible with warnings. Mnemonics, checklists, and worksheets are helpful tools for reducing errors in various product and process evaluation contexts (Peacock & Resnick, 2011). These tools help individuals know what is missing and can inform the communication process. Therefore, research is needed in each specific context to determine how best to communicate critical information to reduce errors. **Critical Thinking Questions:** Considering your courses, identify various errors that arise due to miscommunication. What types of errors are these? How might you enhance the effectiveness of the communications? Now, consider various signs around your campus. How effectively do they communicate the issue? How much pre-existing knowledge is required to understand the sign ("in the world" or "in the head" knowledge)? How would you research the communication process?

Whether human error occurs at the input, process, or output phase, the error potentially affects the success of the system. To have a functional system with minimal error, human factors experts attempt to determine the cause of the human error to ensure successful system functioning. By reducing the likelihood of human error, the users' behaviors become more reliable. We focus on human reliability next.

11.4 HUMAN RELIABILITY

With an understanding of human error, we can now focus on the opposite of human error or human reliability. **Human reliability** is the extent to which the human is error free and will *NOT* make errors (Miller & Swain, 1987). The more reliable the person is, the less likely that person will make errors. As humans function within systems composed of machines, equipment, or other humans, system reliability is a combination of the reliabilities of each component (Miller & Swain, 1987). As you increase the number of components within a system, whether they are additional machines or people, the reliability is likely to decrease because there is a greater likelihood that something will go wrong when there are more parts.

Given the great extent to which engineers have studied machine error, machine reliability is better understood and more clearly determined than human reliability. This is due in part because our understanding of human error, as discussed in the above section, is a bit more complicated than our understanding of machine error given the issue of human cognition. Assuming that machines and equipment have little to no error and are designed appropriately for the task at hand, we need to determine the level of human reliability in order to increase system reliability.

11.4.1 Human reliability analysis

To design safer systems, we need to understand how likely it is there will be no error. As mentioned above, engineers have a deep understanding of how machines function and respond to environmental conditions, whereby they can greatly reduce machine error to a negligible level. Yet, much work is still needed to determine how reliable the worker is in order to increase system safety. The process of identifying the risks and potential errors related to the human operator is **human reliability analysis** (HRA). HRA involves four steps: (1) determine all the actions required by the human and the proper sequence of the actions, (2) identify where error could occur, (3) determine the probabilities of these errors, and (4) estimate the impact on the system (see Table 11.2). After a completed HRA, we generally make suggestions for changes that can reduce error and then repeat steps two and three to determine the new error estimate (Reason, 1990).

Table 11.2 Steps of HRA: a simplified example of a pharmacist filling a prescription

General steps	Example
(1a) Determine all human actions required and the proper sequence of actions	(1) Receives prescription, (2) reads and interprets prescription, (3) prepares drug label, (4) dispenses drug, and (5) cleans the work area.
(1b) Understand expectations of the job and every task to determine performance criteria	Needs to (1) receive prescription in a timely manner for prompt delivery, (2) read and interpret prescription to prepare correct drug and correct dosage, (3) prepare proper label to ensure patient safety (i.e., understanding of drug consumption dosages and times), (4) dispense correct drug to ensure proper treatment and patient safety (i.e., not taking another potentially harmful drug), and (5) clean area to avoid contamination of the next prepared drug.
(2) Identity where error could occur	(1) Does not receive prescription or receives multiple prescriptions at once, (2) confuses drug names or misreads dosage and instructions for taking medicines (i.e., pharmacist could be sleep deprived or interrupted), (3) creates improper label (i.e., incorrect dosage or time to take), (4) confuses drug names or is interrupted and dispenses wrong drug, and (5) unclean area contaminates current drug dispensed (could interact negatively with other drugs or independently have negative effects).
(3) Determine error probabilities	The estimated probabilities are: (1) not receiving the prescription on time, 0.01, (2) misinterpreting drug and/or instructions, 0.10, (3) creating incorrect label 0.07, (4) dispensing incorrect drug, 0.05, and (5), not cleaning area, 0.01.
(4) Estimate system impact	The probability of misinterpreting the drug and/or instructions (0.10), or creating an incorrect label (0.07), or dispensing the incorrect drug (0.05) would be 0.22 (0.10+0.07+0.05) or 22%. Without some type of intervention, patient injury is likely.

The first step in HRA is to understand the expectations of the job and every task, and to determine the performance criteria for these tasks (Miller & Swain, 1987). The process of understanding the job and tasks when investigating human error involves typical research methods (see Chapters 2 and 5) such as observation, questionnaires, experiments, simulations, and case studies (Reason, 1990), as well as task analysis. From the first step, we should understand all the tasks one must perform, the proper sequence of the tasks, and the expected performance criteria (i.e., level of expected performance, see Table 11.2 for examples). We also need to understand the work environment or the **performance shaping factors** (PSFs) such as a worker's level of training or quality of the equipment (Kirwan, 1992). There is also a selected endpoint or point at which there is, or could be, system failure (e.g., the plane crashes, a person takes the wrong medication). This endpoint could be anywhere in a long sequence of behaviors because some behaviors can be corrected. Once the steps, sequences, and PSFs are identified in step 1, we identify potential errors (step 2), their respective probabilities of occurrence (step 3), and the consequences if these errors occur (step 4, Miller & Swain, 1987). See Table 11.2 for examples.

Several methods of HRA exist and continuously evolve as we gain a better understanding of human error (see Setayesh et al., 2022 for a comparison of four HRA models). Much of the earlier work on human error was completed in, or related to, the aviation and nuclear power industries. Therefore, many of the methods for human reliability analysis refer to these situations. Yet, we can apply these processes to other everyday activities. We briefly discuss a couple of these processes below. If you wish to be proficient at conducting HRA, you would require additional training, as ratings of reliability or probability of error often come from *subject matter experts* (SMEs), as well as any probability data available from the literature or databases.

The Technique for Human Error Rate Prediction (THERP) is the oldest and most well-known technique developed by Alan Swain in 1963 as an extension of Luther Rook's 1962 work (Swain, 1963, 1964). Although theoretically strong, THERP is difficult to use (Miller & Swain, 1987). To determine the human reliability, THERP is initiated by selecting a particular point in a sequence and detailing the sequence of events from that point forward in an HRA **event tree** and what happens if something goes wrong for each action identified in the task analysis. This is in contrast to other methods that use an HRA **fault tree**, whereby the sequence of events is described from the point of error backward (see Figure 11.2 as an example). After completing the event tree, the analyst might drop any actions from the model that are not perceived as relevant to the system function; however, one needs to be conservative in this step to avoid eliminating critical actions that would severely impact the reliability analysis (Swain, 1963).

After the sequence of events is identified, the probability of error is predicted for each action or human output in the sequence. The errors identified at this step in THERP include error of omission, error of commission, and any extraneous error (Kirwan, 1992). Human error data are estimated based on experimental data if available and appropriate; however, these data are seldom appropriate given the level of control used to cause error in the laboratory. In addition, an actual operator often has additional tasks requiring attention and much more or more real stress than research participants experience (Swain, 1963). Sometimes the determination of error rate requires a great deal of expert judgment, which reinforces the need for SMEs.

Finally, we use probability theory to combine the error probabilities into a quantitative estimate of human reliability (see an example in Table 11.2, Step 4). As the operator might perform a corrective action after an error, we could create multiple equations that represent numerous behavioral sequences, some of which will lead to ultimate system failure and others will not (Swain, 1963). Therefore, we would have a range of error probabilities

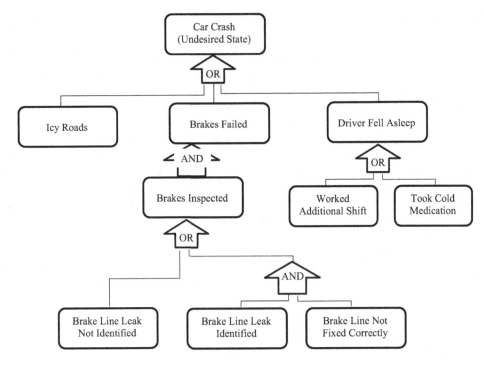

Figure 11.2 An example of a fault tree. (© Nancy J. Stone.)

depending on which sequence an operator follows. Further, these data for the HRA reflect the general rule or "average" and might not apply well to a single individual. It is assumed that the individual operator has sufficient training as well as average motivation to perform the job. As you might guess, variations in any of these variables could impact the final level of system reliability.

Once the THERP analysis is complete, it is possible to make suggested corrections to the system that would reduce error. Then, the cost of these changes in terms of training, personnel, and equipment changes, for example, could be compared to the projected cost (or savings) of the error reduction to determine whether to make these changes. Unfortunately, system changes are often not made due to the expense, even though there is the probability of error.

As described above, THERP depicts or records human error as human output—what the operator did or what actions did or did not happen. Yet, we need to determine how error occurs cognitively or during the process, and possibly the input stage of the system. Other techniques designed to tap cognitive errors are generally based on Rasmussen's skill-, rule-, and knowledge-based (SRK) model of human performance. Although the actual error can appear and have the same impact regardless of the cause of the error, the types of errors could vary depending on when the error occurs along the SRK process (Kirwan, 1992).

James Reason's (1990) Generic Error Modeling System (GEMS) is helpful not only with classifications of errors, but also in human reliability analysis. GEMS determines the cognitive aspects of human error by incorporating the SRK model to explain Norman's slips and mistakes. Specifically, GEMS helps explain when an operator is likely to move into the rule- and knowledge-based functions. As discussed previously, an individual will work at the skill-based level until an error is identified or a slip occurs. At this point, the individual will move into the rule-based level, searching for a standardized solution to the current situation. If there are errors at this stage, then these would be intentional mistakes because the individual is selecting a particular rule to follow. Once these errors create a novel situation, the individual will be required to move into the knowledge-based level of functioning to resolve the problem.

When using GEMS in HRA, the task is to determine the probability that an individual will make an error while monitoring (skill-based error) or will apply the wrong or correct rule at the right or wrong time (rule-based error). At the knowledge-based level, the task is to determine the probability that some type of decision-making bias (see Chapter 7) is likely to occur that prevents the novel thinking required to solve the current problem. Keep in mind that sometimes errors lead to even more errors as the person becomes more exasperated with the situation. In this case, the probability of an error can be double the original probability of the error and the probability of the error could double with each successive behavior (Swain, 1963). These situations require *dynamic fault management* (e.g., Woods, 1994).

Regardless of the type of HRA, one must know the tasks; sequence of the tasks or steps; performance criteria; working environment (external PSFs); and potential errors, their estimated probabilities and their consequences (i.e., importance or impact of error) related to the action to be performed (Miller & Swain, 1987). As can be ascertained from the brief descriptions of the THERP and GEMS HRA techniques, performing an HRA is complicated and requires quite a bit of judgment based on expertise. Other HRAs are listed and briefly described in Table 11.3.

Table 11.3 Descriptions of HRA techniques

HRA technique	Description
Operator-Action Tree (OAT) Method	OAT was developed after THERP. OAT is performed after an accident, focusing on cognitive issues (e.g., reasoning, decision-making) prior to the accident. OAT uses a logic tree to identify three types of cognitive errors related to failures in (1) perceiving the event, (2) diagnosing the event circumstances and corrective actions, and (3) implementing corrective actions in a timely manner. Data are based on experts' judgments or lab experiments. See NUGREG/CR-3010, Hall et al. (1982) for details.
Maintenance Personnel Performance Simulations (MAPPS)	The MAPPS is used to understand quantitatively how behavior changes when maintenance conditions vary for nuclear power plant maintenance personnel. Main variables of interest are quality of performance and performance time, which are assessed relative to variations in personnel, conditions, task, and even the possibility of skipping less relevant tasks when stress levels are high. The model provides information about human reliability under these various conditions and is useful for training if one can anticipate where stress levels will be high and errors more likely. See Siegel et al. (1984) for more details.
Success Likelihood Index Methodology (SLIM)	The purpose of SLIM is to determine a Success Likelihood Index which focuses on performance influencing factors (PIFs), which are similar to performance shaping factors (PSFs). PIFs are more cognitively oriented, as opposed to the behaviorally oriented PSFs assessed in THERP (Reason, 1990). The likelihood of success is determined based on factors related to the individual (e.g., training), the task, or environment. The PIFs are considered most critical for the given situation based on expert judgments. Results can be used to identify and reduce errors that are currently most likely to occur.
Sociotechnical Approach to Assessing Human Reliability (STAHR)	The STAHR method was developed from SLIM. STAHR takes the systems approach to understanding human error. See Phillips et al. (1990) for more details.

(Continued)

Table 11.3 (Continued) Descriptions of HRA techniques

HRA technique	Description
Human Cognitive Reliability (HCR) model	HCR Correlation is a technique used in predicting the amount of time to complete a task when error occurs. When there is error, it is assumed that greater cognitive processing time is required to resolve the situation. This method does not predict the error itself or how often the error is likely to occur (Reason, 1990).
Empirical Technique to Estimate Operators' Errors (TESEO) (Abbreviation from Italian wording)	TESEO applies to processing plants (e.g., chemical, nuclear power). This process predicts the probability of error based on five parameters: (1) activity (routine or not, high attention or not), (2) level of stress, (3) operator qualities or abilities, (4) level of anxiety the activity is likely to induce (e.g., novel situation), and (5) an activity ergonomic factor (microclimate and interface). The technique is simple and results in good predictions similar to expert judgments (Reason, 1990).
Confusion Matrix	Potash, Stewart, Dietz, Lewis, and Dougherty developed the confusion matrix for use in HRA (as cited in Kirwan, 1992). Use of the confusion matrix is relatively simple. A matrix of all possible abnormal events is created. Then, experts judge the level of confusion one event is likely to create with other events (low, medium, or high) and the impact on subsequent actions (low or high). Using this matrix, it is easy to determine where problems (i.e., misdiagnoses) are likely to occur and the likely resultant effects. Such information would be of use to training departments (Kirwan, 1992).
Systematic human action reliability procedure (SHARP)	Given the importance of the human operator in system reliability, this technique helps to combine models and techniques of human reliability analyses with the machine reliability analyses in a system's probability risk assessment (PRA). There is a strong focus on the human–machine interaction functions. There are seven steps: (1) fully describe human interactions comprehensively, (2) select the most critical human interactions for further analysis, (3) identify the tasks and subtasks for each critical human function, (4) model the human interactions to include the various options available, (5) determine the impact of human interactions, (6) quantify the human interactions, and (7) document them. See Hannaman et al. (1984) for additional information.
Systematic Human Error Reduction and Predication Approach (SHERPA)	The goal of SHERPA is to identify where errors are likely to occur. The process is a computerized question and answer system for each step of function identified in the task analysis, whereby this technique might not be used similarly (i.e., systematically) by two different assessors. The types of likely errors are based on the SRK. and GEMS models (Kirwan, 1992).

For more information on human reliability analysis techniques, see Miller and Swain (1987), Kirwan (1996, 1997, 1998a, 1998b), Kirwan et al. (1997), Stanton and Stevenage (1998), Swain (1963, 1990), and Shorrock and Kirwan (2002).

11.5 ERRORS, NEAR MISSES, AND ACCIDENTS

From the previous section, we know that there are various types of errors; however, not all errors lead to accidents. Taking the wrong highway exit is an error, but it is not considered an accident. When we bump into the beverage sitting on our desk, but catch it before it spills, or planes violate the required spacing and get too close to one another, these are near misses. A **near miss** occurs when there is an error with no accident, but there was a

Figure 11.3 Near misses and accidents. The event on the left (By Arpingstone, Public domain, via Wikimedia Commons) is what a near miss error might look like (although this is not an actual near miss). The car incident on the right (By Janne. from Finland, CC BY-SA 2.0 <https://creativecommons. org/licenses/by-sa/2.0>, via Wikimedia Commons) is an accident.

high probability for an accident (see Figure 11.3a). When Sandra's father goes for a walk instead of his appointment, this is an error, but not an accident. When her father takes the wrong medication or the young man in Iowa gets too close to the combine's auger with loose clothing, these errors can lead to accidents. An **accident** occurs when something happens unexpectedly or without intention, and this event leads to some type of consequence such as damage or injury (see Figure 11.3b). We need to clarify the term "intention" in this defini-tion. As some errors are classified as intentional, an accident occurs when someone did not intend to cause the accident or was unaware that this behavior would or could cause damage or injury (Taylor, 1976), even if the behavior was intentional. In summary, errors can cause accidents, but not all errors lead to accidents. Some HF specialists use the Human Factors Analysis and Classification System (HFACS, Shappell & Wiegmann, 2000) to assist with the identification and classification of human error in accidents.

11.5.1 Accident and error reporting

If we report accidents and errors, we can evaluate the preceding events to determine the cause of these errors and accidents. Once the cause or causes are known, we can make efforts to reduce these causes. The problem is, people often get into trouble for mak-ing errors, which can lead to fines, revocation of a license, or other actions associated with errors or accidents. Therefore, we are often reluctant to admit or report errors because the system essentially punishes the reporting behavior. This is problematic (and a system error) when we do not receive notice of error that could help us determine and fix the cause of the errors and accidents. This issue of a culture of blame is discussed later in the section on System Factors.

11.5.2 Error detection and correction

Sometimes we catch our own errors, such as when we take the wrong highway exit and realize we are not in the proper location. We can easily correct this error by getting back on the highway and taking the correct exit. It is important to remember that even though there was no negative consequence due to this error (except perhaps the loss of time), it was still an error. Means to reduce our errors include monitoring ourselves (e.g., self-monitoring, double checking our work), establishing cues within the environment to indicate an error has occurred (e.g., blocking functions that prohibit the next behavior if other behaviors are not completed first), having others review our work (e.g., a quality control system; Reason, 1990), or error reporting systems. Even if we implement various methods of monitoring, cueing, reviewing, or error reporting of our actions, there are other factors that contribute to our inability to identify errors. For us to reduce error, we must be able to correctly identify error.

11.6 FACTORS CONTRIBUTING TO ERRORS

Given that different types of errors possibly occur at the input, process, or output phases of a system, there are different potential causes of errors at each of these stages. Different causes of errors might arise at the individual and system levels (Miller & Swain, 1987). Individual ability, training level, emotional states, personality, and stress level are all possible causes of error that are internal to the individual. Components of the system that might cause errors include the workspace, the environment, task complexity, and shiftwork. For example, if the workspace is too small or the lighting level is inappropriate for the task, the human operator is more likely to make an error. In addition, there are often other contributing factors or events such as a storm that disrupts or eliminates power (Rouse & Rouse, 1983). Next, we will consider some of these individual- and system-level contributing factors.

11.6.1 Individual factors

Individual factors that contribute to error include our personalities and attitudes. In addition, our human limitations in decision-making, information processing, and memory (see Chapters 6 and 7) are contributing factors of error. In turn, decision-making, information processing, and memory are often impacted or affected by additional factors such as level of expertise, sleep deprivation, or stress, which we discuss next.

11.6.1.1 Expertise level

As discussed in Chapter 7, human decision-making is greatly influenced by level of expertise. Individuals with more ability, knowledge, skills, training, or time on the job are more likely to work in an automated fashion and have more resources to draw on when needing to create unique solutions. Essentially, as expertise increases, workload decreases, as the task is well understood. Experts understand the environment, task, and situation well and have the skill-, rule-, and knowledge-based behaviors required for the job. In contrast, novices lack the knowledge and experience to be as efficient as experts and, therefore, are more likely to make errors. Even if the novice performs the task correctly, if the task is performed too slowly, that is still an error. Further, the novice is less likely to have the knowledge-based skills to resolve unique problems requiring creative solutions.

11.6.1.2 Sleep deprivation

Sleep deprivation is another determinant of errors. Researchers often evaluate the effects of prolonged sleep deprivation by not allowing the participants to sleep during the normal sleep cycle. Performance is evaluated over another 10–12 hours of wakefulness, whereby the participants acquire 24–39 hours of sleeplessness. Sleep deprivation tends to reduce our ability to think systematically and has an impact on our memory, perception, concentration, and reaction times. In fact, sleep loss deficits (20–25 hours of wakefulness) are similar to having a blood alcohol content of 0.10% (Orzeł-Gryglewska, 2010). Yet, one does not have to experience total sleep deprivation to demonstrate negative behavioral effects. For example, individuals who repeated a pattern of staying awake for 23 hours and then sleeping fewer than 5 hours over several days demonstrated lower driving performance during a 10-minute driving simulation (Matthews et al., 2012).

Generally, individuals acknowledge quite well their performance decrements as their performance drops. Yet, the participants evaluated themselves relative to other people as more capable of performing well under these conditions (Jones et al., 2006). In addition, younger individuals (early 20s to early 30s) perceive the impact of sleep deprivation on their level of fatigue to be less than older adults between 52 and 63 years of age do. This is a concern given the finding that sleep deprivation had a greater negative impact on younger persons' reaction times (Philip et al., 2004). Although individuals can identify when their performance is likely to drop when sleep deprived, people tend to be overconfident in their ability and this overconfidence is greater for younger individuals (i.e., less than 34 years of age).

As thinking and behaviors are affected by sleep deprivation, which results in errors, perhaps a stimulant such as caffeine could offset these effects and reduce errors. Young Finnish military pilots who were sleep deprived and given a moderate (200 mg) dose of caffeine performed no better on a flight simulation over a 37-hour wakefulness period than individuals who did not receive caffeine. In addition, there was no difference between these two groups in their self-assessments of their level of sleepiness, which increased for everyone. Yet, the pilots given caffeine were more confident in their performance during the simulation (Lohi et al., 2007). The lack of a performance differences might be attributable to the fact that the flight simulation task is an engaging and motivating task for young pilots. Caffeine might also lead to more errors of commission when people have poor sleep (Anderson et al., 2018). Further research is needed to determine if caffeine makes individuals overconfident when they are sleep deprived, especially since young people already tend to be prone to overconfidence in their abilities when sleep deprived.

BOX 11.2:   MEDICAL ERRORS: CAUSES AND CONSEQUENCES

Medical errors are prevalent, especially medication or overdose errors (overdoses) and events such as the amputation of the wrong leg (Kohn et al., 2000). Although we might like to think medical personnel are infallible, they are also human and not perfect. Given that the consequences of medical errors can be catastrophic, it is critical to identify what errors occur and their causes to help reduce these errors. One way to learn about errors is to allow individuals to anonymously report incidents (Dietz et al., 2010). An anonymous error reporting system reduces the punitive culture of reporting errors. One of the potential contributors to error is a lack of sleep. In 2003, the Accreditation Council for Graduate Medical Education (ACGME) guidelines stated that medical residents were restricted to working 80 hours per week and an extended shift could last up to 30 hours. As of 2011, interns were restricted to 16-hour shifts (Fagan, 2013). These changes improved care, leading to reductions in length of hospital stays and total cost of care (Rosenbluth et al., 2013).

Sleep deprivation is not the only contributor to medical errors. Nurse skill level, prescribing the wrong medicine or wrong combination of medicines, the actual dispensing of the medicine (where, how, and when), and the safety culture of the organization also can contribute to elderly acute care (Metsälä & Vaherkoski, 2014). In palliative care, communication, and more generally, the lack of sharing information is often a cause of medical errors. Given their direct and daily contact with patients, nurses are the ones who detect, interrupt, or correct errors (Henneman et al., 2010). Strategies for detecting errors can include knowing the patient, knowing the proper rules and procedures, and double-checking. Interrupting errors includes offering assistance and clarifying. Finally, correcting errors involves offering options and referring to experts. **Critical Thinking Questions:** What type of errors (skill-, rule-, or knowledge-based) are likely when there is a medication overdose or surgery on the incorrect body part? If errors are common in medical practice, what is the likely impact of sleep deprivation? Are sleep-deprived residents more likely to make slips or mistakes? Will sleep deprivation have a greater impact on performance or perceptions of performance? How might you encourage or develop a safety culture?

11.6.1.3 Stress

When we are stressed, we may be more prone to slips such as placing the milk in the cupboard or the cereal in the refrigerator. Although spoiled milk is a bad consequence, stress could lead to human error that has much more serious consequences. For instance, if someone is stressed and fails to notice a stop sign, the consequences could be severe. To address the effects of stress on human performance, we need to understand stress.

Stress generally arises because we perceive too great of a demand on us relative to our ability to cope with these demands. The demands occur due to disturbances in the environment known as **stressors**. Examples of stressors include, but are not limited to, task overload, task complexity, conflict, lighting, time pressure, high traffic flow, noise, glare, sleep deprivation, or fatigue. Stressors are often categorized as environmental, psychological, or temporal. **Environmental stressors** include physical aspects of the environment such as the air quality and temperature. **Psychological stressors** include issues of workload and cognitive appraisal. When considering issues of fatigue, sleep loss or deprivation, and work shifts, these are considered **temporal stressors**. Because of individual differences, **stress** is our individual reaction to a stressor or set of stressors. Not everyone will feel stressed by a particular noise or the loss of sleep for one night, or perceive a challenging task as a threat. As an example, having multiple papers and tests due within the same week motivates some students, whereas other students become overwhelmed.

As stress is a personal interpretation of the stressors, the personal interpretation is influenced by the individual's abilities, available resources, and perception of the importance of the task at hand. Consider Sandra's condition of working in a pharmaceutical lab as well as having the demands of taking care of an elderly parent. If Sandra perceives that she has the skills and resources to handle these demands, she is less likely to experience much stress. On the other hand, if either her work or her father requires skills or resources she lacks, her level of stress will be higher. Similarly, if you have the ability and resources to complete the papers and tests for a course, or if you do not perceive the papers and tests to be important, then your stress level is likely to be lower than someone who lacks the ability or resources, or considers the papers and tests to be extremely important. With enough stressors, exposure to a severe stressor, or chronic stress, everyone's capability to cope can be exceeded. Some tips on how to reduce stress are presented in Table 11.4.

11.6.2 System factors

Recall that Donald Norman proposed that mistakes are often the result of decisions made long before the actual wrong behavior occurs. This is why, as discussed in Chapter 1, human factors and ergonomics (HFE) professionals must understand the role and importance of the

Table 11.4 Some tips on how to reduce stress

* Accept the situation: accept that you cannot control everything and accept the anxiety, which will pass.	* Relax and decompress: take deep breaths, employ relaxation techniques, meditate; take 15–20 minutes for quiet time to reflect.
* Exercise regularly.	* Laugh regularly and heartily.
* Stay positive: have a positive attitude, use positive self-talk.	* Do things you enjoy (are pleasurable) such as listening to music.
* Eat well, eat healthy; some say avoid alcohol.	* Be aware of how stress impacts you physically.
* Get enough sleep.	* Use your networks for social support.
* Manage your time well: this will give yourself more time to complete tasks.	* Get away before stress strikes: when feeling the onset of stress, take a walk.

Information adapted from WebMD (http://www.webmd.com/balance/guide/blissing-out-10-relaxation-techniques-reduce-stress-spot, http://www.webmd.com/balance/stress-management/reducing-stress-tips) and Psychology Today (https://www.psychologytoday.com/blog/finding-cloud9/201308/5-quick-tips-reduce-stress-and-stop-anxiety).

larger system in affecting performance outcomes. This larger system includes the interaction of various components: people, tasks, the environment, organizational structure, and organizational culture. In addition, systems are impacted by external factors that are components of the larger societal system, including laws, regulations, community perspective, and other possible watch-group policies. Given all these interacting components within a system, the feedback loop plays a critical role in helping us determine if the system is working well. For example, without the notification that your recent online purchase was completed, you might submit the order multiple times unnecessarily. With a better system design and a better understanding of all the components of a system and how they interact, we can create a world requiring more of Norman's "in the world" as opposed to "in the head" knowledge.

As systems are composed of people performing various tasks while using an assortment of tools and technology, the individual aspects that contribute to error and safety mentioned above and in Box 11.2 are important; however, system-level aspects such as the organizational culture also can impact safety (Metsälä & Vaherkoski, 2014). Even the most skilled employee may violate standard operating procedures (SOPs) and rely on shortcuts or workarounds if SOPs are seen as inefficient or overly complicated (Carayon et al., 2014). Similarly, the communication processes, and styles of management and leadership, contribute to system function. As the various components of a system are interdependent, sometimes there is a need to correct how communication flows (Henneman et al., 2010) or how management approaches an issue.

Healthcare is a good example of a complex system involving a variety of people (e.g., patients, patients' families, physicians, and nurses) and technology (Carayon, 2006). Working from a systems engineering perspective, Pascale Carayon and colleagues (Carayon et al., 2006; Holden et al., 2013; Carayon et al., 2020) applied systems theory to healthcare to create the Systems Engineering Initiative for Patient Safety (SEIPS) model. The SEIPS model helps practitioners understand the causes of error in healthcare across time and space, as well as potential means to identify and control these errors. As healthcare transitions from the mentality that the doctor makes all the decisions to a more sociotechnical systems collaborative process involving input from all parties (e.g., patients, patients' families, and nurses), the systems perspective or SEIPS model helps us understand these interrelationships. Many of the problems in healthcare are not due to a lack of skills or abilities of the professionals, but rather organizational problems such as poor communication (Carayon et al., 2006).

11.7 REDUCING ERRORS

Selecting or developing human users who are reliable and who are less error prone can help reduce the likelihood of errors. As human error is often a consequence, not a cause, of the situation, we also can apply human factors to the design of system components that might cause errors to reduce the likelihood of error. Equipment, tools, and machinery can be designed better for human use with more "in the world" knowledge. Similarly, the total system can be redesigned for better human use.

11.7.1 Individual factors

11.7.1.1 Expertise

Experts make fewer errors and can resolve problems better. Training increases individuals' knowledge and skill levels, whereby the individual works more efficiently at each of the skill-, rule-, and knowledge-based levels. Expertise reduces the likelihood of errors and the negative consequence of error. Therefore, training is a critical component for decreasing the number of errors an individual displays.

11.7.1.2 Optimal stress and training

As trained individuals (i.e., experts) have a better understanding of the job, they are less stressed and perform better during high demand situations than inexperienced and non-trained individuals (Berkun, 1964). On the other hand, untrained individuals or novices tend to experience stress at lower levels of job demand. This finding relates to the Yerkes–Dodson (1908) law.

Recall from Chapter 10, the **Yerkes–Dodson law** identifies the relationship between arousal and performance as an inverse-U relationship (Yerkes & Dodson, 1908; see Figure 11.4). Performance is low when arousal is either low or high. **Work overload** occurs when the workload is high, requiring more skills and resources than the individual has available. Too low of a workload or **work underload** also can be problematic because the operator becomes bored. Only when there is an optimal level of arousal, or stress, is there optimal performance. The optimal level of arousal and stress is related to one's level of expertise. This explains why trained military personnel performed better in the "real" military situations than novices (see Berkun, 1964). It is likely that the trained military personnel reached an optimal level of arousal (i.e., moderate arousal level) during the high demand situation, whereby they were able to perform more efficiently than the non-trained individuals. Referring to Figure 11.4, the high performance of the expert military personnel most likely

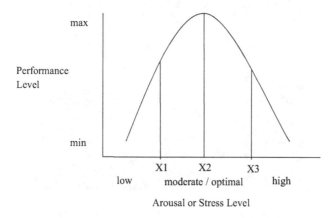

Figure 11.4 An example of the inverse-U relationship. (© Nancy J. Stone.)

reflects the level of arousal represented by X2. For that same level of workload, though, the novices' level of arousal was likely at X3. If the workload were reduced to enhance performance for the novices, the experts' level of arousal would likely reflect the point X1. Training better prepares individuals to handle stressful situations by enhancing an individual's level of expertise, which essentially reduces the task demands required for the job.

Besides training individuals to be experts, another means to reduce stress is to train individuals to develop and use coping mechanisms. For example, developing better time management skills could help the individual better handle situations and reduce stress. On a more personal level, teaching individuals various relaxation techniques also can reduce stress.

11.7.1.3 Checklists

As systems and tasks become more complex, human information processing and memory limitations pose problems. Even the most seasoned operator may experience difficulty recalling complex procedures or conditions where exceptions to standard practice are acceptable (see Chapter 6). Hence, checklists or other performance aids such as written instructions are helpful. Checklists are used by nurses and pilots (see Figure 11.5), as a missed procedure could have extremely dire consequences. Atul Gawande (2010) provides examples of the beneficial impact of simple checklists in the medical field as well as in the construction, financial, restaurant, and aviation fields. Complications and death rates fell over 30% with the implementation of checklists before anesthesia, before incision, and before removing the patient from the operating room (Haynes et al., 2009). In addition, there was an increase in adherence to appropriate antibiotic use during surgery from 56% to 83%, reducing the rates of infections!

Figure 11.5 A pilot completes a checklist. (Adapted photo by J-E Nystrom, Helsinki, Finland. Janke at English Wikipedia, CC BY 3.0 <https://creativecommons.org/licenses/by/3.0>, via Wikimedia Commons.)

BOX 11.3   ENHANCING COMPLIANCE

Although there are numerous situations where people should follow safety precautions, we often do not. For example, people should wear proper eye protection when operating lawn trimmers or mowers, people should use hearing protection when using loud equipment, and people should wash their hands before eating. Yet, many people do not. In the health profession, it is important that individuals use proper hand hygiene, but they often do not. To assist with proper hygiene to reduce the likelihood of infection, adherence engineering was applied to the process of changing a main-line (catheter) dressing (Drews, 2013). A large problem with this procedure is the need to remember all the steps, what materials are needed at each step, and where one left off when an interruption occurred. These conditions are a greater problem for a less experienced healthcare provider. To reduce these errors, the seven aspects of adherence engineering were applied to the main-line dressing change process after completing a task analysis (see Chapter 5). First, *affordances* were created by making the steps visibly identifiable. This was accomplished by using borders and colors to clarify what the current steps were. Next, *task-intrinsic guidance* was provided by creating separate packets of materials for each step and the packets were placed in proper sequence. *Nudging* was the third aspect, whereby only the materials, devices, or equipment needed were included. Fourth, *smart defaults* were provided, which meant the correct number of kits most often used were provided (i.e., don't overload individuals with too many choices). Fifth, *feedback* was easily acquired, as the used packets indicated where someone stopped if there was an interruption. This setup allows for *minimizing cognitive effort*, as the instructions are grouped and identified for each step. Finally, *minimizing physical effort* was possible, as the hand sanitizer (and everything else) was in the packet and the healthcare provider did not need to find the sanitizer somewhere else in the room (Drews, 2013). **Critical Thinking Questions**: How might Norman classify the original process versus the new process (in the head or in the world knowledge—explain)? What types of errors were being made? How is this design better than a checklist?

11.7.2 System factors

At the system level, we can use systems analyses to determine and correct errors (e.g., Carayon et al., 2014). The SEIPS model is an application of human factors and ergonomics to healthcare systems. Although our discussion above was about the systems aspects of healthcare and patient safety, these concepts can be applied to all systems. Not only does the systems perspective help us understand all the interacting (i.e., internal) and impacting (i.e., external) components of a system of interest (e.g., healthcare, work team, or business), the feedback loop is critical in determining how well the actual system output compares with the ideal system output. Any identified discrepancies can be addressed to reduce these errors.

11.7.2.1 Task redesign

One means for decreasing errors at the systems level is to change the task or work considering proper human factors principles (Miller & Swain, 1987). For example, increasing automation of mundane tasks would allow the human operator to focus on tasks leveraging human judgment or insight. Work underload, as discussed above, can be problematic because the operator becomes bored, less attentive to the task, or takes on tasks unrelated to their current task. One of the authors noticed a lifeguard reading a book while on duty. It is understandable that reading can eliminate boredom; however, the lifeguard is likely to miss an emergency such as a distressed swimmer. Therefore, it is important to find the right level of workload for the human so that there is not too much to do as well as too little to do, as both work overload and work underload are stressors. In turn, both of these stressors can lead to problems with human performance.

Task redesign could impact several other factors such as training, communication, and workflow. Additional variables that impact task performance (e.g., environmental design and conditions, improper supervision) and operate outside the individual (i.e., external PSFs; Miller & Swain, 1987) are also considered system factors. Similarly, warnings can impact the type of expertise required of individuals as well as the potential interaction necessitated when a warning needs to be followed after a negative event.

11.7.2.2 Warnings

Warnings come in a variety of forms ranging from fire alarms, bell tones when you leave your keys in the ignition, text instructions that accompany power tools, to the rotten egg smell of a natural gas leak. Warnings supply information about hazards to prevent injury or alert us to changes in our environment. This information should influence safe behaviors as well as serve as a reminder about what dangers or problems exist and what behaviors will help us avoid getting hurt or making an error. People often know what to do, but forget (hence the tone in your car that your keys are in the ignition; Laughery, 2006). When discussing warnings, the first question to address is whether we even notice these warnings. If we do notice warnings, the next question is whether we understand the warning such as the difference between flammable, inflammable, and nonflammable. Even if we understand the warning, do we heed the warning? Often, individuals purchase new equipment and rarely read the assembly instructions let alone the warnings.

The design of effective warnings should occur only after applying good, high-quality human factors design to the product and the design is complemented with other protections such as guards (e.g., a cover over the blade of a circular saw) and procedures (e.g., we must move away from the machine to activate the power switch; Laughery, 2006; Lehto & Salvendy, 1995). As we consider error within a system, other variables such as the selection of the most competent people and the level of training received influence not only the individual's ability to do the task, but the ability to understand and implement the warning (Lehto & Salvendy, 1995). In addition, it is easier for an employer to oversee workers' behavior than it is for a manufacturer to oversee the behavior of the general public. Individuals might use products in many ways not intended. Therefore, there are specific guidelines for effective warnings discussed below.

11.7.2.2.1 Effective warnings

The design of warnings should be informed by the research to increase the likelihood of the warning's effectiveness. Since the 1980s, the use of warnings increased due to an increasingly litigious society (Laughery, 2006; Lehto & Salvendy, 1995). An effective warning is *noticed, understood,* and *implementable* to avoid the error or to remedy the error. For a warning to be noticed, understood, and implementable, one must consider the characteristics of the warning, the intended audience of the message (e.g., children, trained specialists, or general population), and the situation or environment in which the warning is needed (Laughery, 2006).

Kenneth Laughery (2006) reviewed the literature on what makes warnings effective and when people are most likely to comply with the warnings. Although warnings can use any sense (e.g., gustatory, olfaction), warnings are usually visual (e.g., written notices, images) or auditory (e.g., bells, voice messages). Warnings, which are specific types of displays, should reflect the application of good human factors science to the design of displays, as discussed in Chapters 3 and 4. To be noticed, visual warnings need to be accessible by being within the visual field during the event. Where do you store your procedural manuals for various electric and other tools you own? If there is an issue when mowing the lawn, working a combine, or using a blender, can the warning be noticed? Sometimes a warning is on the product, but it might become obscured by wear or objects. Auditory warnings are not as restrictive given the omnidirectional nature of sound. Yet, the messages contained in auditory displays need to be simpler and shorter than verbal written warnings that can be reviewed (see Chapter 4).

11.7.2.2.2 Designing warnings

To notice and encode the information in the warnings, certain variables need to be considered when designing visual warnings: size, location, color, words, and images. Generally, larger visual displays are better. Good warnings are conspicuous (see Chapter 3) and use bold letters, color to create contrast, and borders around the message. When using bold letters, it is critical to consider the width of the letters to ensure the characters are distinguishable. Color is also beneficial as well as animation such as making it flash or blink (Wogalter et al., 2002). The location or placement of warnings is also critical, but poses design challenges. Visual warnings need to be positioned near where the operator is expected to be looking to be seen when needed, and the proximity should reflect consideration of the importance of the warning. The warning message also should provide directions on how to acquire additional information about the warning, if needed. Images can assist in understanding the warning if the individual cannot read, but the image has to be understandable to the user. This will be discussed further below.

The wording on a warning needs to include four components: a *signal word*, the *hazard* (be specific, complete, and brief), the *consequences* of exposure to the hazard, and *explicit instructions* on what to do to avoid the hazard (Wogalter et al., 2002). Wording is best when it is simple and integrated with the task (i.e., the person sees the wording when opening the pill bottle, Hancock et al., 2020). Often the signal word is matched or coded with a specific color enclosed in a triangle with an exclamation mark. Caution is yellow and represents the possibility of minor or moderate injury. Orange signals a warning, which indicates severe injury or death are possibilities. Danger is red and used when there is a certainty of death or severe injury. An example of a warning sign is presented in Figure 11.6. If it is possible to animate the signal word with technology (e.g., a flashing word), this can increase visibility. The format of the written warning should be in a bulleted, outline form. Once the written text is legible (see Chapter 3), add any images to help convey the message.

It is also important to know your audience. Written text is not always understood, and pictorials often do not convey abstract concepts well and must be learned ("in the head" knowledge). A noticed warning is not effective if it is not understood as intended. Literacy levels and age can impact understanding. Prescription labels written at the eighth-grade level or higher are confusing for and not understood correctly by users with low levels of literacy, especially when there are multiple steps (e.g., take before or take after food; Davis et al., 2006). The design of warnings should consider language and cultural fluency, which could impact users' understanding. For literate individuals, images enhance memory of the warning, particularly when paired with effective wording (Hancock et al., 2020). When

Figure 11.6 An example of a warning. The top band is in orange. If the wording were danger or caution, the top bad would be red or yellow, respectively. (© Nancy J. Stone.)

literacy or language levels are lower, well-designed images are helpful because they can be interpreted quickly; however, images also can be misunderstood. Children often misunderstand images. The symbol for poison, skull and crossbones, on a bottle might be interpreted by children as a "fun" pirate drink because of the movies the children watch (Latham et al., 2013). Using multiple modes of learning appears to be one of the best ways to educate children about the true meaning of these images (Latham et al., 2013).

11.7.2.2.3 Warning compliance

Assuming users can understand the warning (i.e., the signal ward, hazard, consequences of the hazard, and actions to take to avoid the hazard), the question remains as to why people often do not follow the warning. Perceptions of the risk and level of familiarity with the risk influence compliance with a warning (Laughery, 2006). If an individual believes there is little risk, the person is less likely to heed the warning. In fact, the person is less likely to look for and read a warning when perceived risk is low. Conversely, if the risk is perceived to be high, a person is more likely to look for and read a warning. Familiarity with a situation also appears to impact perceived risk. The more familiar a person is with a situation, the more comfortable the person becomes with the situation (unless he or she had a bad experience) and the less likely the person will follow, look for, or read a warning. Consider the young man who got caught in the combine's auger in the opening vignette. Perhaps he did not follow the normal procedures for safety because he had a low perception of risk due to high familiarity with the task and an absence of recent accidents.

Compliance with warnings is a cost-benefit decision (Laughery, 2006). When the benefit is perceived to be high (generally when there is a high perceived risk or lack of familiarity), a person is more likely to comply with the warning (i.e., there is a perceived benefit of being safe if the warning is followed). In contrast, compliance is lower if the environment is perceived to be safe and the precautionary behaviors are believed to be unnecessary (i.e., high cost).

To ensure greater compliance, the warning must first be well designed to be noticeable and easily understood, which is enhanced with (understandable) images. When the wording is more serious, individuals perceive warnings to be more believable (Hancock et al., 2020). In addition, warnings should be explicit in (1) what the hazard is (e.g., fire), (2) what the consequences are (e.g., get burned), and (3) proper behavior to avoid an unsafe situation (e.g., remove hand from fire). When reading prescription labels, individuals with low levels of literacy found the warning, "For external use" confusing because no specifics were given on what to do (Davis et al., 2006). Additionally, having other people model the proper behaviors helps increase compliance (Laughery, 2006).

In summary, to reduce errors we first create the best design of the job, task, or environment applying human factors principles. According to Donald Norman, we want to reduce the demands on working memory by reducing the cognitive complexity of the tasks. Standardizing some processes can help with this first issue. In addition, we want to design the environment that leverages "in the world" or intuitive knowledge as much as possible. As we cannot remove all complexity and we assume that errors might occur, we can modify the workforce through selection and training, and support the worker by providing various aides such as checklists (Miller & Swain, 1987). We also need to consider the system, which includes the specific environment (see Chapter 10) as well as the organizational structure (i.e., the sociotechnical view), and how the various system components interact. Lastly, the use of warnings can help reduce the likelihood something will go amiss or reduce the results of an accident to a minimal, acceptable level, avoiding total failure.

BOX 11.4:   CREATING A SAFETY CLIMATE

Good human factors and ergonomics design greatly reduce and deter errors. Effective communication of the potential error and warnings is also important. Yet, communication is not sufficient. It is essential to create a strong safety culture and climate, whereby people are encouraged to be safe and supported when errors arise in order to encourage reporting of incidents. Culture is generally defined at the system level (i.e., over-arching values and beliefs) and climate is at the work group level (i.e., what is thought and practiced by the workers in their own specific contexts). Applying systems theory, Murphy et al. (2014) argued the need for macroergonomics and the integration of organizational principles to the work system design, which applies sociotechnical systems (STS) theory. STS theory is the application of systems theory to organizations and technical systems. Technology is anything that is non-human and includes organizational structure aspects not just machines and equipment (Murphy et al., 2014). At the organizational level, management sets policies for safety; however, at the work group level, supervisors might implement their own informal policies that are not aligned with the formal policies set by management (Murphy et al. 2014). Murphy et al. (2014) also argued for the use of mesoergonomics, the integration of macroergonomics and safety climate at two or more levels (e.g., the worker and management). One method for determining the gap between the worker-level (informal) and organizational-level (formal) safety practices and policies includes SEIPS, which was discussed earlier in the chapter.

To help create a safety culture, Wilson-Donnelly et al. (2004) provided ten tips. These tips included: (1) clearly and unambiguously communicate the message that safety matters, (2) have top-management support of and commitment to the safety attitudes, (3) support error checking and reporting that allows individuals to learn from mistakes, (4) have open communication to allow information to flow appropriately, (5) search for solutions at all levels instead of assigning blame, disciplining, and proceeding as normal, (6) develop a non-punitive means of documenting errors, (7) train workers to appropriately deal with error, (8) continually assess the behaviors to ensure they are safe, (9) be sure you reinforce the desired behaviors not unsafe behaviors, and (10) encourage teamwork to ensure effective communication and coordination. **Critical Thinking Questions**: Consider your university. What are the university's safety policies? How is that culture expressed as climate in your various departments such as in the courses or laboratory work you do? How might you study the gap between the university's culture on the department's climate? How might you implement change? Which tips reflect Normans' slips and mistakes? What is "in the head" and what is "in the world?" Can you answer these questions relative to your job?

Although this discussion on warnings focused on written warnings, auditory warnings are often used. Auditory warnings are signals such as bells, sirens, or app notifications and could be used to get the attention of distracted pedestrians about to cross a busy street. Auditory warnings have the advantage of not requiring a specific location, but must be designed to be distinguishable from the noise within the environment. Also, oral or voice messages cannot be too complex, and people often habituate to these warnings (Wogalter et al., 2002), which means they begin to ignore the warning. For example, in airports, many of the moving walkways have a voice message reminding walkers to watch their steps as they come to the end of the walkway. This is fine for those moving along, but the message can be extremely repetitive and annoying for someone who is waiting for a flight and hears this message over and over.

11.7.3 Safety management

Given that various environments might be more prone to error than others, it is important to be aware of potential problems, take action to avoid error, and record incidents of error to reduce the number of incidents and severity of the consequences of error. As referenced in Chapters 9 and 10, the International Organization for Standards (ISO) provides standards for all types of equipment, environments, and situations. Other organizations and processes provide enforcement of regulations or guidance, which are briefly discussed below.

11.7.3.1 OSHA and NIOSH

The Occupational Safety and Health Association (OSHA), a subunit of the Department of Labor, is the enforcement agency for safety and health regulations and legislation (www.OSHA.gov). Table 11.5 lists some of these regulations. In contrast, the National Institute for Occupational Safety and Health (NIOSH) is a research organization under the Centers

Table 11.5 Various OSHA regulations applicable to human factors

OSHA regulations (Standards 29 CFR)	
Subpart	Title
1910 Subpart D	Walking-Working Surfaces
1910 Subpart E	Exit Routes and Emergency Planning
1910 Subpart H	Hazardous Materials
1910 Subpart I	Personal Protective Equipment
1910 Subpart N	Materials Handling and Storage
1910 Subpart O	Machinery and Machine Guarding
1910 Subpart P	Hand and Portable Powered Tools and Other Hand-Held Equipment
1910 Subpart T	Commercial Driving Operations

Source: Adapted from https://www.osha.gov/laws-regs/regulations/standardnumber/1910.

for Disease Control and Prevention. Individuals at NIOSH conduct research to understand and make recommendations for ways to be free of illness and safer at work (http://www. cdc.gov/niosh/about.html).

11.7.3.2 Product liability

Although OSHA enforces regulations and NIOSH informs the legislation with its research in various work environments, there are many accidents that occur outside of work. In these situations, the issue of liability arises. Product liability and other legal issues are the focus of individuals working in the forensic human factors area. The area of forensic human factors provides the scientific perspective on any litigation issues related to legislation, regulations, or the judicial system (see Mayhorn & Wogalter, 2020 for an introduction to the field). Individuals interested in this forensic human factors should contact the Forensics Professional Group, a technical group within the Human Factors and Ergonomics Society (https://www.hfes.org/Connect/Technical-Groups#Forensics).

BOX 11.5:   WHO'S RIGHT, WHO'S WRONG? DETERMINING FAULT IN ERROR DISPUTES

The role of forensic human factors/ergonomics (HFE) experts in the adversarial environment of resolving lawsuits to determine who is at fault when an error occurs is to convey objectively the specialized and technical information relevant to the case (Cohen & Cohen, 2001). The forensic HFE should not be influenced by the adversarial system, maintaining an ethical code of conduct, and delivering information and/or results that are based on high-quality analysis of the situation (Peacock, 2012).

Unfortunately, there are many causes of error that lead to great harm. Although determining fault might allow someone to cover costs of expenses incurred, it does not restore someone to his or her original condition or bring back family members who died in a mishap. It is often the case that all or most parties are at fault. "Fault" might be due to the nature of human cognition such as attentional demands or decision-making. One case involved the death of a young woman who stepped into the path of an oncoming passenger truck (Nemire, 2011). After a thorough analysis, it was determined that the woman was distracted by conversations on her cell phone and did not look both ways; however, it was unclear as to why the driver did not stop in time. It appeared as though the driver might have been distracted by being late for another appointment and was not attending fully to the task of driving. In another example, a farmer, who had changed many tires, purchased a tire for his truck. Unfortunately, he was sold a 15-in. tire, but he needed a 15.5 in. tire. The tire exploded after overinflating the tire and he broke his arm (Peacock, 2012). In other cases, there are known ways to prevent errors in these everyday situations. For example, a woman used a bottom shelf as a step to reach something on a top shelf. The bottom shelf gave way and she was hurt (Peacock, 2012). A man following a clerk in a pet store slipped on a wet floor, fell, and sprained his ankle (Nemire, 2011). This was not just an issue of his line of sight (he would have been looking at the back of the clerk, not the floor), but also the contrast on the floor and the ability to detect a water spill. **Critical Thinking Questions**: How would you classify the errors described in these cases? How might these errors have been avoided? Consider the different examples presented. What types of design issues might have prevented these errors? How might a warning help (or not) in these situations?

11.8 APPLICATIONS TO SPECIAL POPULATIONS

We can anticipate that individuals are going to use equipment, tools, and environments in ways for which they were not designed. People use screwdrivers as hammers, desks as chairs and ladders (i.e., they sit or stand on them), and machinery without reading the warnings presented in the operating manual. Therefore, the importance of designing out potential error is critical. Regardless of these precautions, as discussed above, children and individuals with lower literacy levels are less likely to understand or are likely to misinterpret how equipment, tools, and environments are supposed to work. Thus, it is important to consider who might acquire a particular tool or function within a particular environment and to design for all potential users. This is an extreme challenge, but we need to design for the limitations of all users such as children, individuals will low levels of literacy, or even Sandra's father, if possible. Similarly, designers need to consider whether individuals who do not speak the native language can understand the environment.

Besides considering the cognitive abilities of individuals, it is important to consider whether an individual will have any physical limitations, which could be within the individual (i.e., a physical disability) or the environment (e.g., a physical constraint preventing the appropriate movement to safely control the equipment, tool, or environment). For example, the use of color in designs can help reduce errors, but not necessarily if the user has some type of color blindness. Regardless of how well an individual knows how to act error-free, if the individual is not able to see or hear the warnings or the product is difficult to operate (i.e., not able to open the pill dispenser), or the desk is too high to properly work on the task, there is a greater potential for error, accidents, and decreased safety.

11.9 SUMMARY

Human error is an intentional or unintentional behavior that leads to an outcome that is not desired, as it does not meet system goals. There are a multitude of ways to classify errors, such as errors of omission or commission; slips or mistakes; and skill-, rule-, and knowledge-based errors. Initially, errors were identified as errors of omission or commission. Donald Norman argued that errors were either slips (an error in the specific behavior performed based on habit or skill that was generally unintentional) or a mistake (an error caused by the design of the environment or process, which was an intentional decision). Much of the current work refers to Jens Rasmussen's work on skill-, rule-, and knowledge-based error.

In contrast to human error, human reliability is the likelihood that the user will be error free. Knowing that human error is generally greater than machine error, it is important to estimate human reliability. Human reliability analysis begins with an understanding of the task and all relevant behaviors, often determined using task analysis. Next, the probability of error for each action is estimated from research data or by SMEs. The level of reliability is a mathematical combination of these error estimates to understand the potential impact on the system. Once the level of human reliability is understood, it is possible to focus on ways to increase human reliability. Several types of HRAs were discussed even though they are generally used in nuclear power industries.

It is important to understand the distinction between errors, near misses, and accidents. An error is an unwanted action or behavior relative to the system goals, but it does not necessarily result in an accident. Near misses are errors that had the potential of being an accident, but did not end in an accident. An accident occurs when there is a negative consequence such as damage to people or items. The reporting of errors, near misses, and accidents is critical in helping us understand where errors arise; however, people are often reluctant to report errors because of the likely negative consequences such as fines or penalties. In the healthcare field, great efforts are being made to develop error reporting systems that allow individuals to report errors anonymously to ensure a better understanding of how things are truly functioning. This is the only way we can make corrections.

The determinants of error can occur at the individual or system level. Some individual determinants include level of expertise, sleep deprivation, and stress. System-level determinants of error include components that bind the system together such as communication, leadership, and coordination of actions. Healthcare providers often know the proper methods, but information is not communicated on time or to the appropriate individual.

To reduce errors, training can increase individuals' level of expertise, which not only reduces error due to ability but also reduces stress by increasing individuals' experience and preparation for high demand situations. At the system level, there is a need to assess the type of leadership, organizational structure, and communication in place to determine the fit and alignment with goals of the system. Regardless of the amount of individual and system preparation, there is often a need for warnings as a backup. Warnings need to be noticed, understood, and implementable. It is critical that one understands the audience of the warning to ensure it can be noticed, encoded, interpreted correctly, and then appropriately enacted.

Because there are numerous accidents as well as ways to reduce error, there are various forms of safety management. OSHA is the enforcement agency, whereas NIOSH conducts research to better understand how to reduce errors and informs various legislative bodies. The field of forensic human factors focuses on informing legal, regulatory, and judicial systems in ways to increase safety.

LIST OF KEY TERMS

Accident

Environmental stressor

Errors of commission

Errors of omission

Event tree

Fault tree

Human error

Human reliability

Human reliability analysis

Knowledge-based behavior

Lapse

Mistake

Near miss

Performance shaping factors

Psychological stressor

Rule-based behavior

Skill-based behavior

Slip

Stress

Stressor

Temporal stressor

Work overload

Work underload

Yerkes–Dodson law

SUGGESTED READINGS

Gawande, A. (2010). *The Checklist Manifesto: How to Get Things Right* (Vol. 200). New York: Metropolitan Books.

This book focuses on the single topic of checklists and their importance in the medical field along with examples from the aviation, financial, and construction industries. This is a good read for individuals interested in how a simple tool can lead to remarkable results.

Hogan and Foster (2013). Multifaceted personality predictors of workplace safety performance: more than conscientiousness. *Human Performance*, 26(1), 20–43. doi:10.1080/08959285.2012.736899

Although conscientiousness is a good predictor of safety, underlying personality facets as how compliant someone is might be better predictors. This article is the first to evaluate personality at the facet as opposed to factor level and its relationship to human error and safety. A composite battery of facet scores is a better predictor of safety than a single dimension in the big five factor model.

Norman, D. A. (1983). Design rules based on analyses of human error. *Communications of the ACM*, 26, 254–258.

This short article provides useful tips on how to design situations to avoid errors.

Norman, D. A. (1988). *The Psychology of Everyday Things*. New York, NY: Basic Books.

In this classic book, Don Norman discusses the everyday things he and others experienced, the slips and mistakes they made, reasons as to why these errors occurred, and suggestions for avoiding these errors. As computers and technology were just gaining momentum, his comments and recommendations were insightful, even prophetic.

Rasmussen, J. (1983). Skills, rules, and knowledge; signals, signs, and symbols, and other distinctions in human performance models. *IEEE Transactions on Systems, Man, and Cybernetics*, 13, 257–266.

This is a foundational article for our understanding of human error. Jens Rasmussen provided us with the different categorizations of human performance used by many for assessing errors.

Reason, J. (1990). *Human Error*. New York, NY: Cambridge University Press.

This book covers the foundational cognitive errors that humans make. It is a good "first book" about how the field started and evolved. James Reason also presents good case studies on various accidents such as Three Mile Island and the gas leak from a pesticide plant in Bhopal India.

Wogalter, M. S., Conzola, V. C., & Smith-Jackson, T. L. (2002). Research-based guidelines for warning design and evaluation. *Applied Ergonomics*, 33, 219–230.

This article is an excellent review of the guidelines for creating effective warnings.

Woods, D. D., Dekker, S., Cook, R., Johannesen, L., & Sarter, N. (2010). *Behind Human Error* (2nd Ed.). Burlington, VT: Ashgate. Retrieved from www.scopus.com

This book gives readers a better understanding of human error as a consequence, not the cause of incidents.

CHAPTER EXERCISES

1. Identify several errors that could occur when driving. Explain how these errors could be:
   a. A slip or a mistake.
   b. An error of omission or commission (and which types).
   c. A skill-, rule-, or knowledge-based error.
   d. Complete (a) through (c) for the following errors.
       i. Missing a deadline.
       ii. Going to the wrong building for your meeting.
       iii. Sending the email to the wrong person(s).
2. Compare and contrast the different ways to categorize errors.
   a. Explain how slips and mistakes are similar to or different from errors of omission and commission.
   b. Explain how slips and mistakes are similar to or different from skill-, rule-, and knowledge-based errors.
   c. Explain how errors of omission and commission are similar to and different from skill-, rule-, and knowledge-based errors.
3. Explain how each classification system helps you understand the underlying problem.
4. Norman talked about knowledge "in the head" and "in the world." Design a process or tool two ways, one way for each of these two different types of knowledge. Explain how these processes or tools work differently because of the focus on the different types of knowledge required.

5. Write out in detail how you would perform the steps of a human reliability analysis for the following situations. You will need to create the situation and estimate probabilities.
   a. A car accident.
   b. Withdrawing the incorrect amount of cash from an ATM.
   c. Completing a semester-long group project for class.
6. Explain whether the following is an accident:
   a. A chemical spill.
   b. Someone loses control of the car and drives over a curb.
   c. The garbage disposal is turned on when a utensil is in it.
7. Consider the following situations. Identify potential individual and system factors contributing to errors.
   a. A driver falls asleep behind the wheel.
   b. A parent forgets to pick up the children after school (yes, this happens!).
   c. Submitted reports are late.
   d. Reports are completed incorrectly.
8. Explain how you might reduce errors by addressing the individual and system factors identified in #7 above.
9. Design warnings.
   a. Identify and describe an example of a bad warning.
      i. Explain how you would make this an effective warning.
      ii. Justify why this modification is good.
   b. Identify and describe a situation that needs a warning. Create a warning to meet this need.
10. Explain how you would test (i.e., research)
    a. The effectiveness of a warning.
    b. The cause of an error.
    c. The best way to reduce error.

REFERENCES

Anderson, J. R., Hagerdorn, P. L., Gunstad, J., & Spitznagel, M. B. (2018). Using coffee to compensate for poor sleep: impact on vigilance and implications for workplace performance. *Applied Ergonomics*, 70, 142–147.

Aucote, H. M., Miner, A., & Dahlhaus, P. (2012). Interpretation and misinterpretation of warning signage: perceptions of rock falls in a naturalistic setting. *Psychology, Health, and Medicine*, 17, 522–529. doi:10.1080/13548506.2011.644247

Beck, L. J., & Cohen, H. H. (1994). Miscommunication and human error: the difference between expectation and reality is the result of communication failure. *Ergonomics in Design*, 2, 16–20. doi:10.1177/106480469400200107

Berkun, M. M. (1964). Performance decrement under psychological stress. *Human Factors*, 6, 21–30. doi:10.1177/001872086400600104

Carayon, P. (2006). Human factors of complex sociotechnical systems. *Applied Ergonomics*, 37, 525–535. doi:10.1016/j.apergo.2006.04.011

Carayon, P., Hundt, A. S., Karsh, B-T., Gurses, A. P., Alvarado, C. J., Smith, M., & Brennan, P. F. (2006). Work system design for patient safety: the SEIPS model. *Quality and Safety in Health Care*, 15(SUPPL. 1), i50–i58. doi:10.1136/qshc.2005.015842

Carayon, P., Wetterneck, T. B., Rivera-Rodriguez, A. J., Hundt, A. S., Hoonakker, P., Holden, R. J., & Gurses, A. P. (2014). Human factors systems approach to healthcare quality and patient safety. *Applied Ergonomics*, 45, 14–25. doi:10.1016/j.apergo.2013.04.023

Carayon, P., Woldridge, A., Hoonakker, P., Hundt, A. S., & Kelly, M. M. (2020). DEIPS 2.0: Human-centered design of the patient journey for patient safety. *Applied Ergonomics, 84*, Article 103033.

Cohen, J., & Cohen, H. H. (2001). Human factors in stairway fall litigation. *Ergonomics in Design, 9*, 19–24. doi:10.1177/106480460100900305

Davis, T. C., Wolf, M. S., Bass, P. F. III, Middlebrooks, M., Kennen, E., Baker, D. W., Bennett, C. L., Durazo-Arvizu, R., Bocchini, A., Savory, S., & Parker, R. M. (2006). Low literacy impairs comprehension of prescription drug warning labels. *Journal of General Internal Medicine, 21*(8), 847–851. doi:10.1111/j.1525-1497.2006.00529.x

Dekker, S. W. A. (2002). The re-invention of human error. Technical report 2002-01. Lund University School of Aviation (pp. 1–16).

Dietz, I., Borasio, G., Schneider, G., & Jox, R. J. (2010). Medical errors and patient safety in palliative care: a review of current literature. *Journal of Palliative Medicine, 13*(12), 1469–1474. doi:10.1089/jpm.2010.0228.

Drews, F. A. (2013). Adherence engineering. *Ergonomics in Design: The Quarterly of Human Factors Applications, 21*, 19–25. doi:10.1177/1064804613497957

Fagan, H. A. (2013). Sixteen hours, education, error, and cost—is enforcing continuity the answer? *Sleep: Journal of Sleep and Sleep Disorders Research, 36*(2), 165–166.

Gawande, A. (2010). *The Checklist Manifesto: How to Get Things Right* (Vol. 200). New York: Metropolitan Books.

Hall, R. E., Fragola, J., & Wreathall, J. (1982). *Post event human decision errors: operator action tree/ time reliability correlation* (NUREG/CR-3010; BNL-NUREG-51601 ON: DE83010152, DOE contract number: AC02-76CH00016), Brookhaven National Lab., Upton, NY. Washington, D.C.: US Nuclear Regulatory Commission. Retrieved from SciTech Connect: https://www.osti.gov/ scitech/servlets/purl/6460666.

Hancock, P. A., Kaplan, A. D., MacArthur, K. R., & Szalma, J. L. (2020). How effective are warnings? A meta-analysis. *Safety Science, 130*, Article 104876.

Hannaman, G. W., Spurgin, A. J., Joksimovick, V., Wreathall, J., & Orvis, D. D. (1984). *Systematic human action reliability procedure (SHARP)*. NP-3583, Research Project 2170-3; NUS Corporation, San Diego, CA, Prepared for Electric Power Research Institute, Palo Alto, CA, Risk Assessment Program, Nuclear Power Division.

Haynes, A. B., Weiser, T. G., Berry, W. R., Lipsitz, S. R., Breizat, A.-H. S., Dellinger, E. P., Herbosa, T., Joseph, S., Kibatala, P. L., Lapitan, M. C. M., Merry, A. F., Moorthy, K., Reznick, R. K., Taylor, B., & Gawande, A. A. (2009). A surgical safety checklist to reduce morbidity and mortality in a global population. *New England Journal of Medicine, 360*, 491–499. doi:10.1056/NEJMsa0810119

Henneman, E. A., Gawlinski, A., Blank, F. S., Henneman, P. L., Jordan, D., & McKenzie, J. B. (2010). Strategies used by critical care nurses to identify, interrupt, and correct medical errors. *American Journal of Critical Care, 19*(6), 500–509. doi:10.4037/ajcc2010167

Hogan, J., & Foster, J. (2013). Multifaceted personality predictors of workplace safety performance: more than conscientiousness. *Human Performance, 26*, 20–43. doi:10.1080/08959285.2012.736899

Holden, R. J., Carayon, P., Gurses, A. P., Hoonakker, P., Hundt, A., Ozok, A., & Rivera-Rodriguez, A. (2013). SEIPS 2.0: a human factors framework for studying and improving the work of healthcare professionals and patients. *Ergonomics, 56*(11), 1669–1686. doi:10.1080/00140139.2013. 838643

Hollnagel, E. (1993). The phenotype of erroneous actions. *International Journal of Man-Machine Studies, 39*(1), 1–32. doi:10.1006/imms.1993.1051

Jones, C. B., Dorrian, J., Jay, S. M., Lamond, N., Ferguson, S., & Dawson, D. (2006). Self-awareness of impairment and the decision to drive after an extended period of wakefulness. *Chronobiology International, 23*, 1253–1263. doi:10.1080/07420520601083391

Kirwan, B. (1992). Human error identification in human reliability assessment. Part 1: overview of approaches. *Applied Ergonomics, 23*, 299–318.

Kirwan, B. (1996). The validation of three human reliability quantification techniques THERP, HEART and JHEDI: Part 1 - technique descriptions and validation issues. *Applied Ergonomics, 27*(6), 359–373.

Kirwan, B. (1997). The validation of three human reliability quantification techniques - THERP, HEART and JHEDI: Part III - practical aspects of the usage of the techniques. *Applied Ergonomics*, 28(1), 27–39.

Kirwan, B. (1998a). Human error identification techniques for risk assessment of high risk systems - Part 1: review and evaluation of techniques. *Applied Ergonomics*, 29(3), 157–177.

Kirwan, B. (1998b). Human error identification techniques for risk assessment of high risk systems - Part 2: towards a framework approach (review). *Applied Ergonomics*, 29(5), 299–318.

Kirwan, B., Kennedy, R., Taylor-Adams, S., & Lambert, B. (1997). The validation of three human reliability quantification techniques — THERP, HEART and JHEDI: Part II — results of validation exercise. *Applied Ergonomics*, 28(1), 17–25.

Kohn, L. T., Corrigan, J. M., & Donaldson, M. S. (Eds.) (2000). *To Err Is Human: Building a Safer Health System*. Washington, D.C.: National Academy Press.

Latham, G., Long, T., & Devitt, P. (2013). Children's misunderstandings of hazard warning signs in the new globally harmonized system for classification and labeling. *Issues in Comprehensive Pediatric Nursing*, 36(4), 262–278.

Laughery, K. R. (2006). Safety communications: Warnings. *Applied Ergonomics*, 37(4), 467–478.

Lehto, M., & Salvendy, G. (1995). Warnings: a supplement not a substitute for other approaches to safety. *Ergonomics*, 38(11), Special Issue: Warnings in research and practice, 2155–2163.

Lohi, J. J., Huttunen, K. H., Lahtinen, T. M., Kilpeläinen, A. A., Muhli, A. A., & Leino, T. K. (2007). Effect of caffeine on simulator flight performance in sleep-deprived military pilot students. *Military Medicine*, 172(9), 982–987.

Matthews, R. W., Ferguson, S. A., Zhou, X., Sargent, C., Darwent, D., Kennaway, D. J., & Roach, G. D. (2012). Time-of-day mediates the influences of extended wake and sleep restriction on simulated driving. *Chronobiology International*, 29(5), 572–579. doi:10.3109/07420528.2012.675845

Mayhorn, C. B., & Wogalter, M. S. (2020, March). Forensic human factors and ergonomics: theory in Practice. *Theoretical Issues in Ergonomics Science*, 21, 259–265.

Metsälä, E., & Vaherkoski, U. (2014). Medication errors in elderly acute care—a systematic review. *Scandinavian Journal of Caring Sciences*, 28(1), 12–28. doi:10.1111/scs.12034

Miller, D. P., & Swain, A. D. (1987). Human error and human reliability. In G. Salvendy (Ed.). *Handbook of Human Factors*, pp. 219–250. New York, NY: John Wiley & Sons.

Murphy, L. A., Robertson, M. M., & Carayon, P. (2014). The next generation of macroergonomics: integrating safety climate. *Accident Analysis and Prevention*, 68, 16–24. doi:10.1016/j.aap.2013.11.011

Nemire, K. (2011). Cognitive human factors in litigation. *Ergonomics in Design*, 19, 16–20. doi:10.1177/1064804611400988

Norman, D. A. (1981). Categorization of action slips. *Psychological Review*, 88, 1–15.

Norman, D. A. (1983). Design rules based on analyses of human error. *Communications of the ACM*, 26, 254–258.

Norman, D. A. (1988). *The Psychology of Everyday Things*. New York, NY: Basic Books.

Orzeł-Gryglewska, J. (2010). Consequences of sleep deprivation. *International Journal of Occupational Medicine & Environmental Health*, 23(1), 95–114. doi:10.2478/v10001-010-0004-9

Peacock, B. (2012). Forensics cases highlight human variability in product use. *Ergonomics in Design*, 20, 19–22. doi:10.1177/1064804612445007

Peacock, B., & Resnick, M. (2011). The six is: an ergonomics approach to enhancing product and process evaluations. *Ergonomics in Design*, 19, 25–29. doi:10.1177/1064804611408016

Phillips, L. D., Humphreys, P., Embrey, D., & Selby, D. L. (1990). *A socio-technical approach to assessing human reliability*. In R. M. Oliver, & J. Q. Smith (Eds.). *Influence Diagrams, Belief Nets and Decision Analysis* (pp. 253–276). Wiley series in probability & statistics. West Sussex, UK: Wiley & Sons Ltd.

Philip, P., Taillard, J., Sagaspe, P., Valtat, C., Sanchez-Ortuno, M., Moore, N., Charles, A., & Bioulac, B. (2004). Age, performance and sleep deprivation. *Journal of Sleep Research*, 13(2), 105–110. doi:10.1111/j.1365-2869.2004.00399.x

Rasmussen, J. (1983). Skills, rules, and knowledge; signals, signs, and symbols, and other distinctions in human performance models. *IEEE Transactions on Systems, Man, and Cybernetics*, 13, 257–266.

Rasmussen, J. (1985). Trends in human reliability analysis. *Ergonomics, 28*, 1185–1195.

Rasmussen, J., & Jensen, A. (1974). Mental procedures in real-life tasks: a case study of electronic trouble shooting, *Ergonomics, 17*, 293–307

Reason, J. (1990). *Human Error.* New York, NY: Cambridge University Press.

Rosenbluth, G., Fiore, D. M., Maselli, J. H., Vittinghoff, E., Wilson, S. D., & Auerbach, A. D. (2013). Association between adaptations to ACGME duty hour requirements, length of stay, and costs. *Sleep: Journal of Sleep & Sleep Disorders Research, 36*(2), 245–248.

Rouse, W. B., & Rouse, S. H. (1983). Analysis and classification of human error. *IEEE Transaction on Systems, Man, and Cybernetics, SMC-13*(4), 539–549.

Setayesh, A., Di Pasquale, V., & Neumann, W. P. (2022). An inter-method comparison of four human reliability assessment models. *Applied Ergonomics, 102*, Article 103750.

Shappell, S. A., & Wiegmann, D. A. (2000). The Human Factors Analysis and Classification System--HFACS. Retrieved from https://commons.erau.edu/publication/737

Shorrock, S. T., & Kirwan, B. (2002). Development and application of a human error identification tool for air traffic control (review). *Applied Ergonomics, 33*(4), 319–336.

Siegel, A. I., Bartter, W. D., Wolf, J. J., Knee, H. E., & Haas, P. M. (1984). The maintenance personnel performance simulation (MAAPS) model. *Proceedings of the Human Factors and Ergonomics Society, 28*, 247–251. doi:10.1177/154193128402800311

Stanton, N. A., & Stevenage, S. V. (1998). Learning to predict human error: issues of acceptability, reliability and validity. *Ergonomics, 41*, 1737–1756.

Swain, A. D. (1963). *A Method for Performing a Human-Factors Reliability Analysis* (Sandia Corporation Monograph, SCR-685). Albuquerque, NM: Sandia Corporation.

Swain, A. D. (1964). Some problems in the measurement of human performance in man-machine systems. *Human Factors, 6*, 687–700. doi:10.1177/001872086400600611

Swain, A.D. (1990). Human reliability analysis: need, status, trends and limitations. *Reliability Engineering and System Safety, 29*(3), 301–313.

Taylor, D. H. (1976). Accidents, risks, and models of explanation. *Human Factors, 18*(4), 371–380.

Wilson-Donnelly, K. A., Priest, H. A., Burke, C. S., & Salas, E. (2004). Tips for creating a safety culture in organizations. *Ergonomics in Design, 12*, 25–20. doi:10.1177/106480460401200407

Wogalter, M. S., Conzola, V. C., & Smith-Jackson, T. L. (2002). Research-based guidelines for warning design and evaluation. *Applied Ergonomics, 33*, 219–230.

Woods, D. D. (1994). Cognitive demands and activities in dynamic fault management: abduction and disturbance management. In N. Stanton (Ed.). *Human Factors of Alarm Design*, pp. 88–107, Chapter 5, pp. 63–92. London: Taylor & Francis.

Woods, D. D., Dekker, S., Cook, R., Johannesen, L., & Sarter, N. (2010). *Behind Human Error* (2nd Ed.). Burlington, VT: Ashgate. Retrieved from www.scopus.com

Yerkes, R. M., & Dodson, J. D. (1908). The relation of strength of stimulus to rapidity of habit-formation. *Journal of Comparative Neurological Psychology, 18*, 459–482.

Chapter 12

Future trends in human factors

Chapter Vignette

Doctor Smith was excited to use the new surgical robot. Dr. Smith received training over a weekend after the system was delivered and scheduled his first surgery the next week to remove a patient's prostate gland. On the day of the surgery, Dr. Smith turned the machine on—the electronic rumble brought the machine to life, with its octopus-like arms conducting their startup dance. The patient was prepped and brought into surgery and the first 30 minutes went swimmingly. Suddenly, during the removal of one of the cancerous lobes of the patient's prostate, the sound of an electrical discharge was heard. The patient's heart stopped, and alarms began to sound. Although an expert resuscitation team responded within moments, the patient still died on the table just a few minutes after the event. Dr. Smith was unsure of what happened, but he was certain that if he still had his medical license after this event, he'd never use the robot again.

Consider another scenario, where Sue is in her home preparing to meet a friend when a chime on her cell phone reminds her to take her medicine. Using voice commands, she instructs her wheelchair to take her to her bedroom. The wheelchair wheels around and heads toward her bedroom deftly avoiding a robotic sweeper cleaning the floor. In the bedroom, she instructs her robotic assistant consisting of a robotic arm mounted on the wheelchair to prepare her daily insulin shot. The robotic assistant retrieves the insulin from a dispenser and injects her arm with the glucose. Her computer politely reminds her of her next meeting by highlighting her schedule on a display that doubles as a work surface on her wheelchair. Completing preparations, she commands the wheelchair to take her to her car where she instructs it to take her to the site of her next meeting. The car backs out of her garage and heads toward the restaurant while seamlessly navigating the traffic, stopping at traffic lights and stop signs and waiting for pedestrians before turning. Arriving early, she asks the car information system to let her granddaughter know she has arrived. The robotic assistant aids her with eating, and the computer system automatically monitors her incoming communication (email, text messages, etc.) and alerts her to priority messages when there are breaks in the conversation.

The vignettes above highlight recent trends including the recognition of the relevance of human factors (HF) principles and methods to medical settings as well as the design of emerging technologies that may impact you or someone you know soon. Over the last decade, technological advances seem to occur at an ever-faster rate and concepts that only recently seemed in the realm of science fiction are now more readily attainable. In this chapter, we will highlight the growing interface between HF and medicine, with research being conducted across the care spectrum, from the design of electronic medical records (EMRs), team training, and communication protocols to the re-engineering of an entire hospital system.

DOI: 10.1201/9781003515463-12

This chapter also provides a brief, high-level overview of important issues concerning impact of sophisticated automation in transportation, as well as emerging technologies that may allow operators to interact with tools using Brain Control Interfaces, which holds promise for user populations including the disabled.

12.1 CHAPTER OBJECTIVES

After this chapter you should be able to:

- Understand HF in the context of up-and-coming future research areas with a particular focus on medical and automation industries.
- Identify some of the major issues facing human factors research and applications as the science advances into the 21st century.
- Apply knowledge that cuts across previous chapters to understand how to solve high-risk, real-life HF/E problems facing the world today.
- Understand why HF/E is needed, and will continue to be needed, in high-risk and complex socio-technical systems.
- Human factors issues associated with the use of automation.
- Introduces the emerging field Neuroergonomics.
- Highlights other emerging trends including Brain Control Interfaces.
- Highlights how advances in technology can have a positive impact on quality of life for special populations.

12.2 WHAT ARE FUTURE TRENDS IN HUMAN FACTORS?

With the rapid advancement of technology and specializations, medicine is more complicated than it has ever been, and the pressure on providers to deliver perfect care in an imperfect system is mounting. The interface between human factors and medicine is growing rapidly, with research being conducted across the care spectrum, from the design of electronic medical records (EMRs), team training, and communication protocols to the re-engineering of an entire hospital system. Below we provide a brief, high level overview of where HF/E research fits within the medical world to provide the reader a better understanding of the future state of these sciences.

Recent research shows medical errors are the third leading cause of death in the United States (Makary & Daniel, 2016). Since the Institutes of Medicine (IOM) report in 1999, the number has steadily increased—from less than 100,000 medical error related deaths to almost half a million (Makary & Daniel, 2016). The reason behind this increase in errors is still unclear. The simplest explanation is that more errors are occurring. If we use our human factors way of thinking, it's more likely that instead we may be getting better at measuring errors, or we are uncovering more errors as we investigate them in this setting. Regardless, the integration of human factors into healthcare has begun and seems to be here to stay. Human can be applied to the medical setting in a multitude of ways—from the design of the electronic devices and interfaces that practitioners need to interact with daily, to the redesign of workflow, team practices, protocols, and checklists. Due to the need for error mitigation and prevention (see Chapter 11), human factors is being applied across the medical setting, including inpatient and outpatient settings. Although HF/E has a multitude of approaches at its disposal, there appear to be only four ways in which HF/E can intervene in this setting—changing the equipment, changing the training, changing the procedures, or changing the organization (Stanton, 2005). Below we will discuss the various ways HF/E can be applied to the medical domain with these four intervention strategies as the basis for application.

12.3 MEDICINE—A MODERN EXAMPLE OF THE NEED FOR HUMAN FACTORS

As technology advances at an ever-increasing pace, there continues to be a need for human-systems integration. This need becomes further necessary as we integrate advanced technologies into complex and high-risk systems. Nowhere has this become more pertinent than applications in healthcare, where new technologies can easily lead to potentially devastating consequences for patients. As an example, the Da Vinci robotic surgical system has had rapid and widespread adoption—with approximately a quadrupling in use between 2008 and 2009 (from 114,000 robotic assisted surgeries to 367,000; "Surgical Robot da Vinci Scrutinized by FDA After Deaths, other Surgical Nightmares", 2013).

This adoption, like that of any technology, does not come without consequences— through 2012, there were approximately 500 reports of problems. These problems are the same as with any complex systems including failures in both the machinery and the human operators understanding of the system capabilities and knowledge of how to react during failures (see Chapter 11). As evidence, a Cornell university study demonstrated that out of 10,624 adverse events with the robotic system, 144 led to death, 1,391 led to injuries, and 8,061 to machine malfunctions— "These adverse events included burnt or broken pieces of the robotic instruments falling into the patient (14.7 percent), electrical arcing of the instruments (10.5 percent), and the ever-dreaded system error (5 percent)." (Alemzadeh et al., 2016, p. 1).

Further, although the system has been made more robust over the past few years, the training users receive on such machines is limited—with some reports stating that surgeons had only used the device five times across a 3-day training prior to conducting a surgery on patients (Rabin, 2013). Considering that the device is arguably as complex as a plane, could you imagine flying on an aircraft where the pilot only has 3 days of training, and all that training solely in simulation!? Also, there appears to be no form of validated post-training assessment to confirm a user's understanding of the system or their competence in using device.

What this demonstrates is a clear need for human factors in medicine. Although medicine, in general, has accepted that human factors is needed after the Institutes of Medicine (IOM) report "To Err is Human" (Kohn et al., 2000), it seems that the current state is one of a reactionary approach rather than a preemptive approach. The following sections will discuss the potential hazards of various domains where technology integration is inevitable. These areas will be those that you, as a student and future practitioner of human factors, may very well be involved if you maintain a career in the field.

12.3.1 Sociotechnical systems in medicine

One concept that has aided in integrating HF/E into the medical domain is the application of the sociotechnical systems models. One of the most prominent in recent years is Pascale Carayon's Systems Engineering in Patient Safety (SEIPS) model (see Figure 12.1). As discussed in Chapter 11, this model demonstrates the intricacies of the medical system across providers, patients, tasks, devices, and organizational variables. This model highlights the important components of any medical task or procedure and aids practitioners in uncovering potential pitfalls and pinch-points within the medical system to improve and create targeted interventions. Although some argue that medical systems are extremely complex, and therefore, unmeasurable and unknowable in their entirety (Dekker, 2011), it is important that we attempt to intervene as best we can. Carayon's model utilizes a common organizational structure for its basis, specifically the Input-Process-Output-Input (IPOI) model

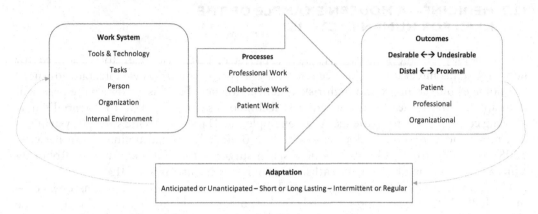

Figure 12.1 An adaptation of the SEIPS 2.0 model demonstrating the work system, processes, and patient, provider, and organizational level outcomes. (Adapted from Holden et al., 2013.)

(Ilgen et al., 2005). The IPOI model is a closed-loop system (see Chapter 1). This type of model captures variables as both aspects of the system that are pre-states (e.g., hospital culture), the various processes that occur throughout operation of the system (e.g., patient care), and the various outcomes associated with organizational success (e.g., hospital mortality rate). Further, the outcome of the first three stages has implications for future states, as demonstrated by a feedback loop from outcomes back to inputs. The model focuses on three principles: systems orientation, person centeredness, and design-driven improvements (Holden et al., 2013). This model can be used as a framework for guidance on where a new system has potential for failure but can also demonstrate how changes in any one part of a system can have profound effects on the entire system.

SEIPS can be used to aid organizations in thinking deeply about systematic issues that arise in their units. As noted in Chapter 11, it is easy to blame humans when they commit errors (Dekker, 2011), but humans often falter due to system constraints and poor design. Therefore, it's pertinent for organizations, practitioners, and scientists to understand all potential precursors to an error, and the SEIPS model provides the groundwork for this type of guided thinking. For instance, imagine a scenario where a provider gives a patient a drug, they happen to have an allergy to. In an older way of thinking, they might very well lose their job, be demoted, or face loss of medical licensure for making such a mistake— especially if it led to patient death. If we utilize the SEIPS model in this instance, we quickly realize that the work system and current process might have set this provider up for the kind of error they committed. Questions arise such as: Was the provider informed that the patient had this allergy? Was this information salient in the EMR, or on the patient's wristband? Was there stop-gaps present in the retrieval and use of the medication that remind providers to check for allergies? Were policies in place that trained or guided providers to be careful concerning medication administration? Was the drug properly labeled to make allergies a salient concern?

All these questions point away from the human and toward the work systems, process, and organizational policies, the modern way of thinking with human factors that we've learned throughout this text and the SEIPS model provides a robust model for this type of thinking.

The mindset behind these types of systematic approaches is a drastic cultural change from current practice in medicine. The SEIPS model is one way to understand the medical system as a set of extremely complex, inter-related processes. In recognizing this, the fault of medical errors no longer rests solely on the shoulders of practitioners but is also a function of organizational norms and culture, equipment training and usability, as well as process efficiency and control. The future should be brighter as we move away from persecuting the failures of practitioners, and instead re-engineer the healthcare system to capture errors before they propagate and instead support human performance.

BOX 12.1   ORGANIZATIONAL POLICY VERSUS PRACTICAL NEEDS: MAKING CRITICAL DECISIONS TO SAVE PATIENTS

An EMS agent kneels down beside a victim of a hit and run on the side of a busy interstate. It was getting late when this call came in, and having hardly slept the night before, the EMS agent had asked his partner to drive—normally a duty he was unwilling to relinquish. As the agent stooped over his patient, he noticed that she was wheezing. She needed to be intubated immediately to ensure she was receiving enough oxygen to make it to the hospital alive. Unfortunately, her jaw was clenched shut, and the agent needed to administer a muscle relaxant to unclench her jaw and move forward with the orally inserted tube that would bring oxygen to her lungs. Usually, the agent needs to call a local hospital to get approval from an MD before using this drug. Due to time constraints and the patient's labored breathing and rapidly fading pulse, he believed that calling this one in would cost the patient her life, so he moved forward without approval. Fumbling around in the darkness, the agent found the correct vial, measured out the dose, and injected the relaxant into the patient—unclenching her jaw and clearing the airway for intubation. Although this is the ultimate reward, he was dreading the inevitable meeting with his medical director about breaking protocol. **Critical Thinking Questions**: This is clearly a representation of the contradictory goals of a healthcare organization. Think about how this relates to the idea of a sociotechnical system—specifically, what may be the best approach to dealing with this type of incident? What role could human factors play in making this process more efficient and safer? How might you apply the SEIPS model to better understand this system? What aspects of the SEIPS would help describe what happened? Where does the SEIPS possibly break down?

12.3.2 Transitions of care—communication, teamwork, and handovers

Healthcare's approach to medical errors has changed drastically over the last decade and will continue to change as HF/E is further integrated into medical practice. Not too long ago, medical errors were solely attributed to providers, with an error being an obvious result of incompetence on the provider's part. As described above, this mindset has changed drastically, with a much deeper understanding of organizational and technological constraints being accountable for adverse outcomes. With the changing mindset and approach to blame, medical errors have become more heavily studied and this has led to the emergence of themes, one being questioning the very nature of an "error." One of the most common areas of medical errors is during transitions of care when a provider or team or providers gives a patient over to another provider or team. These transitions have been associated with upward of 70% of medical errors (Sutcliffe et al., 2004), evidence that they are a critical and risky moment in patient care. Broadly defined, transitions of care are when responsibility of a patient's care is passed from one provider or team of providers to another. This could be as simple as the one-to-one exchange that occurs between doctors from day shift to night shift, or as complex as two team's passing a surgical case from surgery to the Intensive Care Unit. In all instances of transitions of care, teamwork, and specifically, communication and coordination, is paramount to ensuring that the information being passed to the incoming provider or team of providers is up to date, complete, but also succinct and clear, and that both the sending and receiving team have a shared set of goals, expectations, and action plans in regard to the patient.

12.3.2.1 Developing handoff protocols

There are many ways the field of human factors can aid the medical community concerning transitions of care. One growing area of research is specifically focused on the development and evaluation of protocols for transitions of care (Keebler et al., 2016). Just like checklists used by pilots to ensure they execute flight procedures in the correct order without missing any steps (see Chapter 11), protocols aid providers in structuring their communication during transitions of care to ensure the most important information about a patient case is passed to the incoming provider or team. Although some argue that EMRs might solve this problem, to date EMRs have been shown to be unreliable, difficult to use, and in many ways an impediment to the day-to-day workings of medical practice (Mador & Shaw, 2009).

12.3.3 Teams, teamwork, and team training

Although protocols have been shown to enhance performance for the transition of patients, the knowledge, skills, and attitudes (KSAs) of teamwork are still an all-encompassing aspect of excellent provider performance (Figure 12.2). The literature has defined a *team* as "a set of two or more individuals that adaptively and dynamically interact through specified roles as they work toward shared and valued goals" (Salas et al., 2009, p. 40). This dynamic, inter-active, and usually inter-dependent action(s) toward a goal is *teamwork*. Finally, *team training*, the training of team-related competencies, usually referred to as the KSAs of teamwork, which are necessary for good teamwork but differ from individual expertise or *taskwork*. Usually, knowledge refers to cognitive aspects of teamwork, while skills refer to behavioral aspects and attitudes toward dispositional aspects. Due to this synonymous meaning, KSAs are also often referred to as ABCs in the literature—attitudes, behaviors, and cognitions. Holding the appropriate KSAs for one's taskwork is part of a professional's expertise, but holding the KSAs for teamwork is a fresh area for research and application. Therefore, the next section will describe what leads to good acquisition of team KSAs and how this relates to provider performance and medical outcomes.

Good teamwork leads to good *team performance yet* requires training that is separate from individual level expertise training. For instance, training to be an expert doctor or pilot does not necessarily entail expertise in teamwork. Teams in healthcare are complex, fluid, and filled with experts in various professions. This leads to difficulty in individuals knowing who is on their team, and sometimes clouds perceptions of leadership and goals. As an example, a recent meta-analysis (Hughes et al., 2016) has shown that team training in healthcare is effective at enhancing provider performance and patient outcomes. One remedy to enhance **teamwork** in the medical domain is through team training. One of the most widespread team training programs in healthcare is the DoD/AHRQ's TeamSTEPPS program. TeamSTEPPS was developed based on the existing theories of teamwork and is a program of team training, integrated with protocols, checklists, and other materials that aid medical workers in learning the fundamental aspects of teamwork. Figure 12.3 depicts the major constructs of teamwork that TeamSTEPPS assesses.

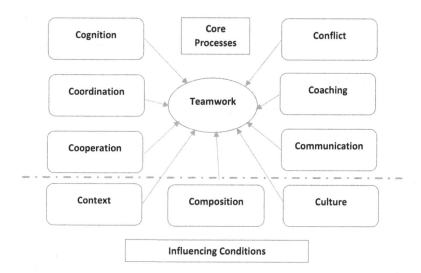

Figure 12.2 An example of major teamwork competencies. (Adapted from Salas et al., 2014.)

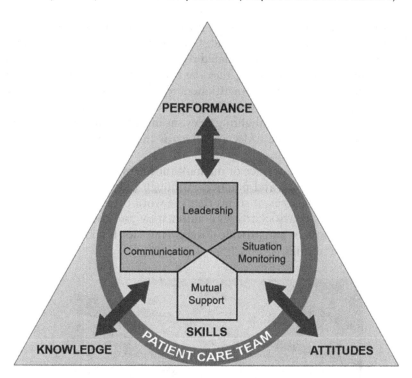

Figure 12.3 Team competencies in the TeamSTEPPS framework.

12.3.4 Telemedicine and HRI

Another emerging technology field within medicine known as **telemedicine** or **telehealth** utilizes remote technology to have patients and providers interact over longer distances. This could be as simple as a phone conversation with a psychiatrist, to a robotic asset that allows doctors to monitor highly acute patients while keeping a sterile environment, to surgery across an ocean (Marescaux et al., 2001)! Telemedicine is an area ripe for the application of human factors science due to its inherent reliance on technology. One area that is still poorly understood is teamwork in telemedicine. As this chapter has spoken at length about teamwork, it's important to note that very few studies have empirically evaluated how telemedical assets affect the patient-provider team or looked at the effects of telemedical technology on patient and provider outcomes. With the advent of more useful and lower cost robots, more powerful communication technologies, and an aging population living in rural areas with limited access to medical professionals, we are certain to see an explosion of the use of telemedicine.

12.3.5 Electronic medical records

The days of using paper-based tools to record patient information are quickly disappearing. Instead, they are being replaced by EMRs, which are integrated electronic systems known that can track a patient across time and keep this information in a system accessible by any of the patient's provider. The science that studies the information around patient health is called **bioinformatics**, and it is a growing field involving designers, human factors engineers, and medical practitioners. EMRs (Figure 12.4) have become popularized over the last decade, with a mandate from the Joint Commission stating that all medical units must be using an EMR by the year (Chassin et al., 2010). As with any technology, EMRs bring both benefits and costs to the medical world. On one hand, they provide computer access to a patient's entire history and can be updated regularly—far surpassing the capabilities of paper-based records. On the other hand, they are often criticized for being difficult to use, being cluttered, and forcing medical practitioners to spend more time on their computers and less time with patients. This leaves a large research gap where HF/E can be applied. Principles for better design, for highlighting relevant information, and for providing training on updates and changes to the EMR are all areas that will need to be explored in the future.

We have barely scratched the surface of the multitude of ways that human factors can be integrated into medical research and practice. Although there is much to still discover in the cross-over between these areas, it seems that the work being done is fruitful. Our next section will unpack the ideas associated with automation and robotic systems—a growing industry that will supplant jobs and provide novel avenues for human factors applications in industrial and domestic settings.

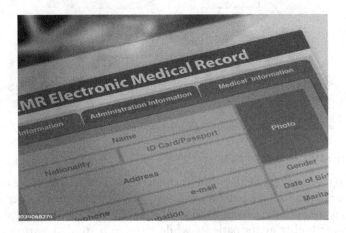

Figure 12.4 Example of an electronic medical record made by EPIC.

12.4 AUTOMATION

Advances in software and integrated sensors allow a wide range of tasks to be automated from the sweeping of your apartment floor to driving your car to assisting with medical needs. Automatic floor sweepers are commercially available as are cars with automatic features such as parallel parking and hazard detectors initiate actions including slowing down or avoiding a hazard. Waymo, owned by Google has made headlines with the introduction of their autonomous taxi service that has navigated the congested roads of California.

The trends cited above highlight the increasing reach of automated systems that offer users a wide range of benefits. First, they free the human operator to perform other tasks where humans excel like being creative and problem solving rather than performing monotonous, repetitive, and dangerous tasks that automated systems can perform with greater precision, speed, and safety. The wider application of automation is due to advances on multiple fronts including more sophisticated software algorithms, lighter, smaller, cheaper, and powerful electronic sensors.

Until recently, most applications of sophisticated automation addressed safety critical systems in heavily regulated industries such as energy production (i.e., coal, natural gas, and nuclear power plants), industrial processes (i.e., petrochemical industry), and aviation where automation found early applications in autopilot and navigation systems (Wiener, 1988). Automation is used to monitor and control complex processes to improve efficiency and reduce the potential for human error. Additionally, these applications involved user populations consisting mostly of highly trained and carefully selected operators. Increasingly, recent applications of automation like the Tesla cars have targeted systems used by the public.

Many of the human factors issues concerning the use or misuse of automation by experts in highly regulated industries are just as pertinent to the use of automation in "consumer electronics." Researchers (Dekker & Woods, 2002) have identified how automation changes the nature of the operator's work. When systems are automated, the operator gives up direct control of the system and is expected to monitor or supervise the actions of the automation, intervening when they detect a system malfunction, or they believe the systems response is not optimal. Consider the second vignette from the opening of the chapter where Sue relies on a driverless vehicle to deliver her safely to her destination. If there is an emergency, Sue cannot be expected to take over control of the vehicle. Is this scenario realistic? How much trust must operators have in an automated system that they would not continuously monitor its performance or behavior? In modern automated applications, operators are expected to monitor the system to identify system malfunctions, yet the extensive vigilance literature (see Chapter 6) has documented how operators are prone to errors in monitoring. Is this a reasonable expectation for the lay public? It is ironic that complex automation which relieved human operators of performing certain tasks now requires the human to supervise and monitor its operation. This expectation is inconsistent with the purported benefits of automated cars—namely that it frees the driver use of the time in the vehicle more productively by reading, replying to emails, or chatting with other passengers.

Automation also often requires the user to learn skills that were not necessary before. For instance, the surgeon using a robotic surgical system must now monitor rather than perform the surgery and must also learn to use the systems software to program and execute the surgery. Similarly, a consumer using a new thermostat that can be controlled remotely must learn to use the application software to set the time and temperature settings for their home. These new "programming" skills create opportunities for new types of errors (see Chapter 11)—such as entering the wrong information or forgetting how to program the system—that were not possible before. These issues highlight the importance of well-designed software that makes programming the system more intuitive and provides

Table 12.1 Table levels of automation

Level of automation	Description
Level 0: No-automation	The driver is in complete and sole control of the primary vehicle controls—brake, steering, throttle, and motive power—at all times.
Level 1: Function-specific	Automation at this level involves one or more specific control functions. Example/comment: Examples include electronic stability control or pre-charged brakes, where the vehicle automatically assists with braking to enable the driver to regain control of the vehicle or stop faster than possible by acting alone.
Level 2: Combined function	This level involves automation of at least two primary control functions designed to work in unison to relieve the driver of control of those functions. Example/comment: An example of combined functions enabling a Level 2 system is adaptive cruise control in combination with lane centering.
Level 3: Limited self-driving automation	Vehicles at this level of automation enable the driver to cede full control of all safety-critical functions under certain traffic or environmental conditions and in those conditions to rely heavily on the vehicle to monitor for changes in those conditions requiring transition back to driver control. The driver is expected to be available for occasional control, but with sufficiently comfortable transition time. Example/comment: The Tesla and Google cars are an example of limited self-driving automation.
Level 4: Full self-driving automation	The vehicle is designed to perform all safety-critical driving functions and monitor roadway conditions for an entire trip. Such a design anticipates that the driver will provide destination or navigation input but is not expected to be available for control at any time during the trip. This includes both occupied and unoccupied vehicles. Example/comment: Google has a prototype car without a steering wheel, gas or brake pedal which would presume level 4 automation.

feedback in ways that enables the user to recognize errors, as well as training that facilitates the development of a mental model of how the system works.

The use of automation is not an all or none decision, but rather often is a matter of degree, as is readily evident in Table 12.1. The National Highway Traffic Safety Agency (NHTSA, 2013) defines five levels of automation for vehicles as shown in the table. The levels range from 0 where the vehicle has no automation to 5 where the car is fully self-driving, and the driver cannot retake control of the car. The vehicles produced by Tesla and Google correspond to a level 3 and 4 of automation, respectively. Google has also publicly stated a goal of producing a car without a steering wheel, and gas and brake pedals for driving around town. Its top speed would be capped at 25 mph (40.23 kph) for safety reasons.

A report on autonomous vehicle technology (Anderson et al., 2014) notes that level 3 automation could significantly reduce crashes and injuries due to driver error and those involving motorcycles, pedestrians, and cyclists since the automation is not prone to distraction, impaired driving, or recklessness. Also, level 4 could importantly reduce accidents and fatalities related to alcohol if drivers would use an autonomous vehicle to get home.

While technology can be a boon for many users, it can also pose considerable risk if misused. We have all experienced how the passenger accompanying us on a road trip promptly falls asleep while we drive. In an autonomous vehicle, the driver is along for the ride, how long might we expect them to stay awake once the car starts moving? How will novice/young or older drivers respond to automation? What is the potential for misuse? How will users maintain their skills if they rarely get to practice them? An example of a crash involving a Tesla car is reviewed in Box 12.2.

BOX 12.2   FIRST FATALITY ASSOCIATED WITH AN AUTOMATED SELF-DRIVING CAR

On May 7, 2016, Joshua Brown was driving on a Florida Highway in his Tesla Model S using the automated driving system and was playing a DVD on the cars infotainment system, when a semi-trailer turned in front of his vehicle. The automated driving system employs a variety of sensors, including a camera, lidar (acronym for Light Detection and Ranging), radar, and GPS allowing the car to manage its speed, stay in its lane, and change lanes when the driver activates the turn signal failed to detect the white tractor-trailer against the bright sky. Mr. Brown died instantly.

Like Thomas E. Selfridge and Henry H. Bliss, Joshua Brown will forever be associated with a first—in this case, the first fatality associated with a self-driving car. Thomas Selfridge and Henry Bliss were also firsts—the first fatality in an airplane and the first fatality in a motorized vehicle, respectively. These inauspicious "firsts" shadow important technological developments in many areas including transportation (shipping, aviation, rail and road), nuclear power, space travel, and advances in medicine. However, we may be better prepared to anticipate some of the negative consequences that follow these achievements if we attend to some lessons from our recent past.

Tesla touted the autopilot system used by Mr. Brown as an "assist feature" that can be turn on by the driver only after explicitly acknowledging that the technology is new and in beta phase testing, implying that the system is under development but can be used "experimentally" by drivers. Drivers must keep their hands on the steering wheel, which are monitored by sensors on the wheel; otherwise, the car will slow until the drivers' hands are again on the steering wheel. Mr. Brown's accident was one of three accidents involving Tesla vehicles that occurred over a 4-week period between May 7 and July 8, 2016, and all involved drivers using the autopilot system. One of the accidents occurred late at night on an undivided mountain road, and the driver did not have his hands on the steering wheel, nor did they respond to system warnings instructing them to do so. Language may have played a contributing role as the audio warnings are in English and the driver was a non-native English speaker. **Critical Thinking Questions**: What role can human factors play in advancing the development of this type of technology? What types of errors may be more common or unique to this type of application? How the driver's attention to the driving scene by can constructively engaged so that when they need to take over they are better prepared to do so?

The accident highlighted in Box 12.2 illustrates the continuing relevance of issues related to automation trust, operator monitoring and supervision of the automation, and operator training and understanding of systems limitations. These issues came to the forefront during investigations of accidents involving the use of the first generation of automated systems (Wiener, 1988). Engendering the proper level of user trust in an automated system may be difficult given that all the environmental and situational factors that reduce the system's ability to detect an obstacle may not be known by the driver or the engineers who designed the system. Thus, a gap will exist between the perceived capabilities of self-driving cars and their real capabilities. This is illustrated by another incident involving an autonomous Google car that was involved in an accident with a transit bus (della Cava, 2016).

The Google car was in a right lane planning to turn right when it detected an obstacle. To avoid the obstacles which were sandbags placed near a storm drain, it maneuvered toward the left predicting that a bus approaching from behind would yield to it. A Google test driver was in the car, observing the unfolding events, but did not intervene because he also assumed that the bus would yield to the car. The collision of the bus and car was unexpected and demonstrated a weakness of the cars algorithms to handle a situation that humans also find challenging. The failure of the human driver to intervene also serves as a caution of the limitations of human oversight of automation.

Likewise expecting drivers to maintain a high level of alertness and preparedness to take over the vehicle in advent of an unanticipated behavior is unrealistic given what we know about human abilities and behavior (see Chapters 6–8). Alertness waxes and wanes depending on a wide range of factors including the time of day, driver age, and the driving environment, and this is true even when a driver is in full control of the vehicle. While providing the user with information detailing the limitations and capabilities of a system is necessary, it is also important to recognize that operators often do not read the users manuals of many complex and potentially dangerous devices, they use daily—cars notwithstanding.

Another concern with automated systems is how proficient the user will be in taking over manual control of a system should the automation fail. The issue of operator proficiency in assuming manual control (see Chapter 8) of the system was brought to the fore by Air France flight 447 that crashed into the Atlantic Ocean while on route from Rio de Janeiro Brazil to Paris France (BEA, 2012). The aircraft had a sophisticated auto pilot and flight management system that was used for a significant proportion of most flights. Prior to the mishap, pilots' groups have expressed concerns that spending so much time monitoring the automated systems can lead to declines in pilot proficiency in manually flying the aircraft. Consequently, the pilot might be ill prepared to fly the aircraft should unusual circumstances arise. In this case, aviation experts raised concerns that the pilots may not have known how to regain control when flying at high altitude in turbulence (BEA, 2012). Similar concerns are pertinent for autonomous cars. How will novice young drivers acquire the skills needed to take over should automation fail if they rarely drive the car in manual mode? Also, how will experienced drivers maintain proficiency if they are not driving very often?

12.4.1 Automation trust or mistrust

As mentioned earlier, the propensity of an individual to use automation depends on the level of comfort or trust they have in the automation. Evidence indicates that users don't always use automation appropriately and may fail to use it when it would aid performance or efficiency (Parasuraman & Riley, 1997). Alternatively, other users may rely blindly on it and fail to monitor or question the actions of the automation. Automation trust has several facets and has been defined as "the attitude that an agent will help achieve an individual's goals in a situation characterized by uncertainty and vulnerability" (Lee & See, 2004, p. 50). Users will use and rely on automation that they trust and fail to utilize automation that they don't. Both misuse and disuse of automation represent incongruence between the perceived capabilities of the system and its actual capabilities. Calibration between a system's capability and the degree of user trust underlies the correct use of automation. Users who have an unrealistic belief in the capabilities of the automation are more prone toward misuse of the system, whereas users who underestimate system capabilities are prone toward underutilization of automation. It should be readily apparent from this discussion that "trust" develops

over time as the operator uses the automation and gains experience that informs their understanding of the systems capabilities and limitations, as well as situational factors that affect
the systems performance.

12.4.2 Automation as an enabler

Despite the potential challenges, the development of sophisticated automation could be
a boon for individuals with disabilities. Autonomous technologies offer the promise of
greater independence and the opportunity to live richer productive lives for individuals
with disabilities. As the scenario at the start of the chapter illustrated, automation has the
potential of expanding human abilities. The autonomous car allows greater independence,
the ability to stay socially engaged in the community, and to be employed, thus empow-
ering the disabled. Perhaps you have seen the YouTube video (https://www.youtube.com/
watch?v=peDy2st2XpQ) featuring the google autonomous car and the blind passenger. In
the video, the car takes the passenger to the dry cleaner and in route successfully negotiates
the drive-through lane allowing the blind passenger to order fast food. Automation would
enable persons with physical and sensory limitations to engage in many activities that they
can't currently perform. This however would also necessitate changes to existing rules that
require the user of self-driving vehicles to take control in the case of an emergency or a
malfunction.

12.5 EMERGING TECHNOLOGIES

Advances in technology are altering the way we interact with many of the devices we use
daily. These developments have potential benefits for a broad range of users including individuals with different types of disabilities. In most cases, we interact with computer systems
indirectly through keyboards and input devices like mice and trackballs using our hands.
Multi-touch screens on cellphones are replacing physical keyboards providing greater flexibility in the use of the screen. Using touch, the systems recognize gestures like pinching or
spreading fingers to magnify or minimize, and sweeping fingers to scroll up-down, left or
right. The SIRI® voice command system for Apple cellular phones and other software applications which convert voice to text are now widely available to the public and could find
important applications in systems for paraplegics. The challenges faced by disabled users
have also sparked an interest in using signals generated by the nervous system to control
mechanical devices or to interact with computers and software.

The methods and techniques described in Chapter 5 can be used to evaluate the effectiveness of these new interaction methods. Some of the human factors issues include the
accuracy of these systems in identifying human voice commands or speech under noisy conditions, their response times, ease of use, and the cognitive demands they place on the user.

The Wii game system is illustrative of an emerging class of systems that can identify physical gestures made by an operator remotely. These types of systems eliminate the need for
the user to interact directly with a keyboard or mouse, but allow the computer system to
recognize gestural movements like those made by the character played by Tom Cruise in the
movie titled Minority Report. In the movie, Tom Cruise stands in front of a virtual monitor
making a series of hand movements that allow him to find, select, and manipulate images
shown on a virtual screen. Game players including the Kinect © can recognize movements,

which hold promise as another means of interacting with devices without having to physically touch the interface. This frees the operator's hands from the keyboard or work surface. They could be used in medical settings where minimizing physical contact with other surfaces, such as keyboards or touch screens, reduces the risk of contamination. Doctors and nurses could use hand movements to search through a patients EMR, request and review X-rays and other medical images, or enter patient data.

12.6 APPLICATIONS TO SPECIAL POPULATIONS

For many years, a few dedicated researchers worked on developing technologies that allowed users suffering from paralysis due to spinal cord injuries to use computers to communicate. Early interactive systems designed for the disabled allowed paraplegics to spell messages by fixating on individual letters displayed on a computer screen. New technologies allow users to control devices using brainwaves thus not requiring any neuromuscular control (McFarland & Wolpaw, 2011). Instead of monitoring muscle activation, these devices rely on cues provided by the pattern of brain activity associated with certain behavioral goals to identify specific signatures associated with an object, number, or letters on the screen selected by the user (Grigorescu et al., 2012).

12.6.1 Brain control interfaces (BCIs)

The development of **Brain Control Interfaces** (BCIs) is motivated by the needs of individuals that suffer from conditions including locked-in syndrome, a condition where the person cannot physically move but is otherwise cognitively intact, and military veterans who have lost the use of a limb due to a combat injury (Nicolas-Alonso & Gomez-Gil, 2012). Through a BCI, individuals can control a wheelchair, surf the web, and play games.

BCIs are not mind readers, that is, they do not identify the user's thoughts or goals, but rather it detects a specific pattern of brain activity that the user has learned to reliably produce. The computer monitors brain activity via sensors placed in contact with the head and is programmed to execute a set of actions when it detects a particular pattern of brain activity (Grigorescu et al., 2012). In cases like the control of a prosthetic limb, the BCI is designed to monitor brain activity associated with the activation of somatosensory and motor control areas of the brain. When a user imagines grasping an object, specific areas of the somatosensory and motor areas of the brain associated with generating efferent signals to the limbs are activated. This pattern of activity can be monitored and used to control a robotic hand or arm (Grigorescu et al., 2012). BCIs require distinct patterns of activity that can be produced reliably by the participant on demand, and which are large enough that they can be detected among the other electrical activity produced by the brain and nervous system. Applications of BCIs include the activation of environmental controls, a speller that allows a patient to write text messages, play games, control a wheelchair or robotic assistant, and grasp objects.

Importantly, many of the human factors issues discussed in earlier chapters apply equally well to this domain. For instance, how large must a target be to support accurate target acquisition when the cursor is controlled by brain activity (see Chapter 8)? How fast can a participant move and acquire a target, then disengage and acquire another? How much training is required to be proficient in using the system? Does the user interface match the capabilities of users with different types of physical limitations?

BOX 12.3   HUMAN FACTORS CHALLENGES OF BCIS

BrainGate is one example of a BCI. BrainGate sensor consists of 100 thin electrodes in a pill size piece of silicon. The sensor is surgically implanted in the part of the brain that generates motor movement. Using BrainGate, two patients who could not speak or move their limbs as a result of strokes learned to control a robotic arm (Hochberg et al., 2012). One patient who had suffered a stroke 15 years before was able to grasp a water bottle and take a sip using the robotic arm. While promising, there remain significant hurdles before the system would be ready for day-to-day use. For instance, the control and movements of the robotic arm can be frustrating slow and inaccurate. In some trials, the participants were only able to successfully grasp the foam target on roughly half the trails, and the action took approximately 10 seconds to complete. This might be unacceptable to users with experience making fast and accurate motor actions. This does not diminish the significance of the scientific advancements made by the research teams. They acknowledge the importance of the "Human Factors" considerations—the focus of this book—including accuracy, response time, ease of use, and training to user acceptance and adoption of this technology. Feedback from the patients is being used to refine the design of the system. **Critical Thinking Questions:** What other human factors considerations besides the ones listed above might be important to consider in the design of a BCI? Are human factors considerations that may be more important or relevant to disabled users?

BCIs may also benefit users without disabilities. Researchers are investigating how brain activity can be monitored non-invasively to detect different mental states that can in turn be used to control tools or devices or to interrupt users to present them certain information or delay presenting them with information. This area of research has been called neuroergonomics.

12.6.2 Neuroergonomics

The term neuroergonomics was coined to draw attention to the confluence of several areas of research including human factors, ergonomics, and the neurosciences (Parasuraman, 2003 & Wilson, 2008). With the advent of smaller, portable, non-invasive tools and techniques for observing, monitoring, and recording brain activity, it is possible to identify the neural correlates of persons engaged in perceptual and cognitive tasks as well as when they perform physical tasks like grasping, lifting, or moving objects. Some of these scientific tools often referred to by a soup of acronyms, including fMRI (Functional Magnetic Resonance Imaging), TCD (Transcranial Doppler sonography, and EGG (Electroencephalography). Each tool uses different technologies to localize brain activity, monitor changes in this activity across time, and thereby record the neural signature of people performing work. This emerging area of study called "**neuroergonomics**" is concerned with the activity of the brain when an operator is performing work so that we can "understand the neural bases of such functions as seeing, attending, remembering, deciding, and planning in relation to technologies and settings in the real world" (Parasuraman & Wilson, 2008, p. 468).

The monitoring of neural activity while an operator works would provide a basis for the development of new systems that dynamically reallocate tasks depending on situational demands. For instance, it could be possible to monitor an operator's neural activity and determine their current cognitive workload. Consequently, like the second scenario described at the start of the chapter, alerts, warnings, text message, emails, or other tasks might be prioritized so that only the most critical are presented to the operator, or the pacing of the tasks might be altered allowing the operator to perform optimally. Alternatively, the monitoring of human cognitive activity may indicate whether the operator is aware of an alarm or warning and instruct the system to repeat the warning or increase its intensity to ensure detection of the signal. Additionally, improvements in neural signal processing and recognition might support identification of different types of cognitive work enabling the system to anticipate the user needs and make available information without the user's explicit request. Developments in the field of neuroscience indicate that it may be possible to identify what an individual is attending to (Kamitani & Tong, 2005), what information they are storing in working memory (Harrison & Tong, 2009), or even what they are seeing or imagining (Nishimoto et al., 2011).

Advances in technology will continue to create new ways for users to interact with technology, new ways to monitor the physical activity, and physiological and cognitive states of the user, thereby enabling the creation of novel tools to support their work. Human Factors will play an increasingly important role in the development of these tools to ensure that they are designed to enable, facilitate, and maximize human performance.

12.7 SUMMARY

The early impetus for the field of human factors can be found in manufacturing and military applications where researchers sought to address issues related to work productivity and efficiency, human error, safety, personnel selection and training. More recently, the computer revolution had a major influence on field of human factors shifting the focus of much research to the design of graphical computer interfaces, input devices, and automation. In this chapter, we have sought to highlight emerging trends including the increased acceptance of human factors principles in medicine and foreshadowed technical advances that will change the way we interact with the tools we use. In addition, we highlighted how advances in technology continue to create new ways for users to interact with technology, new ways to monitor the physical activity, physiological, and cognitive states of the user thereby enabling the creation of novel tools to support their work. These technological achievements allow individual with disabilities greater independence and participating in a broader range of work and social activities. Human factors will play an increasingly important role in the development of these tools to ensure that they are designed to enable, facilitate, and maximize human performance.

LIST OF KEY TERMS

Automation
Automation trust and mistrust
Bioinformatics
Brain Control Interfaces
Electronic medical record
Neuroergonomics
SEIPS model
Sociotechnical systems models
Taskworks

Team
Team performance
TEamSTEPPS
Team training
Teamwork
Telehealth
Telemedicine

SUGGESTED READINGS

Dekker, S. W., & Woods, D. D. (2002). MABA-MABA or abracadabra? Progress on human–automation co-ordination. *Cognition, Technology & Work*, 4(4), 240–244.
   The authors of this paper discuss the challenges of choosing when and how to functions that may automated in a complex system.
Parasuraman, R., & Wilson, G. F. (2008). Putting the brain to work: neuroergonomics past, present, and future. *Human Factors: The Journal of the Human Factors and Ergonomics Society*, 50(3), 468–474
   This paper provides an informative overview of the emerging field of Neuroergonomics and its potential applications.

CHAPTER EXERCISES

1. Practicing your literature review skills—conduct a medical HF literature review
   a. Preparation—students will need access to a computer and hopefully a school library although scholar.google.com can suffice.
   b. Students should decide on a topic of interest in the area of HF/E in medicine and find 10 relevant articles from no later than 10 years ago.
   c. Students should read each article and create a code sheet using excel or a similar program. The spreadsheet should contain the following columns: (1) The article's title and authors. (2) The article's overarching purpose. (3) Whether it's a theoretical, review, or empirical paper. (4) What the major contributions of the paper are to the field. (5) A summary of next steps for research given the articles findings.
   d. The students should create two reference lists—one in APA format and one in AMA format. APA rules can be found here: http://www.easybib.com/guides/students/writing-guide/iv-write/a-formatting/apa-paper-formatting/
      AMA rules can be found here: https://www.lib.jmu.edu/citation/amaguide.pdf
2. Teamwork in Medicine—two teamwork models are provided in Chapter 12. Noticeably, they both have only a few constructs/factors that they focus on. Teamwork is much more complex than the models presented, yet both of these models are important and adopted in the field. Why do you think these models are simplistic? What are positive and negative ramifications of this simplification? What could be done to capture more information in the model without making it over-complex in regards to medical practice?
3. Thinking about standardization—usability versus flexibility—medical professionals are inundated with rules, regulations, and mandates from all levels of their organization, their professional societies, and the government. We know from the chapter on usability that most tools have a usability flexibility tradeoff. In general, the more flexible a tool is, the less usable it is and vice versa. In recent years, there has been a mandate for medical professionals to use standardized tools when handing patient's over during care transitions (i.e., passing one patient to another provider or unit for the next step in their care). This is often accompanied by a set of checklists or protocols that help guide providers in their information exchange. What kind of issues could tools like this face when being implemented? What role does technology play, if any, in this sharing of information? It is speculated that approximately 70% of adverse medical events occur due to "failures in communication" —how can we address this problem using HF/E methods?
4. The problem with medical errors—preparation—the instructor should provide the following two articles to students if available, and their abstracts if not:

    a. Makary, M. A., & Daniel, M. (2016). Medical error—the third leading cause of death in the US. *BMJ*, 353, i2139.

    b. Stokowski, L. A. (2016). Who believes that medical error is the third leading cause of hospital deaths. Medscape, May 26.

    Medical errors appear to be a huge problem in modern day medicine, but their measurement and evaluation are still not standardized and validated. Have students read Makary's and Stowkowski's papers and compare and contrast the two articles in regards to medical error rates. Which article seems to be more accurate? Is there any bias in either article that could be affecting their findings? Is there a way to synthesize the two disparate views into a more comprehensive understanding of the medical error space?

5. Error thought problem—discuss the philosophy of an error—when does an error actually become an error? Often times we estimate an error's presence by the outcomes, we see accidents, injuries, and deaths. But can errors exist without a negative outcome? Can negative outcomes exist without errors? How does a "systems approach" affect the way we look at errors and react to them?

6. Provide four examples of automated system that you interact with during your daily activities.

7. Using Table 12.1 as an example, identify the levels of automation employed by the following systems: vehicle cruise control, thermostats, microwave, and Roomba?

8. List the pros and cons of requiring human operators to monitor an automated system. Considering your list of pros and cons, what recommendation would you make regarding the role of a human operator in supervising the activity of an automated system?

9. Describe the types of interfaces you use during your daily activities (i.e., physical keyboards, voice, etc.). Are there particular applications that are better suited for certain modes of interaction (i.e., touch, voice, gesture recognition, etc.)? What are some of the challenges or problems you have experienced using each of these types of interfaces?

10. Describe one application where BCIs may be of use to the broader population users either in the workplace or at home.

REFERENCES

Alemzadeh, H., Raman, J., Leveson, N., Kalbarczyk, Z., & Iyer, R. K. (2016). Adverse events in robotic surgery: a retrospective study of 14 years of FDA data. *PloS One*, *11*(4), e0151470.

Anderson, J. M., Nidhi, K., Stanley, K. D., Sorensen, P., Samaras, C., & Oluwatola, O. A. (2014). *Autonomous Vehicle Technology: A Guide for Policymakers*. Rand Corporation.

Bureau d'Enquêtes et d'Analyses pour la sécurité de l'aviation civile (2012). Safety Investigation Following the Accident on 1ST June 2009 to the Airbus A300-203, Flight AF 447.

Chassin, M. R., Loeb, J. M., Schmaltz, S. P., & Wachter, R. M. (2010). Accountability measures—using measurement to promote quality improvement. *New England Journal of Medicine*, *363*(7), 683–688.

Dekker, S. (2011). *Patient Safety: A Human Factors Approach*. Boca Raton, FL: CRC Press.

Dekker, S. W., & Woods, D. D. (2002). MABA-MABA or abracadabra? Progress on human–automation co-ordination. *Cognition, Technology & Work*, *4*(4), 240–244.

della Cava, M. (February 29, 2016). Google car hits bus, first time at fault. *USA Today*, volume number. Retrieved from https://www.usatoday.com/story/tech/news/2016/02/29/google-car-hits-bus-first-time-fault/81115258/

Feder, B. J. (2008). Prepping robots to perform surgery. *The New York Times*. Retrieved from https://www.nytimes.com/2008/05/04/business/04moll.html?_r=0

Grigorescu, S. M., Lüth, T., Fragkopoulos, C., Cyriacks, M., & Gräser, A. (2012). A BCI-controlled robotic assistant for quadriplegic people in domestic and professional life. *Robotica*, *30*(03), 419–431.

Harrison, S. A., & Tong, F. (2009). Decoding reveals the contents of visual working memory in early visual areas. *Nature*, *458*(7238), 632–635.

Hochberg, L. R., Bacher, D., Jarosiewicz, B., Masse, N. Y., Simeral, J. D., Vogel, J., Haddadin, S., Liu, J., van der Smagt, P., & Donoghue, J. P. (2012). Reach and grasp by people with tetraplegia using a neurally controlled robotic arm. *Nature*, *485*(7398), 372–375.

Holden, R. J., Carayon, P., Gurses, A. P., Hoonakker, P., Hundt, A. S., Ozok, A. A., & Rivera-Rodriguez, A. J. (2013). SEIPS 2.0: a human factors framework for studying and improving the work of healthcare professionals and patients. *Ergonomics*, *56*(11), 1669–1686.

Hughes, A. M., Gregory, M. E., Joseph, D. L., Sonesh, S. C., Marlow, S. L., Lacerenza, C. N., Benishek, L. E., King, H. B., & Salas, E. (2016). Saving lives: a meta-analysis of team training in healthcare. *Journal of Applied Psychology*, *101*(9), 1266–1304.

Ilgen, D. R., Hollenbeck, J. R., Johnson, M., & Jundt, D. (2005). Teams in organizations: from input-process-output models to IMOI models. *Annual Review of Psychology*, *56*, 517–543.

Kamitani, Y., & Tong, F. (2005). Decoding the visual and subjective contents of the human brain. *Nature Neuroscience*, *8*(5), 679–685.

Keebler, J. R., Lazzara, E. H., Patzer, B. S., Palmer, E. M., Plummer, J. P., Smith, D. C., Lew, V., Fouquet, S., Raymond Chan, Y., & Riss, R. (2016). Meta-analyses of the effects of standardized handoff protocols on patient, provider, and organizational outcomes. *Human Factors: The Journal of the Human Factors and Ergonomics Society*, *58*, 0018720816672309.

Kohn, L. T., Corrigan, J. M., & Donaldson, M. S. (Eds.). (2000). *To Err Is Human: Building a Safer Health System* (Vol. 6). National Academies Press, Washington DC.

Lee, J. D., & See, K. A. (2004). Trust in automation: designing for appropriate reliance. *Human Factors: The Journal of the Human Factors and Ergonomics Society*, *46*(1), 50–80.

Mador, R. L., & Shaw, N. T. (2009). The impact of a Critical Care Information System (CCIS) on time spent charting and in direct patient care by staff in the ICU: a review of the literature. *International Journal of Medical Informatics*, *78*(7), 435–445.

Makary, M. A., & Daniel, M. (2016). Medical error—the third leading cause of death in the US. *BMJ*, *353*, i2139.

Marescaux, J., Leroy, J., Gagner, M., Rubino, F., Mutter, D., Vix, M., Butner, S. E., & Smith, M. K. (2001). Transatlantic robot-assisted telesurgery. *Nature*, *413*(6854), 379–380.

McFarland, D. J., & Wolpaw, J. R. (2011). Brain-computer interfaces for communication and control. *Communications of the ACM*, *54*(5), 60–66.

NHTSA. (2013). *Preliminary Statement of Policy Concerning Automated Vehicles*. Publication NHTSA 14-13. U.S. Department of Transportation, Washington, D.C.

Nicolas-Alonso, L. F., & Gomez-Gil, J. (2012). Brain computer interfaces, a review. *Sensors*, *12*(2), 1211–1279.

Nishimoto, S., Vu, A. T., Naselaris, T., Benjamini, Y., Yu, B., & Gallant, J. L. (2011). Reconstructing visual experiences from brain activity evoked by natural movies. *Current Biology*, *21*(19), 1641–1646.

Parasuraman, R. (2003). Neuroergonomics: research and practice. *Theoretical Issues in Ergonomics Science*, *4*(1–2), 5–20.

Parasuraman, R., & Riley, V. (1997). Humans and automation: use, misuse, disuse, abuse. *Human Factors: The Journal of the Human Factors and Ergonomics Society*, *39*(2), 230–253.

Parasuraman, R., & Wilson, G. F. (2008). Putting the brain to work: neuroergonomics past, present, and future. *Human Factors: The Journal of the Human Factors and Ergonomics Society*, *50*(3), 468–474.

Rabin, C. R. (2013). Questions about robotic hysterectomy. *The New York Times*. Retrieved from https://well.blogs.nytimes.com/2013/02/25/questions-about-a-robotic-surgery/

Rabin, C. R. (2013). Salesmen in the surgical suite. *The New York Times*. Retrieved from https://www.nytimes.com/2013/03/26/health/salesmen-in-the-surgical-suite.html?pagewanted=all

Salas, E., Rosen, M. A., Burke, C. S., & Goodwin, G. F. (2009). The wisdom of collectives in organizations: an update of the teamwork competencies. In E. Salas, G. F. Goodwin, & C. S. Burke (Eds.). *Team Effectiveness in Complex Organizations. Cross-Disciplinary Perspectives and Approaches*, pp. 39–79. Routledge/Taylor & Francis Group, New York, NY.

Salas, E., Shuffler, M. L., Thayer, A. L., Bedwell, W. L., & Lazzara, E. H. (2014). Understanding and improving teamwork in organizations: a scientifically based practical guide. *Human Resource Management*, *54*(4), 599–622.

Stanton, N. A. (2005). Human factors and ergonomics methods. In N. A. Stanton, A. Hedge, K. Brookhuis, E. Salas, & H. Hendrick (Eds.). *Handbook of Human Factors and Ergonomics Methods*, pp. 27–38. Boca Raton, FL: CRC Press.

Stokowski LA. (2016). Who believes that medical error is the third leading cause of hospital deaths? Medscape Medical News; Available from: https://www.medscape.com/viewarticle/863788. Accessed May 26, 2024.

Sutcliffe, K. M., Lewton, E., & Rosenthal, M. M. (2004). Communication failures: an insidious contributor to medical mishaps. *Academic medicine*, *79*(2), 186–194.

Surgical robot Da Vinci scrutinized by FDA after deaths, other surgical nightmares (2013, April). *Daily News*. Retrieved from https://www.nydailynews.com/life-style/health/surgical-robot-scrutinized-fda-deaths-nightmares-article-1.1311447

U.S. Department of Transportation. (2013). U.S. Department of Transportation Releases Policy on Automated Vehicle Development. National Highway Traffic Safety Administration. 30 May 2013.

Wiener, E. L. (1988). Cockpit automation. In E. L. Wiener, & D. C. Nagel (Eds.), pp. 433–461. *Human Factors in Aviation*. San Diego: Academic Press.

Index

Note: **Bold** page numbers refer to tables and *italic* ones to figures.

Printed in the United States
by Baker & Taylor Publisher Services